MATHEMATICAL
PHYSICS

D1225467

LICENSE, DISCLAIMER OF LIABILITY, AND LIMITED WARRANTY

By purchasing or using this book (the "Work"), you agree that this license grants permission to use the contents contained herein, but does not give you the right of ownership to any of the textual content in the book or ownership to any of the information or products contained in it. *This license does not permit uploading of the Work onto the Internet or on a network (of any kind) without the written consent of the Publisher.* Duplication or dissemination of any text, code, simulations, images, etc. contained herein is limited to and subject to licensing terms for the respective products, and permission must be obtained from the Publisher or the owner of the content, etc., in order to reproduce or network any portion of the textual material (in any media) that is contained in the Work.

MERCURY LEARNING AND INFORMATION ("MLI" or "the Publisher") and anyone involved in the creation, writing, or production of the companion disc, accompanying algorithms, code, or computer programs ("the software"), and any accompanying Web site or software of the Work, cannot and do not warrant the performance or results that might be obtained by using the contents of the Work. The author, developers, and the Publisher have used their best efforts to insure the accuracy and functionality of the textual material and/or programs contained in this package; we, however, make no warranty of any kind, express or implied, regarding the performance of these contents or programs. The Work is sold "as is" without warranty (except for defective materials used in manufacturing the book or due to faulty workmanship).

The author, developers, and the publisher of any accompanying content, and anyone involved in the composition, production, and manufacturing of this work will not be liable for damages of any kind arising out of the use of (or the inability to use) the algorithms, source code, computer programs, or textual material contained in this publication. This includes, but is not limited to, loss of revenue or profit, or other incidental, physical, or consequential damages arising out of the use of this Work.

The sole remedy in the event of a claim of any kind is expressly limited to replacement of the book, and only at the discretion of the Publisher. The use of "implied warranty" and certain "exclusions" vary from state to state, and might not apply to the purchaser of this product.

MATHEMATICAL PHYSICS

An Introduction

Derek Raine, PhD et al.

MERCURY LEARNING AND INFORMATION
Dulles, Virginia
Boston, Massachusetts
New Delhi

Copyright © 2019 by MERCURY LEARNING AND INFORMATION LLC. All rights reserved.

Original title and copyright: *Mathematical Methods for Physical Science* by Derek Raine et al. Copyright © 2018 Pantaneto Press. All rights reserved.

This publication, portions of it, or any accompanying software may not be reproduced in any way, stored in a retrieval system of any type, or transmitted by any means, media, electronic display or mechanical display, including, but not limited to, photocopy, recording, Internet postings, or scanning, without prior permission in writing from the publisher.

Publisher: David Pallai
MERCURY LEARNING AND INFORMATION
22841 Quicksilver Drive
Dulles, VA 20166
info@merclearning.com
www.merclearning.com
1-(800)-232-0223

D. Raine et al. *Mathematical Physics*.
ISBN: 9781683922056

The publisher recognizes and respects all marks used by companies, manufacturers, and developers as a means to distinguish their products. All brand names and product names mentioned in this book are trademarks or service marks of their respective companies. Any omission or misuse (of any kind) of service marks or trademarks, etc. is not an attempt to infringe on the property of others.

Library of Congress Control Number: 2018949983

181920321 This book is printed on acid-free paper in the United States of America.

Our titles are available for adoption, license, or bulk purchase by institutions, corporations, etc. For additional information, please contact the Customer Service Dept. at (800) 232-0223 (toll free).

All of our titles are available in digital format at *authorcloudware.com* and other digital vendors. The sole obligation of MERCURY LEARNING AND INFORMATION to the purchaser is to replace the book, based on defective materials or faulty workmanship, but not based on the operation or functionality of the product.

CONTENTS

PREFACE

The book differs from other introduction to mathematical methods at this level in several important areas.

First, it does not follow the usual presentation of a description of the theory followed by examples and exercises. Rather we use examples to introduce the theory. This approach is not new; it goes back to the methods by which the scribes of Ancient Babylon learned mathematics: by example problems and, to judge from the numbers of surviving cuneiform tablets, lots of them!

Second, to help the reader digest the text, it is broken up into quite short sections (often a page or so) followed by exercises. It may be tempting to skip the exercises, (especially if one is used to doing only a selection of "end-of-chapter" problems) on the grounds that one can get through the book more quickly that way. This is true, in the same sense that watching a film speeded up x8 will get to the end more quickly, but it will be without much understanding of the plot. There are optional additional exercises at the end of each chapter; the ones in the text represent the minimum we think you need.

Third, however, we have tried to avoid too many "plug-and-chug" exercises, that is, exercises which you solve by following the text but substituting some different numbers ("pattern matching"). These are useful to reinforce memory, but they are not very useful to develop or test understanding. Rather, we have tried to make the exercises diagnostic in the sense that they do test understanding, that is, they test the ability to use what has been learned in a slightly different context. An instructor can therefore use these to target support for students.

Finally, and related to the previous point, while we hope it is perfectly possible to use this book for self-study, it was not designed for that purpose. It is intended for use as a course text. In this regard it might be useful to say a little about the background to the writing of the book - especially if you are intrigued to know why there are so many authors. We would also like to thank contributions to various versions of the text from Paul Abel, Mike Dampier, Andrew King, and Tim Yeoman.

About forty years ago, it was agreed that our conventional presentation of mathematical methods for our physics students - lectures, marked homework,and examinations - was not as effective as we might have hoped. So, instead of spending lecture time going through theory and exercises on the board, we produced a text as, in effect, the lecture notes, and refocused class time on weekly workshops and small group tutorials. Lectures were restricted to a weekly one-hour introduction to the topics for that week. We made the examinations harder, by requiring passes separately in the major topics (calculus of one variable, many variables, linear algebra, differential equations, vector calculus) and the pass rates soared. The initial text has been refined over the years (hence the number of authors) and this book is another, more outward-facing version, which we are pleased to have the opportunity to share with you.

D. J. Raine (lead author)
and the teaching team:
G. A. Wynn
S. Vaughan
M. Roy
R. O. Davies
E. J. Bunce

Leicester, 2017

CHAPTER 1

DERIVATIVES AND INTEGRALS

In this chapter we quickly cover the basic and most commonly used rules for differentiation and integration of simple functions and combinations of functions. We assume that you are familiar with basic algebra, the concept of a function and the idea of a limit. Much of the chapter will be familiar to you, but perhaps not all of it. And it is important to practice these procedures as much as possible – by working through the exercises, even if they look simple – to prepare you for the more advanced material that will build upon these ideas.

1.1. DEFINITION OF A DERIVATIVE

The derivative of a function is the slope (or gradient) of the tangent to the graph of the function at any point; i.e. it is the slope approached by the line between two points on the graph as they get closer (Figure 1.1). By the "slope" we mean (as usual) the change in height divided by the horizontal displacement. To use this definition to calculate the slopes of graphs we put it into more formal language.

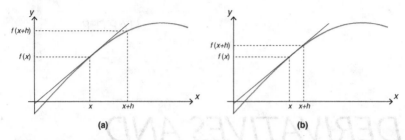

FIGURE 1.1: The tangent at an arbitrary point x as the limit of line joining points $(x, f(x))$ and $(x + h, f(x + h))$ as h gets smaller.

More formally, the *derivative* of a function $y = f(x)$ is the function $f'(x)$ or dy/dx *defined by*

$$f'(x) = \lim_{h \to 0} \frac{f(x+h) - f(x)}{h}. \tag{1.1}$$

Equivalently, in a commonly used alternative notation, we write δx in place of h to emphasize that we are making a small increment in x. We also sometimes write $y = y(x)$ instead of $y = f(x)$ to indicate that y is a function of x. The definition is then written

$$\frac{dy}{dx} = \lim_{\delta x \to 0} \frac{y(x + \delta x) - y(x)}{\delta x}. \tag{1.2}$$

For a function of the form $y = f(x)$ we use the notation $f'(x)$, y' and dy/dx interchangeably. The definition expresses in an exact manner the fact that the derivative at x is the slope of the tangent at x. In order to use it we need a way of taking the limit. The following example shows how this is done for a simple function such as a power of x.

Example 1.1 Use the definition from equation (1.1) to find $f'(x)$ if $f(x) = x^3$.

To use the definition we first need $f(x + h)$.

$$f(x) = x^3, \quad \text{and} \quad f(x+h) = (x+h)^3.$$

Next, expanding out the parenthetical, we have,

$$f(x + h) = x^3 + 3x^2h + 3xh^2 + h^3,$$

therefore

$$\frac{f(x+h)-f(x)}{h} = 3x^2 + 3xh + h^2,$$

and

$$f'(x) = \lim_{h \to 0}\left(3x^2 + 3xh + h^2\right),$$
$$= 3x^2 + 0 + 0,$$
$$= 3x^2.$$

Exercise 1.1 Use the definition of the derivative to prove that if $f(x) = x^4$ then $f'(x) = 4x^3$.

Example 1.2 Use the definition of a derivative to find the derivative of the function $y = x^n$ (with n a positive integer).

We begin with the function, and write out the necessary limit

$$y = x^n, \quad \text{and} \quad \frac{dy}{dx} = \lim_{h \to 0}\frac{(x+h)^n - x^n}{h}.$$

Then, using the binomial expansion (which you may be familiar with and will find discussed in Chapter 2),

$$(x+h)^n = x^n + nx^{n-1}h + \frac{1}{2}n(n-1)x^{n-2}h^2 + \ldots + h^n,$$

and

$$(x+h)^n - x^n = nx^{n-1}h + \frac{1}{2}n(n-1)x^{n-2}h^2 + \ldots + h^n,$$

and so

$$\frac{(x+h)^n - x^n}{h} = nx^{n-1} + h\left[\frac{1}{2}n(n-1)x^{n-2} + \ldots + h^{n-2}\right].$$

Taking the limit as $h \to 0$ gives

$$\lim_{h\to 0}\frac{(x+h)^n - x^n}{h} = nx^{n-1},$$

and therefore

$$\frac{dy}{dx} = nx^{n-1}.$$

In principle, the derivative of any function can be worked out from the definition. In practice, we memorize a few simple examples (Section 1.2, Table 1.1) and we learn to build up more complicated derivatives from these by a variety of methods (Sections 1.3 – 1.5).

1.2. SOME BASIC DERIVATIVES

You should already know (or be willing to take on trust for the present) the derivatives in Table 1.1.

Note: the trigonometric functions are introduced in section A.4 of Appendix A and the logarithm in Section A.3. Throughout this course $\ln(x)$ will be written for $\log_e(x)$, the "natural logarithm." It is common to use $\log(x) = \log_{10}(x)$, and the use of any other base is indicated explicitly, as in $\log_2(x)$. The derivatives in the table may be taken as given.

TABLE 1.1: Table of some standard derivatives.

Function $y(x)$	Derivative dy/dx
a, a constant	0
x^a	ax^{a-1}
$\sin(x)$	$\cos(x)$
$\cos(x)$	$-\sin(x)$
$\tan(x)$	$\sec^2(x)$
$\ln(x) = \log_e(x)$	$1/x$
e^{ax}	ae^{ax}

The derivative of the sum of two functions is the sum of the derivatives. This means that if $f(x)$ and $g(x)$ are both functions of x then

$$\frac{d}{dx}(f+g) = \frac{df}{dx} + \frac{dg}{dx} \tag{1.3}$$

Exercise 1.2 What is dy/dx if

(i) $y = x^n + c$ where c is a constant and n an integer

(ii) $y = x^{-3} + x^4$

(iii) $y = 2\sin(x) = \sin(x) + \sin(x)$

(iv) $y = \sin(x) + \cos(x)$.

We also learn rules for the derivative of a "function of a function" and of a product, which enable us to construct the derivatives of quite complicated expressions from the table without having to return to the basic definition.

1.3. "FUNCTION OF A FUNCTION" (CHAIN RULE)

Let's consider the function $y = (1 + 2x)^2$. This can be thought of as the composite of two functions: $u = 1 + 2x$ and $y = u^2$ because

$y(u(x)) = (1 + 2x)^2$. What is the relationship between the derivatives of these three functions?

$$y = (1+2x)^2 = 4x^2 + 4x + 1, \quad \text{so} \quad \frac{dy}{dx} = 8x + 4,$$

and, alternatively

$$y = u^2, \quad \text{so} \quad \frac{dy}{du} = 2u = 2(1+2x),$$

$$u = 1 + 2x, \quad \text{so} \quad \frac{du}{dx} = 2.$$

We can see that the following holds:

$$8x + 4 = 2(1 + 2x) \times 2.$$

And so the relation in this case is

$$\frac{dy}{dx} = \frac{dy}{du}\frac{du}{dx}. \tag{1.4}$$

This is a general rule that can be proved from the definition of a derivative: the derivative of a composite of two functions (a "function of a function") is the product of their derivatives. In words: the rate of change of $y(u(x))$ with respect to x is the rate of change of y with respect to u times the rate of change of u with respect to x. This is usually called the *chain rule*. So if $y = f(u(x))$ then we use equation (1.4) to compute its derivative as the product of dy/du and du/dx.

Example 1.3 If $y = \cos(x^2)$ what is dy/dx?

We use the chain rule, choosing u to simplify the process. Let $u = x^2$ so $y = \cos(u)$ and

$$dy/du = -\sin(u),$$
$$du/dx = 2x.$$

Then, using the chain rule,

$$\frac{dy}{dx} = \frac{dy}{du}\frac{du}{dx},$$
$$= -\sin(u)\cdot 2x,$$
$$= -2x\sin(x^2).$$

Alternatively, and more economically, you can write down the chain rule directly:

$$\frac{d}{dx}\{\cos(x^2)\} = \frac{d\cos(x^2)}{dx^2}\cdot\frac{dx^2}{dx} = \left(-\sin(x^2)\right)\cdot 2x.$$

Exercise 1.3 What is dy/dx if

(i) $y = \sin(2x)$

(ii) $y = \sin(x^2)$

(iii) $y = \cos^3(x)$

(iv) $y = \sin(ax)$

(v) $y = (1 + ax)^n$

(vi) $y = (1 + \cos(x))^{1/2}$?

Exercise 1.4 What is dy/dx if $y = \log_{10}(x)$? (Hint: see section A.3 to convert this to the derivative of $\ln(x)$ which you can get from the table.)

Exercise 1.5 If $y = \ln[x + (1 + x^2)^{1/2}]$ show that $dy/dx = (1 + x^2)^{-1/2}$. (Hint: look at what you are seeking to prove in order to see how to simplify your initial expression.)

Exercise 1.6 If $y = 1/f(x)$ show that $dy/dx = -f'/f^2$ (where, as usual, f' stands for df/dx).

Note that it is a matter of choice whether you commit the result of this last exercise to memory and use it in specific cases, or use the chain rule directly as needed.

1.4. PRODUCT (AND QUOTIENT) RULE

Suppose now that we want the derivative of a product of two functions in 1.1, $y(x) = u(x)v(x)$, say. We might guess that this is $\dfrac{du}{dx}\dfrac{dv}{dx}$. You can readily check that this is wrong by applying it to a simple example, say $y = x^2 = x \times x$. The correct derivative of the product of two functions, like $y(x) = u(x)v(x)$, is the sum of two terms as follows:

$$\frac{dy}{dx} = u\frac{dv}{dx} + v\frac{du}{dx}, \quad \text{equivalently} \quad (uv)' = uv' + vu'. \quad (1.5)$$

This is known as the *product rule* or the Leibniz rule. An alternative way to remember the product rule is as follows:

$$d(uv) = udv + vdu.$$

Example 1.4 If $y = x \sin(x)$, what is dy/dx?

Let $u = x$, $v = \sin(x)$. Then we have that

$$\frac{du}{dx} = 1, \quad \frac{dv}{dx} = \cos(x).$$

Therefore, by equation (1.5),

$$\frac{dy}{dx} = x \cdot \cos(x) + \sin(x) \cdot 1 = x\cos(x) + \sin(x).$$

Example 1.5 If $y = x^2 \cos^3(2x)$, what is dy/dx ?

This combines the chain rule and product rule. First break it down as a product of two terms $y = uv$:

$$u = x^2, \text{ and } v = \cos^3(2x).$$

We can see that

$$\frac{du}{dx} = 2x.$$

As v is a function of a function we use the chain rule to find its to find its derivative:

$$v = w^3 \text{ with } w = \cos(2x),$$

$$\frac{dw}{dx} = -2\sin(2x),$$

$$\frac{dv}{dx} = \frac{dv}{dw}\frac{dw}{dx} = (3w^2)(-2\sin(2x)),$$

$$= -6\cos^2(2x)\sin(2x).$$

Putting it all together

$$\frac{dy}{dx} = u\frac{dv}{dx} + v\frac{du}{dx},$$

$$= -6x^2\cos^2(2x)\sin(2x) + 2x\cos^3(2x).$$

Eventually you should be able to do some of the intermediate steps in your head. So your working might look something like this:

$$(x^2\cos^3(2x))' = x^2(\cos^3(2x)^3)' + 2x\cos^3(2x),$$
$$= x^2 \cdot 3\cos^2(2x)(\cos(2x))' + 2x\cos^3(2x),$$
$$= x^2 \cdot 3\cos^2(2x)\sin(2x)(-2) + 2x\cos^3(2x),$$
$$= -6x^2\cos^2(2x)\sin(2x) + 2x\cos^3(2x).$$

You should cultivate the habit of working like this, resorting to introducing u and v explicitly only if you get stuck.

If you are familiar with proof by induction, you may like the following example.

Example 1.6 Assuming that $\frac{dx}{dx} = 1$ prove by induction that $\frac{dx^n}{dx} = nx^{n-1}$ when n is a positive integer.

We start by assuming that for integers up to some n we have $\dfrac{dx^n}{dx} = nx^{n-1}$. Then

$$\frac{dx^{n+1}}{dx} = \frac{d}{dx}xx^n = 1 \cdot x^n + x \cdot nx^{n-1} = (n+1)x^n.$$

Thus, if the result is true up to n it is true for $n + 1$. Since it is true for $n = 1$ it is true for all n.

Exercise 1.7 What is dy/dx if

(i) $y = x \ln(x)$

(ii) $y = x^2 \sin(x)$

(iii) $y = \dfrac{\sin(2x)}{x^3}$? (Treat this as $x^{-3}\sin(2x)$.)

The derivative of the quotient of two functions, like $y(x) = u(x) / v(x)$, has its own rule:

$$\frac{dy}{dx} = \frac{\dfrac{du}{dx}v - \dfrac{dv}{dx}u}{v^2}, \quad \text{equivalently} \quad (u/v)' = \frac{u'v - v'u}{v^2}. \quad (1.6)$$

Exercise 1.8

(i) Derive the quotient rule: By putting $f = u$, $g = 1/v$ show that if $y = fg = u/v$ then $dy/dx = (u'v' - v'u)/v^2$.

(ii) Use the quotient rule to find dy/dx if $y = \sin(2x)/x^3$ and check your result with exercise 1.7(iii).

(iii) Use the quotient rule to find dy/dx if $y = \tan(x)$. (Recall that $\tan(x) = \sin(x)/\cos(x)$.)

Rather than learning the quotient rule it is often easier to calculate $(u/v)'$ using the product rule for $(uv^{-1})'$. Alternatively, you can memorize the rule as

$$d(u/v) = \frac{v\,du - u\,dv}{v^2}.$$

1.5. IMPLICIT DIFFERENTIATION

Sometimes it is convenient to find the derivative dy/dx from an equation like $f(y) = g(x)$ without first solving for y as a function of x. This is little more than an application of the chain rule.

Example 1.7 Find dy/dx if $y = \sin^{-1}(x)$. (Note that $\sin^{-1}(x)$ is the inverse sine function, also called $\arcsin(x)$; it is *not* the same as $1/\sin(x)$.)

We begin by transforming to something more familiar. If $y = \sin^{-1}(x)$,

$$x = \sin(y)$$

and differentiating both sides with respect to x and using the chain rule we have

$$1 = \frac{d}{dx}\big(\sin(y)\big) = \frac{dy}{dx}\frac{d}{dy}\big(\sin(y)\big),$$

$$= \frac{dy}{dx}\cos(y).$$

Rearranging for dy/dx gives

$$\frac{dy}{dx} = \frac{1}{\cos(y)}.$$

We want the answer as a function of x so we have to get $\cos(y)$ in terms of x given that $\sin(y) = x$. We first substitute for $\cos(y)$ using $\sin^2(y) + \cos^2(y) = 1$. Then,

$$\frac{dy}{dx} = \frac{1}{\left(1 - \sin^2(y)\right)^{1/2}} = \frac{1}{\left(1 - x^2\right)^{1/2}}.$$

The method used in this example is referred to as *implicit differentiation* because we differentiate the equation $\sin(y) = x$ which gives y implicitly as a function of x, in contrast to $y = \sin^{-1}(x)$ where y is given explicitly.

Exercise 1.9 Find dy/dx if

(i) $y = \cos^{-1}(x)$

(ii) $y = \tan^{-1}(x)$

(iii) $y = x^x$. (Hint: begin by taking ln of both sides. The answer is not $x \cdot x^{x-1}$ because the power to which x is raised is not a constant!)

1.6. PIECEWISE DIFFERENTIABLE FUNCTIONS

Many useful functions cannot be described fully using a single formula, but can be described by different formulae that apply under different circumstances. A *piecewise* function $f(x)$ defines a function in terms of formulae that apply for different ranges of x.

Two useful examples of piecewise functions are $|x|$ (pronounced "mod x") and sgn(x) (pronounced "sign x" or "signum x"). We treat them in turn.

First, $f(x) = |x| = \text{mod}(x)$ is defined by

$$f(y) = \begin{cases} x & \text{if } x \geq 0 \\ -x & \text{if } x < 0. \end{cases} \tag{1.7}$$

This function is plotted in Figure 1.2(a).

Example 1.8 Evaluate

(i) $|-3|$

(ii) $|7|$

The function is defined separately for $x \geq 0$ and $x < 0$. Since -3 is less than 0, look at the second part of the definition: for $x = -3$, $y = -x$ gives $y = -(-3) = +3$.

Next, we have 7 so we want the definition for $x \geq 0$: for $x = 7$, $y = +x$, so $y = +7$.

Obviously $|x|$ is simply the absolute value of x (i.e. its magnitude regardless of sign), but looking at the definition of a quantity will often provide the first step in the answer to a question.

Second, $f(x) = \text{sgn}(x)$. This is defined by

$$f(x) = \begin{cases} +1 & \text{if} \quad x > 0 \\ 0 & \text{if} \quad x = 0 \\ -1 & \text{if} \quad x < 0. \end{cases} \qquad (1.8)$$

This function is plotted in Figure 1.2(b).

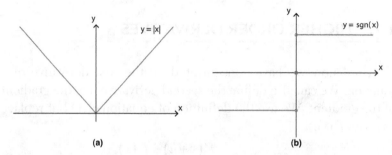

FIGURE 1.2: (a) The graph of the "modulus" function $y = |x|$. (b) The "sign" function $y = \text{sgn}(x)$.

Exercise 1.10 What is

(i) $\text{sgn}(1)$

(ii) $\text{sgn}(-5)$?

Example 1.9 If $y = \text{sgn}(x)$ what is dy/dx?

Look at the definition. If $x > 0$, then $y = +1$, so $dy/dx = 0$. If $x < 0$, then $y = 1$, so $dy/dx = 0$ (because $y = $ constant there). The derivative does not exist at $x = 0$. (The graph, shown in Figure 1.2, obviously does not have a tangent there). Looking at the graph confirms that the slope is indeed zero everywhere else.

Notice that any real number can be expressed as the product of its sign and its modulus

$$x = \text{sgn}(x) \cdot |x|.$$

These two functions $f(x)$ are *piecewise linear* functions because each piece of the graph is a straight line. For mod(x), the graph is continuous at $x = 0$, but its derivative is discontinuous at this point: there is an instantaneous change in the gradient. For sgn(x) the graph has a discontinuity at $x = 0$. We shall discuss continuous and discontinuous functions more in section 1.16.

Exercise 1.11 If $y = |x|$ what is dy/dx? (Hint: start from the definition of the mod function!)

1.7. HIGHER ORDER DERIVATIVES

Until now we have concentrated on the first derivative of a function. We can also define the second derivative, i.e. the gradient of the gradient. We use the definition of equation (1.1) but replace $f(x)$ with $f'(x)$:

$$f''(x) = \lim_{h \to 0} \frac{f'(x+h) - f'(x)}{h}, \qquad (1.9)$$

or, equivalently,

$$\frac{d^2y}{dx^2} = \lim_{\delta x \to 0} \frac{y'(x + \delta x) - y'(x)}{\delta x}. \qquad (1.10)$$

Similarly, we can define the third, fourth and higher order derivatives. Often the nth derivative of a function $y = f(x)$ is written as $f^{(n)}(x)$, or $d^n y/dx^n$. Note the parentheses around the superscript (n) in $f^{(n)}(x)$ to distinguish it from $(f(x))^n$. In practice we continue to differentiate the result after each differentiating (tidying up the expressions as we go) until we get to the required derivative.

Example 1.10 If $y = x^n$ calculate d^2y/dx^2.

If $y = x^n$ then $dy/dx = nx^{n-1}$. Next,

$$\frac{d^2y}{dx^2} = \frac{d}{dx}\left(\frac{dy}{dx}\right) = \frac{d}{dx}\left(nx^{n-1}\right)$$
$$= n(n-1)x^{n-2}.$$

Exercise 1.12 Find all the derivatives (first, second etc.) of the function $f(x) = x^3 + 6x^2 - 9x + 12$ until the derivative is zero.

Exercise 1.13 Find derivatives of the first five orders of the function $f(x) = \sin(x)$.

1.8. STATIONARY POINTS

A common application of differentiation is to find the maxima or minima of functions.

Let $y = f(x)$. A point at which $dy/dx = 0$ is called a *stationary point* of the function $f(x)$. Equivalently, at a stationary point of $f(x)$, we have $f'(x) = 0$. See Figure 1.3.

- A stationary point of $y = f(x)$ is a *minimum* if, at this point, $d^2y/dx^2 > 0$;

- A stationary point of $y = f(x)$ is a *maximum* if, at this point, $d^2y/dx^2 < 0$;

- A stationary point of $y = f(x)$ is (usually) a *stationary point of inflection* if, at this point, $d^2y/dx^2 = 0$ (i.e. the derivative has a maximum or minimum).

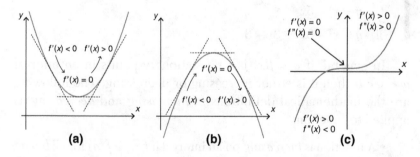

FIGURE 1.3: (a) At a minimum the slope increases as x increases, so $d(y')/dx = y'' > 0$. (b) At a maximum, the slope decreases as x increases, so $y'' < 0$. (c) At a stationary point of inflection the slope is zero and the rate of change of slope is zero.

Example 1.11 Find the stationary points of the function $y = 2x^3 + 3x^2 - 36x + 24$ and determine whether they are maxima, minima, or points of inflection.

We begin by finding dy/dx:

$$dy/dx = 6x^2 + 6x - 36,$$
$$= 6(x^2 + x - 6),$$
$$= 6(x + 3)(x - 2).$$

Then we set $dy/dx = 0$ and solve for x, which gives $x = -3$ and $x = 2$. So the stationary points are $x = -3$, $x = 2$. Now we calculate the second derivative,

$$d^2y/dx^2 = 6(2x + 1),$$

and evaluate it at the stationary points.

At $x = -3$, $d^2y/dx^2 < 0$, so this is a maximum,
at $x = 2$, $d^2y/dx^2 > 0$, so this is a minimum.

Exercise 1.14 Find the stationary points of the following functions, and determine whether they are maxima, minima, or points of inflection.

(i) $y = x^5 - 5x$

(ii) $y = x^3 - 3x^2 + 3x + 3$

In general, if $y = f(x)$ has no stationary points in an interval $a < x < b$ then it is either increasing or decreasing. The following are the mathematical definitions of *increasing* and *decreasing* as applied to functions:

• A function is *increasing* on an interval if $f(b) \geq f(a)$ for all $b > a$ and a, b in the interval.

• A function is *decreasing* on an interval if $f(b) \leq f(a)$ for all $b > a$ and a, b in the interval.

Notice that we make no assumption that the functions are continuous.

In the case of a continuous function, for which the derivative is well defined at each point in an interval, we can use the first derivative to establish whether the function is increasing or decreasing.

- A function is *increasing* on an interval if $f'(x) \geq 0$ for all x in the interval.

- A function is *decreasing* on an interval if $f'(x) \leq 0$ for all x in the interval.

Notice here that, mathematically speaking, "increasing" and "decreasing" include being constant (i.e. "non-decreasing" and "non-increasing," respectively). If we want to exclude this possibility we say a function is strictly increasing $(f'(x) > 0)$ or strictly decreasing $(f'(x) < 0)$.

Example 1.12 Show that the function $f(x) = x - x^2$ is (strictly) increasing in $0 < x < \frac{1}{2}$. Hence show that $x > x^2$ for $0 < x < \frac{1}{2}$.

To find out if the function is increasing ("going uphill") we look at the slope of the tangent: $f'(x) = 1 - 2x$. Now we determine where the slope is positive. In this case $f'(x) > 0$ if $1 - 2x > 0$, hence the slope is positive over the stated region $\left(0 < x < \frac{1}{2}\right)$. So $f(x)$ is increasing for $0 < x < \frac{1}{2}$; the function is indeed increasing in this interval.

This means that $f(b) > f(a)$ if $b > a$ in this interval. Take $a = 0$ and $b = x$. Then $f(x) > f(0)$ implies $x - x^2 > 0 - 0$, which in turn implies $x > x^2$ for $0 < x < \frac{1}{2}$.

Exercise 1.15 Show that $x - \sin(x)$ is (strictly) increasing for $x > 0$. Deduce that $\sin(x) < x$ for $x > 0$.

Exercise 1.16 Show that $y = \ln(1 + x) - x$ is (strictly) decreasing for $x > 0$. Deduce that $\ln(1 + x) < x$ in this range.

1.9.   INTEGRALS OF ELEMENTARY FUNCTIONS

Integration is the inverse of differentiation. Thus, $F(x)$ is the *indefinite integral* of $f(x)$ if $f(x)$ is the derivative of $F(x)$, i.e.

$$F(x) = \int f(x)\,dx \quad \text{if} \quad f(x) = \frac{dF(x)}{dx}. \qquad (1.11)$$

In some books $F(x)$ is called the *anti-derivative* of $f(x)$.

Example 1.13 Find $\int x^3 dx$

We see that $\int x^3 dx = \frac{1}{4}x^4 + c$ (where c is a constant) because

$$x^3 = \frac{d}{dx}\left\{\frac{1}{4}x^4 + c\right\}.$$

Notice the "$+c$", where c is any constant. There's always a "plus c" in an indefinite integral because anti-derivatives are defined only up to an arbitrary additive constant.

Example 1.14 Find the indefinite integral $\int bx^m\,dx$, where b and m are constants.

Begin by writing down a relevant derivative. For $y = ax^n$ we have $dy/dx = nax^{n-1}$. This gives an integral of the right form,

$$\int nax^{n-1} dx = ax^n + c.$$

Compare with the integral we want. Put $na = b$, $n - 1 = m$. Hence $n = m + 1$ and $a = b/(m + 1)$, provided $m \neq -1$. Then,

$$\int bx^m\,dx = \frac{b}{(m+1)}x^{m+1} + c. \quad \text{(for } m \neq -1\text{)}$$

In practice we memorize this particular result, usually as "add one to the power and divide by the new power." But it is worth remembering the method as well: to find an inverse we can guess a result of the right form and then tidy up the details by differentiation and substitution. (If we guess wrong this won't work, so it's a fail-safe method.)

The case $m = -1$ is special: we have

$$\int x^{-1}\, dx = \int \frac{dx}{x} = \ln|x| + c.$$

Note the absolute value of x on the right.

Example 1.15 Find the indefinite integral $\int \cos(2x)dx$.

Start from a relevant known integral. We know $\int \cos(x)dx = \sin(x) + c$ because $\frac{d}{dx}\{\sin(x)\} = \cos(x)$. Then we check that the integral of $\cos(2x)$ is proportional to $\sin(2x)$. We have,

$$\frac{d}{dx}\{\sin(2x)\} = 2\cos(2x)$$

so that,

$$\cos(2x) = \frac{d}{dx}\left\{\frac{1}{2}\sin(2x)\right\}$$

and therefore,

$$\int \cos(2x)dx = \frac{1}{2}\sin(2x) + c.$$

This shows you again how to go about obtaining a new integral from a similar known one. But there is no need to go through all the working once you know the result. The following exercise can be done by comparison with the previous examples. Always check by differentiating your answer.

Exercise 1.17 What are the indefinite integrals of

(i) $\sin(3x)$

(ii) a/x?

TABLE 1.2: Table of some standard integrals; c is an arbitrary constant.

Function y(x)	Indefinite integral $\int y\,dx$
x^a	$\dfrac{x^{a+1}}{a+1}+c$ ($a \neq -1$)
$1/x$	$\ln(x)+c$
$\sin(x)$	$-\cos(x)+c$
$\cos(x)$	$\sin(x)+c$
$\tan(x)$	$-\ln|\cos(x)|+c$
$\ln(x)$	$x\ln(x)-x+c$
e^{ax}	$a^{-1}e^{ax}+c$ ($a \neq 0$)
$f'(x)/f(x)$	$\ln(f(x))+c$

In order to integrate we need to have memorized a repertoire of derivatives and to use systematic guesswork. In practice, the problem is often to reduce an integral to some standard form which can be looked up in tables of integrals (for example Table 1.2). The examples in this chapter give some ways of doing this. The methods of Sections 1.11 and 1.12 are fundamental in their own right and must be learned.

1.10. INTEGRALS OF COMBINATIONS OF FUNCTIONS

Sometimes integrals which look complex are actually easy to do, as long as we know the correct rule or can apply some intuition. For example, consider any function $f(x)$. Then,

$$\int \frac{f'(x)}{f(x)}dx = \ln(f(x))+c.$$

We can show this must be true by differentiating the right hand side using the chain rule. If we let $f(x) = u$, then

$$\frac{d}{dx}\{\ln(f(x))+c\} = \frac{d}{dx}(\ln(u)),$$

$$= \frac{1}{u}\frac{du}{dx},$$

$$= \frac{1}{f(x)}f'(x).$$

Integration of functions like this – which can be written as a fraction where the numerator is the derivative of the denominator – is often called *logarithmic integration*.

Exercise 1.18 Show that $\int f'(x)(f(x))^n\, dx = \frac{1}{n+1}(f(x))^{n+1}+c$.
(Remember that to prove a given integral you differentiate the anti-derivative.)

1.11.   INTEGRATION BY SUBSTITUTION

It is often possible and useful to make a substitution that turns a complicated integral into a simpler integral or a "standard" integral that is given in charts like Table 1.2. Knowing which substitution to use comes with experience and trial and error (and also guesswork!).

Example 1.16 By using a suitable substitution calculate
$$\int \frac{dx}{ax+b}.$$

First, we look for an explicit change of variable such that

$$\int \frac{dx}{ax+b} = \frac{1}{a}\int \frac{dt}{t}.$$

We can then evaluate the integral, $\int dt/t$, using a standard result from Table 1.2.

As we want to manipulate the integral into a form where the new denominator involves t instead of $ax+b$, we start by trying

$t = ax + b$. Then $dt = adx$ so $dx = dt/a$. Next we substitute for x and dx to give

$$\int \frac{dx}{ax+b} = \frac{1}{a}\int \frac{dt}{t}.$$

Finally, from the standard integral in Table 1.2, $\int dt/t = \ln(t)$, so that

$$\int \frac{dx}{ax+b} = \frac{1}{a}\ln(t) + c,$$

$$= \frac{1}{a}\ln(ax+b) + c.$$

Example 1.17 Find an explicit change of variable such that

$$\int \frac{dx}{a^2 + x^2} = K\int \frac{dt}{1+t^2}$$

where K is a constant.

We want to replace the a^2 with a 1 in the denominator, so we take out factor a^2 to get

$$\int \frac{dx}{a^2 + x^2} = \frac{1}{a^2}\int \frac{dx}{1+x^2/a^2}.$$

We now want t^2 for x^2/a^2 so we try $t = x/a$, and $dt = dx/a$ or $dx = a\,dt$. Then

$$\int \frac{dx}{a^2 + x^2} = \frac{1}{a^2}\int \frac{a\,dt}{1+t^2},$$

$$= \frac{1}{a}\int \frac{dt}{1+t^2}.$$

Exercise 1.19

(i) Find an explicit change of variable that converts

$$\int \frac{dx}{b^2 - (x-a)^2} \quad \text{into} \quad \frac{1}{b}\int \frac{dt}{1-t^2}$$

(ii) Similarly, simplify

$$\int \frac{dx}{\left(1 + b^2 (x+a)^2\right)}$$

In harder questions where we are not given the answer, the best approach is often to try to make the integrand as simple as possible. There is usually more than one way to start and, if you do not succeed the first time, try again!

Example 1.18 Find $I = \int \dfrac{dx}{\left(1-x^2\right)^{1/2}}$.

It is tempting to try $t^2 = 1 - x^2$ to get rid of the square root. Then, by differentiation $2tdt = -2xdx$ and therefore $dx = -tdt/(1-t^2)^{1/2}$. So

$$I = -\int \frac{1}{t} \cdot \frac{t\,dt}{\left(1-t^2\right)^{1/2}} = -\int \frac{dt}{\left(1-t^2\right)^{1/2}}.$$

This is back where we started – so the substitution $t^2 = 1 - x^2$ does not work and we need some other way of removing the square root.

If we had $1 - \cos^2(\theta) = \sin^2(\theta)$ in the denominator we could extract the square root. So we try $x = \cos(\theta)$, then $dx = -\sin(\theta)\,d\theta$ and we have

$$I = -\int \frac{-\sin(\theta)\,d\theta}{\left(1-\cos^2(\theta)\right)^{1/2}},$$

$$= -\int \frac{-\sin(\theta)\,d\theta}{\sin(\theta)},$$

$$= -\int d\theta,$$

$$= -\theta + c.$$

Finally, we must substitute back for x to give

$$I = - \cos^{-1}(x) + c$$

Note that we could have chosen $x = \sin(\theta)$ to get $I = \sin^{-1}(x) + k$ (where k is a different constant). These two results are the same because $\sin^{-1}(x) = \pi/2 - \cos^{-1}(x)$.

Exercise 1.20 Find

$$\int \frac{dx}{\left(9 - x^2\right)^{1/2}}.$$

Example 1.19 Find $I = \int \sin(x)\cos(x)\, dx$.

This will simplify by letting either $t = \sin(x)$, then $dt = \cos(x)\, dx$, or $t = \cos(x)$, then $dt = - \sin(x)$. We can choose either of these. Using $t = \sin(x)$ we have

$$I = \int t\, dt,$$

$$= \frac{1}{2}t^2 + c,$$

$$= \frac{1}{2}\sin^2(x) + c.$$

With practice we can set out our working more compactly as follows:

$$I = \int \sin(x)\cos(x)\, dx = \int \sin(x)\, d\left(\sin(x)\right) = \frac{1}{2}\left(\sin(x)\right)^2 + c.$$

Note that in the final step the integral is of the form $\int f\, df = \frac{1}{2}f^2 + c$, where f here happens to be $\sin(x)$.

Exercise 1.21 Find

(i) $\int \sin^2(x)\cos(x)\, dx$

(ii) $\int \cos^n(x)\sin(x)\, dx\, (n > 0)$

(iii) $\int \cot(x)\, dx$

(iv) $\int (1 + x^2)^5 x\, dx$

1.12. INTEGRATION BY PARTS

This technique is one of the most important in mathematical physics. You will probably have met the basic formula in one of two equivalent guises, either

$$\int u\, dv = uv - \int v\, du, \tag{1.12}$$

or

$$\int uv'\, dx = uv - \int vu'\, dx. \tag{1.13}$$

The formula is derived by integrating the rule for differentiating a product (Section 1.4). Use whichever form you are familiar with or learn whichever you prefer.

Example 1.20 Find $I = \int x \sin(x)\, dx$.

Solution 1: First using equation (1.12): we define $u = x$, and $dv = \sin(x)dx$. Then,

$$du = dx,$$
$$v = -\cos(x).$$

Next, using equation (1.12), we find

$$I = \int x\, d(-\cos(x)),$$
$$= x(-\cos(x)) - \int (-\cos(x))\, dx,$$
$$= -x\cos(x) + \sin(x) + c.$$

Note: the integration constant in $v = \int v'dx$ is arbitrary. It will always cancel in the final answer, so we set it to zero for convenience.

Example 1.21 Find $I=\int x\sin(x)\,dx$.

Solution 2: We could also use equation (1.13). First we define $u = x$, and $v' = \sin(x)$, then,

$$u' = 1, \quad \text{and} \quad v' = -\cos(x),$$
$$I = uv - \int vu'dx,$$
$$= x(-\cos(x)) - (-\cos(x)).1\,dx,$$
$$= -x\cos(x) + \sin(x) + c.$$

Note that we chose $u = x$ because differentiating removes the awkward factor x from the integrand. The choice of the correct approach is a matter of experience, and trial and error.

Finally, with some experience, we can set out the working more compactly,

$$I = \int x\sin(x)\,dx = -\int x\,d(\cos(x))$$
$$= -x\cos(x) + \int \cos(x)\,dx = -x\cos(x) + \sin(x) + c.$$

Example 1.22 Find $\int x(1 + x)^4\,dx$ using integration by parts.

We have

$$\int x(1+x)^4\,dx = \frac{1}{5}\int x\,d(1+x)^5,$$
$$= \frac{1}{5}x(1+x)^5 - \frac{1}{5}\int(1+x)^5\,dx,$$
$$= \frac{1}{5}x(1+x)^5 - \frac{1}{30}(1+x)^6 + c.$$

Exercise 1.22 Find

(i) $\int x\cos(x)\,dx$

(ii) $\int \ln(x)\,dx$. (Hint: this is $\int u\,dv$ where $u = \ln(x)$ and $v = x$.)

1.13.   INTEGRATION OF RATIONAL FUNCTIONS

A rational function is one of the form

$$f(x) = \frac{\text{polynomial in } x}{\text{another polynomial in } x}$$

You should review Section A.5 on how to divide polynomials and Section A.6 on partial fractions.

Example 1.23 Find

$$I = \int \frac{2x^3 + 9x^2 + 4x - 20}{2x + 5}\,dx$$

Divide the numerator by the denominator so as to leave a remainder of smaller degree than the denominator.

$$I = \int \left(x^2 + 2x - 3 - \frac{5}{2x + 5} \right)dx.$$

Then, integrating each term,

$$I = \frac{x^3}{3} + x^2 - 3x - \frac{5}{2}\ln|2x + 5| + c.$$

Exercise 1.23 Find $\displaystyle\int \frac{2x^2 - 5x + 10}{x - 4}\,dx$

Exercise 1.24 Find the following integrals. (Hint: first express the integrand in partial fractions.)

(i) $\displaystyle\int \frac{dx}{(x+2)(x-1)}$

(ii) $\displaystyle\int \frac{x\,dx}{(x+4)(x+1)}$

(iii) $\displaystyle\int \frac{x^2 + 6x + 4}{(x+4)(x+1)}\,dx$

1.14. DEFINITE INTEGRALS

The definite integral of a function $f(x)$ between limits $x = a$ and $x = b$ is evaluated from the indefinite integral $F(x)$ at the limits of integration

$$\int_a^b f(x)\,dx = F(b) - F(a) = [F]_a^b. \qquad (1.14)$$

The square brackets on the right are simply a convenient notation for the difference between the value of the function (in this case $F(x)$) evaluated at each of the two stated values (in this case $x = b$ and $x = a$).

Example 1.24 Find $\int_{-1}^{1} x^2\,dx$ (the definite integral of x^2 between -1 and 1) and also find $\int_1^{-1} x^2\,dx$.

$$\int_{-1}^{1} x^2\,dx = \left[\frac{x^3}{3}\right]_{-1}^{1} = \frac{1}{3} - \left(-\frac{1}{3}\right) = \frac{2}{3}.$$

And similarly

$$\int_1^{-1} x^2\,dx = \left[\frac{x^3}{3}\right]_1^{-1} = -\frac{1}{3} - \left(\frac{1}{3}\right) = -\frac{2}{3}.$$

In general

$$\int_a^b f(x)\,dx = -\int_b^a f(x)\,dx.$$

Exercise 1.25 Find

(i) $\int_1^2 x^3\,dx$

(ii) $\int_0^{\pi} \sin^2(x)\,dx.$ (Hint: $\sin^2(x) = \frac{1}{2}[1 - \cos(2x)]$.)

It is useful to memorize how to find this last integral.

1.15. AREA UNDER A GRAPH

The area "under" the graph of a function (i.e. the area between the graph and the x-axis, with areas below the axis counted as negative) is given by the definite integral of the function.

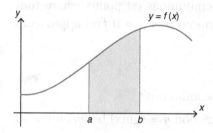

FIGURE 1.4: The definite integral of a function between limits a and b gives the area "under" the graph of the function.

Example 1.25 Find the area under the graph of $\sin(x)$

    (i) between 0 and π

    (ii) between 0 and 2π

First

$$\int_0^\pi \sin(x)\,dx = \left[-\cos(x)\right]_0^\pi = -(-1)-(-1) = 2.$$

And second, using different limits of integration,

$$\int_0^{2\pi} \sin(x)\,dx = \left[-\cos(x)\right]_0^{2\pi} = -1-(-1) = 0.$$

Note that the area below the x-axis is counted as negative.

Exercise 1.26 What is the area under the graph of

  (i) $y = x^2$ between -1 and 1

  (ii) $y = \sin^2(x)$ between 0 and π?

Exercise 1.27 Using the result of Example 1.20, find the area under the graph of $y = x\sin(x)$ between 0 and π.

1.16. CONTINUOUS AND DISCONTINUOUS FUNCTIONS

A *continuous function* is one which has a graph with no breaks. For example, all elementary functions (polynomials, logarithms, trigonometric functions etc.) are continuous (at points where they are finite). Formally, $f(x)$ is continuous at $x = a$ if $f(x)$ approaches $f(a)$ as x approaches a.

Exercise 1.28

   (i)   At what point is sgn(x) discontinuous?

   (ii) Draw the graph of the function $y = $ sgn(x) between $x = -1$ and $x = 2$.

   (iii) Find the area under the graph (as defined in Section 1.15).

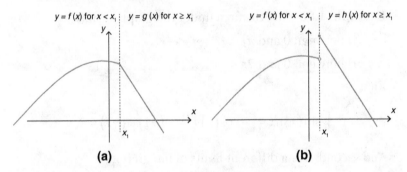

FIGURE 1.5: (a) An example of a function defined piecewise that is continuous. In the region $x < x_1$, the function is $y = f(x)$, and in the region $x \geq x_1$ the function is $y = g(x)$. The two functions are equal at x_1, and continuous across x_1. But notice the derivative is not continuous; there is a sudden change in the gradient at x_1. (b) Now with a discontinuity at the point x_1. The graph is no longer continuous and the derivative is not defined at x_1.

The function sgn(x) (Section 1.6) is an example of a discontinuous function. To integrate a discontinuous function you have to split up the range into a union of intervals in which the function is continuous. This works because of a property of integrals

$$\int_a^c f(x)\,dx = \int_a^b f(x)\,dx + \int_b^c f(x)\,dx$$

where b is any number. We can break up the range of integration into subranges that might be easier to work with.

Example 1.26 Consider the following piecewise function.

$$f(t) = \begin{cases} C - t & \text{if } t \le 3 \\ \exp(-t/5) & \text{if } t > 3 \end{cases}$$

Find the value of C that makes the function continuous over the range $t = 0$ to $t = 10$, and then compute the definite integral of $f(t)$ over this range. See Figure 1.6.

The change in the definition of the function occurs at $t = 3$, so we examine this point. Inserting this value into each formula we get $\exp(-3/5)$ and $C - 3$. If the function is continuous over this point, these two must be equal. Therefore

$$\exp(-3/5) = C - 3 \Rightarrow C = \exp(-3/5) + 3.$$

In order to integrate the above function we split the range of integration $(t = 0 - 10)$ into two parts, and integrate each definition of function within its own range:

$$\int_0^{10} f(t)\,dt = \int_0^3 (C - t)\,dt + \int_3^{10} \exp(-t/5)\,dt.$$

We know how to integrate each piece.

$$\int_0^3 (C - t)\,dt = \left[Ct - t^2/2 \right]_0^3 = 3\exp(-3/5) + 9/2 = 6.1464...$$

$$\int_3^{10} \exp(-t/5)\,dt = \left[-5\exp(-t/5) \right]_3^{10} = 5\left[\exp(-3/5) - \exp(-10/5) \right]$$

$$= 2.0674...$$

The final result is the sum of these,

$$\int_0^{10} f(t)\,dt = (6.1464...) + (2.0674...) = 8.214...$$

Exercise 1.29 The function $f(x)$ is defined by

$$f(x) = \begin{cases} Ax + 1 & \text{for } x \le 1 \\ 2x & \text{for } x > 1. \end{cases}$$

(i) Show that the function is continuous at $x = 1$ if $A = 1$

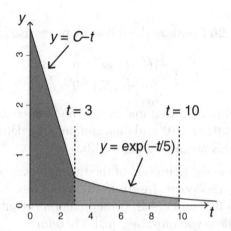

FIGURE 1.6: The piecewise function $y = f(t)$ of example 1.26. The range of integration is split into two subranges $t = 0 - 3$ and $t = 3 - 10$ and the area under the graph over $t = 0 - 10$ is the sum of these two areas.

(ii) If $A = 1$

    (a) sketch the function $f(x)$ between $x = 0$ and $x = 2$

    (b) calculate the area under the graph between $x = 0$ and $x = 2$

Note that if $A = 1$ in the above exercise then $f(x)$ is continuous at $x = 1$, but its derivative is discontinuous at this point.

1.17. ESTIMATES OF INTEGRALS

Sometimes we cannot evaluate a *definite* integral explicitly but we can estimate it approximately. This is achieved by *integrating an inequality*.

Example 1.27 Show that

$$\int_0^{2\pi} x^p \sin^2(x)\, dx \le \frac{(2\pi)^{p+1}}{p+1} \quad \text{for} \quad p > -1.$$

The difficult part is usually to find a suitable inequality that can be integrated. We do this by trial and error in order to come up with an approximate integrand that can be integrated and that gives a useful bound. In the present example we could note that $\sin^2(x) \leq 1$ and, over the range 0 to 2π, $x^p \leq (2\pi)^p$ so

$$x^p \sin^2(x) \leq (2\pi)^p.$$

Integrating both sides gives

$$\int_0^2 x^p \sin^2(x) \leq (2\pi)^{p+1}.$$

A better approximation is obtained if we get rid of just the troublesome $\sin^2(x)$ from the integrand using the fact that $\sin^2(x) \leq 1$. We then have that

$$x^p \sin^2(x) \leq x^p$$

for $0 \leq x \leq 2\pi$ since $\sin(x) \leq 1$. Then, integrating both sides,

$$\int_0^{2\pi} x^p \sin^2(x)\,dx \leq \int_0^{2\pi} x^p dx$$

$$\leq \left[\frac{x^{p+1}}{p+1} \right]_0^2$$

$$\leq \frac{(2\pi)^{p+1}}{p+1}.$$

Note that the answer is not unique – the usefulness of the inequality you obtain will depend on how you choose the inequality to integrate.

Exercise 1.30 From $\sin(x) \leq x$ for $0 \leq x \leq 1$ show that

$$\int_0^1 (\sin(x))^p\,dx \leq \frac{1}{p+1}, \quad \text{for} \quad p \geq 0.$$

1.18. DERIVATIVES OF INTEGRALS

Let

$$F(x) = \int_a^x f(t)\,dt$$

where a is a constant. Then, since integration is the inverse of differentiation, we have

$$\frac{dF}{dx} = f(x).$$

We can check this as follows. First note that $F(a) = \int_a^a f(t)dt = 0$.
Then, changing the integration variable from x to t for clarity,

$$\int_a^x \frac{dF}{dt} dt = F(x) - F(a) = F(x) = \int_a^x f(t)dt.$$

Alternatively, it is sometimes useful to go back to the definition of the derivative:

$$F(x+h) - F(x) = \int_x^{x+h} f(t)dt \approx hf(x).$$

Dividing through by h and taking the limit as $h \to 0$ gives the result.

Example 1.28 Let $f(t) = t^2$; then

$$F(x) = \int_0^x f(t)dt = \int_0^x t^2\, dt = \left[\frac{t^3}{3}\right]_0^x = \frac{x^3}{3},$$

so

$$\frac{dF}{dx} = \frac{1}{3}\frac{d}{dx}\{x^3\} = x^2 = f(x).$$

It should be obvious that in practice there is no need to carry out the integration explicitly to obtain the answer. We've just done it to show how it would work. Now suppose

$$F(x) = \int_a^{g(x)} f(t)dt$$

where $g(x)$ is some given function of x. Then,

$$\frac{dF}{dx} = f(g(x))\frac{dg}{dx}. \tag{1.15}$$

The proof is as follows: F is a function of $g(x)$, i.e. a function of a function, so

$$\frac{dF}{dx} = \frac{dF}{dg}\frac{dg}{dx} = f\left(g(x)\right)\frac{dg}{dx}. \qquad (1.16)$$

Example 1.29 Calculate dF/dx if $F(x) = \int_0^{x^2} e^{-t}\,dt$

Note that $f(g(x))$ is obtained by replacing t in $f(t)$ by $g(x) = x^2$. We have $g(x) = x^2$, and $f(t) = e^{-t}$. So

$$dg/dx = 2x,$$

$$f(g(x)) = e^{-x^2}.$$

Then, using equation (1.15),

$$\frac{dF}{dx} = e^{-x^2} \cdot 2x.$$

Exercise 1.31

(i) If $F(x) = \int_1^{e^{2x}} \left(\ln(t)\right)^{1/2} dt$ what is dF/dx?

(ii) If $G(x) = \int_{e^{2x}}^{1} \left(\ln(t)\right)^{1/2} dt$ what is dG/dx?

(Do not attempt to compute the integrals explicitly! For (ii) begin by finding the relation between G and F.)

1.19. REDUCTION FORMULAE

Sometimes an integration cannot be performed directly but can be related to a simpler integral which can be evaluated explicitly. Reducing the problem in this way leads to a *reduction formula*.

Example 1.30 Find $\int_0^{\pi/2} \cos^3(\theta)d\theta$.

Reduction formulae are usually obtained by integration by parts. The key is to write the integrand as a suitable product. Here we write (with the benefit of hindsight) $\cos^3(\theta) = \cos^2(\theta)$ $\cos(\theta)$, then

$$I = \int_0^{\pi/2} \cos^3(\theta)d\theta = \int_0^{\pi/2} \cos^2(\theta)\cos(\theta)d\theta.$$

Next, integrate by parts with $u = \cos^2(\theta)$ and $v' = \cos(\theta)$, to give

$$= \left[\cos^2(\theta)\sin(\theta)\right]_0^{\pi/2} + \int_0^{\pi/2}(2\cos(\theta)\sin(\theta))\sin(\theta)d\theta.$$

As $\cos(\pi/2) = 0$ and $\sin(0) = 0$ we have

$$= 0 + \int_0^{\pi/2} 2\cos(\theta)\sin^2(\theta)d\theta.$$

Then, removing the sine terms using a standard identity,

$$= \int_0^{\pi/2} 2\cos(\theta)\left(1 - \cos^2(\theta)\right)d\theta$$

$$= 2\int_0^{\pi/2} \cos(\theta)d\theta - 2\int_0^{\pi/2} \cos^3(\theta)d\theta$$

Notice that the last term above is twice the integral we started with, I. So,

$$I = 2\int_0^{\pi/2} \cos(\theta)d\theta - 2I.$$

Finally we can collect all the I terms on the left to give a *reduction formula* expressing the integral of $\cos^3(\theta)$ in terms of the integral of a reduced power (here $\cos(\theta)$) which we can evaluate,

$$3I = 2\int_0^{\pi/2} \cos(\theta)d\theta = 2, \Rightarrow \int_0^{\pi/2} \cos^3(\theta)d\theta = \frac{2}{3}.$$

The *reduction formula* (sometimes also called a recurrence relation) expresses the required integral in terms of a similar integral, but with a reduced value of some parameter. Reduction formulae are very often obtained by integration by parts. You may have to experiment a bit to get the integrand in a suitable form. Reduction formulae of the form found in Example 1.31 occur in many other contexts as a way of specifying an infinite number of terms.

Exercise 1.32 Use a similar method to find a reduction formula for $\int_0^{\pi/2} \cos^5(x)$ and hence evaluate the integral.

Example 1.31 Let

$$C_n = \int_0^{\pi/2} \cos^{2n-1}(\theta)\, d\theta, \quad \text{for} \quad n = 1,2,3,\dots$$

Show that for $n > 1$,

$$C_n = \frac{(2n-2)}{2n-1} C_{n-1}.$$

We begin by splitting the integrand into a product $\cos^{2n-2}(\theta)\cos(\theta)$ so we can use integration by parts. Then we write

$$u = \cos^{2n-2}(\theta), \quad \text{and} \quad v' = \cos(\theta)$$

Then

$$u' = (2n-2)\cos^{2n-3}(\theta) \times (-\sin(\theta)), \quad \text{and} \quad v = \sin(\theta),$$

$$C_n = \left[\cos^{2n-2}(\theta)\sin(\theta)\right]_0^{\pi/2} + \int_0^{\pi/2}(2n-2)\cos^{2n-3}(\theta)\sin(\theta)\sin(\theta)\, d\theta.$$

Since C_n involves only $\cos(\theta)$ we next remove the sine terms using $\sin^2(\theta) = 1 - \cos^2(\theta)$. Then,

$$C_n = 0 + \int_0^{\pi/2}(2n-2)\cos^{2n-3}(\theta)\left(1-\cos^2(\theta)\right) d\theta,$$

$$= (2n-2)\left[\int_0^{\pi/2}\cos^{2n-3}(\theta)\, d\theta - \int_0^{\pi/2}\cos^{2n-1}(\theta)\, d\theta\right].$$

The integrals can now all be written in terms of C_n and C_{n-1}, where C_{n-1} is obtained from C_n by replacing n by $n-1$. We have

$$Cn = (2n-2)Cn - 1 - (2n-2)Cn.$$

Finally, collecting all the C_n terms on the left, we obtain the reduction formula or *recurrence relation* for C_n,

$$C_n = \frac{(2n-2)}{(2n-1)}C_{n-1}.$$

Example 1.32 Following from Example 1.31 show that

$$C_n = \frac{(2n-2)(2n-4)\dots(2)}{(2n-1)(2n-3)\dots(3)}.$$

We begin with

$$C_n = \frac{(2n-2)}{(2n-1)}C_{n-1}.$$

Then, replacing n with $n-1$,

$$C_{n-1} = \frac{(2n-4)}{(2n-3)}C_{n-2},$$

and comparing these two equations we see that

$$C_n = \frac{(2n-2)(2n-4)}{(2n-1)(2n-3)}C_{n-2}.$$

If we continue until we get C_1 on the right hand side, we find that

$$C_n = \frac{(2n-2)(2n-4)\dots(2)}{(2n-1)(2n-3)\dots(3)}C_1.$$

At this point we can stop because we can integrate C_1 explicitly and also the reduction formula no longer applies. We have

$$C_1 = \int_0^{\pi/2} \cos(\theta)\,d\theta = 1$$

and so, as required,

$$C_n = \frac{(2n-2)(2n-4)\dots(2)}{(2n-1)(2n-3)\dots(3)}.$$

The result of example 1.32 is an explicit expression for the integral of an odd power of $\cos(\theta)$ which is sometimes useful. You should remember that such a formula exists! The formula for the integral of an even power in Exercise 1.34 is obtained in exactly the same way. Expressions for the integrals of powers of $\sin(\theta)$ can also be found.

Exercise 1.33 Find a reduction formula for $\int_0^{\pi/2} \cos^7(x)\,dx$ and hence evaluate the integral.

Exercise 1.34 Let

$$D_n = \int_0^{\pi/2} \cos^{2n}(\theta)\,d\theta, \quad \text{for} \quad n = 1,2,\dots$$

Show that for $n > 1$,

$$D_n = \frac{2n-1}{2n} D_{n-1}$$

and hence that

$$D_n = \frac{(2n-1)(2n-3)\dots(1)}{(2n)(2n-2)\dots(2)}\frac{\pi}{4}.$$

Exercise 1.35 The gamma function is defined for positive integer arguments $(n \geq 1)$ by

$$\Gamma(n) = (n-1)\,\Gamma(n-1),$$

with $\Gamma(1) = 1$. Show that $\Gamma(n) = (n-1)(n-2)(n-3)\dots 1$, and hence that $\Gamma(6) = 120$.

Revision Notes

After completing this chapter you should be able to

- Know the derivatives of elementary functions including powers as well as log and trigonometric functions
- Differentiate any combination of elementary functions using the *chain rule, product rule* or by *implicit differentiation*

- Find and identify the *stationary points* of a given function

- Recall the integrals of simple functions including powers, sin and cos

- Know the following methods of integration: *substitution,* use of *trigonometric identities, partial fractions, integration by parts, reduction formulae*

- Calculate the definite integral of a discontinuous function

- Estimate the value of an integral by integrating an inequality

- Differentiate an integral with respect to a limit

- Be able to obtain *reduction formulae* by integration by parts

Further Questions

The sets of additional questions ending chapters divide into two groups. Although they are not necessarily straightforward, the first set comprises exercises that usually illustrate single concepts or techniques from the text. The second set contains more difficult problems and requires a degree of insight usually drawing on or illustrating a range of ideas, possibly using material from earlier chapters.

1.20. EXERCISES

1. A continuous graph $y = f(x)$ consists of the infinite straight lines

$$y = \begin{cases} 3x - 2 & \text{for} \quad x \geq 1 \\ -3x + 4 & \text{for} \quad x \leq 1. \end{cases}$$

Sketch the graph and find a formula, involving modulus signs, for $f(x)$.

2. Solve the equation $2x - 4 = |x - 3|$.

3. Find the derivatives with respect to x of x^x and $x^{\left(x^x\right)}$.

4. Simplify $\dfrac{x^5 + x^3 + 5x^2 - 6x - 10}{x^2 - 2}$

5. Verify that the derivative of the function

$$g(t) = \tan^{-1} t - \frac{3t}{3 + t^2}$$

is equal to $4t^4/(1 + t^2)(3 + t^2)^2$, and deduce that for $t > 0$

$$(3 + t^2)\tan^{-1} t > 3t.$$

6. Show that $f(x) = 3x^3 - 3x^2 + x$ is an increasing function of x and explain why the equation $f(x) = 1/2$ must have a unique solution between $x = 0.8$ and $x = 0.9$.

7. Verify that $\dfrac{d}{dx}\left\{\tan^{-1}\left(\dfrac{\sin x}{2 - \cos x}\right)\right\} = \dfrac{2\cos x - 1}{5 - 4\cos x}$.

8. Show that the function $f(x) = x^3 - 6x^2 + 12x - 6$ is increasing for all real x. For what values of x is its inverse function g (such that $g(f(x)) = x$) defined?

9. Find the stationary points of $y = 3x^4 - 4x^3 - 72x^2 + 1$ and determine whether they are maxima, minima, or points of inflection.

10. Verify that if the approximate integration formula

$$\int_{-1}^{1} f(x)\,dx \approx \frac{1}{\mu}\left[f(-1) + \lambda f(0) + f(1)\right]$$

gives the correct answer when f is a constant then $\mu = 1 + \lambda/2$, and that if it also gives the correct answer when $f(x) = x^2$ then $\lambda = 4$ and $\mu = 3$.

11. Find the function the derivative of which is $2/(1 - 6x + 11x^2 - 6x^3)$ and which vanishes when $x = 2/3$.

12. Find $\int 2\left(1 + x^2\right)^4 x\,dx$.

13. With the help of suitable changes of variables, find

(i) $\int x^5 \left(2 - 5x^3\right)^{2/3} dx$

(ii) $\int x^2 (1-x)^{1/2}\, dx$

(iii) $\int \dfrac{\sin x \cos^3 x}{1+\cos^2 x}\, dx$

(iv) $\int \dfrac{\tan^{-1} x}{1+x^2}\, dx$

(v) $\int \dfrac{x^{11}}{x^8 + 3x^4 + 2}\, dx$

14. Find $\int \dfrac{dx}{x^{2/3} - x^{4/3}}$.

15. Indicating the method used, obtain the values $A = 1$, $B = 2$ for the constants in the partial fraction identity

$$f(x) = \frac{A}{6-x} + \frac{B}{7-x}, \quad \text{where } f(x) = \frac{5-x}{(6-x)(7-x)}.$$

At which points P, Q does the curve $y = f(x)$ meet the axes Ox, Oy? Show that the area of the region bounded by OP, OQ and the arc PQ of the curve is $\ln(49/24)$.

16. In the partial fraction expression

$$\frac{2t}{(1+t)(1+3t)} = \frac{A}{1+t} + \frac{B}{1+3t},$$

you are given that $A = 1$; find the value of B. Hence show that

$$\int_0^1 \frac{2t^2}{(1+t^2)(1+3t^2)}\, dt = \frac{\pi}{4} - \frac{\pi}{3\sqrt{3}}.$$

17. Use integration by parts to show that

$$\int u e^u\, du = u e^u - e^u + C,$$

(C an arbitrary constant) and with the help of a suitable change of variable find $\int_0^0 e^{\sin(t)} \sin(2t)\, dt$.

18. With the help of an integration by parts, show that

$$\int_0^\pi x \cos(x)\, dx = -2.$$

19. Use integration by parts to show that

$$\int 2s\ln(s)\,ds = \left(\ln(s)\right)s^2 - \frac{1}{2}s^2 + C$$

(C a constant) and with the help of a suitable change of variable deduce that

$$\int_{\pi/4}^{\pi/2} \sin(2t)\ln(\sin(t))\,dt = \frac{1}{4}\ln\left(\frac{2}{e}\right).$$

20. Show that $\displaystyle\int_{\pi/2}^{\pi} \ln\left(\frac{y}{\pi}\right)d\big(y\cos(y)\big) = 1.$

21. It is known that

$$\frac{4x}{(x+1)(x-1)^2} = \frac{A}{x+1} + \frac{B}{(x-1)^2} + \frac{C}{(x-1)}$$

where A, B and C are constants. By multiplying both sides by $x + 1$ and then putting $x = -1$ in the resulting equation show that $A = -1$. Similarly, show that $B = 2$. Finally, by putting $x = 0$ in the original equation, or otherwise, obtain the value of C.

22. Find the range of values of x for which $\dfrac{1}{1+x^2} \leq 1 - \dfrac{1}{2}x^2$ and by integrating the inequality over this range deduce that

$$\pi \leq \frac{10}{3}.$$

23. A continuous function $f(x)$ is defined by

$$f(x) = \begin{cases} (x-1)^2 & \text{for } x \leq 1 \\ -x^3 + 4x^2 - 5x + C & \text{for } x > 1. \end{cases}$$

Find the value of the constant C and show that $f(x)$ is differentiable at $x = 1$.

24. Given that the sequence a_n is generated by the recurrence relation

$$a_n = \frac{(n+1)(n-3)}{(n-1)!}a_{n-1}$$

with $a_0 = 1$, find a_1, a_2, a_3, a_4 and a_{2001}. Note that $0! = 1$ – see the discussion of factorials in Section 2.1.

1.21. PROBLEMS

1. A continuous graph $y = f(x)$ consists of the straight line
    segment joining the points $(1, 8)$ and $(2, 1)$ together with
    infinite straight lines of respective slopes $+5$ and -5 over
    the intervals $(-\infty, 1)$ and $(2, +\infty)$. Find a formula for $f(x)$
    of the form $f(x) = \pm|ax - b| \pm| cx - d |+h$, where $a, b, c, d,$ and
    h are constants.

2. Divide the polynomial $x^3 - 7x^2 + 36$ by $x - 6$.
    Let $f(x) = (7x^2 - 36)^{1/3}$. Determine the ranges of values of x
    for which the inequality $f(x) > x$ is true, and state the values
    of x for which $f(x) > 6$.
    Sketch the graph of $y = f(x)$ for $x > 0$, indicating in particu-
    lar the points at which the graph crosses the x-axis and the
    lines $y = x$ and $y = 6$.
    Show, on the basis of this graph, that if a sequence of num-
    bers $x_1, x_2, x_3,...$ is defined (from a given x_1) by $x_{n+1} = f(x_n)$ for
    $n = 1, 2,...$ then for $x_1 > 6$ the sequence (x_n) is decreasing,
    with limit 6, while on the other hand if $0 < x_1 < 3$ then x_2 is
    further away from 3 than x_1. What happens if $3 < x_1 < 6$?

3. For $y = xe^x$ calculate y', y'' and y'''. Hence write down a
    formula for the nth derivative $y^{(n)}$.
    Deduce the value of A in the formula for the nth derivative
    of a product

    $$(uv)^{(n)} = uv^{(n)} + au'v^{(n-1)} + ...$$

    If $z = x \sin(x)$ what is the value of d^8z/dx^8 at $x = 0$?

4. Let $I_k = \int_0^\infty (1 + 2x)^{2k+1} e^{-x-x^2} dx$. Find the value of I_0, obtain a
    reduction formula for I_k and deduce that $I_3 = 493$.

5. Evaluate $\int_0^1 \dfrac{dx}{1 + x + x^2}$. By expanding the integrand, written
    in the form $(1 - x)(1 - x^3)^{-1}$, in a series of powers of x, and
    integrating term by term, show that

    $$1 - \frac{1}{2} + \frac{1}{4} - \frac{1}{5} + \frac{1}{7} - \frac{1}{8} + ... = \frac{\pi}{3\sqrt{3}}.$$

6. The potential energy of a system is given by

$$V(x) = x^4 - 14x^2 + 24x.$$

Find and classify the equilibrium points, i.e. stationary points of $V(x)$, and sketch the general shape of the graph; show that the minimum at $x = 2$ is not global (i.e. that $V(x)$ takes values less than 8). Find an approximate expression for $V(x)$ near the point $x = 2$, as a quadratic in $x - 2$.

7. A function $f(t)$ is defined by the formula

$$f(t) = \frac{t^2}{(1+t)(2+t)^2}.$$

(a) Obtain an expression for $f(t)$ in the form

$$f(t) = \frac{A}{1+t} + \frac{B}{(2+t)^2}$$

where A, B are real constants, showing the method used.

(b) Deduce that for $x > -1$

$$\ln(1+x) - \frac{x}{1+x/2} = \int_0^x f(t)\,dt.$$

(c) The left side is therefore positive for $x > 0$. Deduce that $\ln(7/5) > 1/3$, and hence that $e < 2.744$.

(d) By any method, and quoting any standard series you may use, find the constant term and coefficients of x, x^2, x^3 in the expansions in ascending powers of x for

(i) $f(x)$, (ii) $g(x) = \ln(1 + x) - x/(1 + x/2)$, and state for what values of x these expansions are valid.

8. Let $f(x)$ be a differentiable function of x increasing on the interval $a < x < b$. What condition is satisfied by the derivative $f'(x)$? If $f(x)$ is an increasing function show that $1/f(x)$ is decreasing, except at points where $f(x) = 0$. Show that

$f(x) = \dfrac{x^2 - 1}{x}$ is increasing for $x \neq 0$ and deduce that $\dfrac{x}{x^2 - 1}$

is decreasing except at $x = \pm 1$. Sketch a graph of $\dfrac{x}{x^2 - 1}$.

Deduce the range (or ranges) of x for which $\dfrac{x}{x^2 - 1} > 3x$.

9. Let $J_m = \displaystyle\int_0^{\pi/2} \cos^m(\theta)\, d\theta$. Calculate the value of J_0

and J_1. Writing $J_m = \displaystyle\int_0^{\pi/2} \cos^{m-1}(\theta)\, d\sin(\theta)$, show that

$J_m = \dfrac{m-1}{m} J_{m-2}$ where m is a positive integer. Deduce that

$$J_{2n} = \frac{2n-1}{2n}\frac{2n-3}{2n-2}\cdots\frac{1}{2}J_0 \quad \text{and} \quad J_{2n+1} = \frac{2n}{2n+1}\frac{2n-2}{2n-1}\cdots\frac{2}{3}J_1$$

where n is a positive integer, and hence that

$$\frac{\pi}{2} = \left[\frac{2.4.6...2n}{3.5...(2n-1)}\right]^2 \frac{1}{2n+1}\frac{J_{2n}}{J_{2n+1}}.$$

Show that $J_{2n-1} \geq J_{2n} \geq J_{2n+1}$ for any n, and use the relation

$$\frac{J_{2n-1}}{J_{2n+1}} = \frac{2n+1}{2n}$$

to deduce that

$$\lim_{n\to\infty}\frac{J_{2n-1}}{J_{2n+1}} = \lim_{n\to\infty}\frac{J_{2n}}{J_{2n+1}} = 1.$$

Hence show that

$$\frac{\pi}{2} = \lim_{n\to\infty}\left(\frac{2}{1}\cdot\frac{2}{3}\cdot\frac{4}{3}\cdot\frac{4}{5}\cdot\frac{6}{5}\cdots\frac{2n-2}{2n-1}\cdot\frac{2n}{2n-1}\cdot\frac{2n}{2n+1}\right).$$

10. Let

$$I_{n,m} = \int_0^1 x^n (1-x)^m\, dx$$

where n and m are positive integers. Show that

$$I_{n,m} = \frac{n}{m+1} I_{n-1,m+1}$$

and hence that

$$I_{n,m} = \frac{n!m!}{(n+m+1)!}.$$

Verify this formula by direct integration in the case $n = 1$, $m = 2$.

ELEMENTARY FUNCTIONS

In the first part of this chapter we look at the representation of an arbitrary function $f(x)$ as a power series, $f(x) \approx a_0 + a_1x + a_2x^2 + \ldots$ where $a_0, a_1, a_2 \ldots$ are constants. We call this representation the "expansion" of the function (or more precisely, the power series expansion of the function about the point $x = 0$). The expansion is intended to be exact if we include the infinite number of terms of the whole series or an approximation if we cut off the series after a finite number of terms. Power series expansions are very useful if we want to know the form of $f(x)$ for small x, and this is often the case in physical problems where systems operate close to equilibrium. We shall find that power series expansions are particularly useful near to a stationary point or where the solution of a problem depends on a small parameter. Later on (in Fourier series) we shall meet other types of expansion which will be useful for other purposes.

In the second part of the chapter we introduce the exponential and related functions and their power series expansions. Together with polynomials, the trigonometric functions and their inverses and combinations, these make up what are called the *elementary functions*.

2.1. BINOMIAL EXPANSION

You are probably familiar with the fact that a function like $(1 + x)^n$, where n is a positive integer, can be expanded in powers of x. For example, for $n = 1, 2, 3$,

$$(1 + x)^1 = 1 + x,$$
$$(1 + x)^2 = (1 + x)(1 + x) = 1 + 2x + x^2,$$
$$(1 + x)^3 = (1 + x)(1 + 2x + x^2) = 1 + 3x + 3x^2 + x^3.$$

We can see there is a pattern in the terms: there are $n + 1$ terms in the expansion, and there is a symmetry to the coefficients in front of each term (e.g. 1, 3, 3, 1 for $n = 3$).

In fact, the general expression can be written as a sum of $n + 1$ terms called the *binomial expansion*,

$$(1+x)^n = 1 + nx + \frac{n(n-1)}{2 \cdot 1}x^2 + \frac{n(n-1)(n-2)}{3 \cdot 2 \cdot 1}x^3 + \ldots + x^n \qquad (2.1)$$

Example 2.1 Expand $(1 + x)^5$.

We apply the definition of equation (2.1),

$$(1+x)^5 = 1 + 5x + \frac{5 \cdot 4}{2 \cdot 1}x^2 + \frac{5 \cdot 4 \cdot 3}{3 \cdot 2 \cdot 1}x^2 + \frac{5 \cdot 4 \cdot 3 \cdot 2}{4 \cdot 3 \cdot 2 \cdot 1}x^4 + \frac{5 \cdot 4 \cdot 3 \cdot 2 \cdot 1}{5 \cdot 4 \cdot 3 \cdot 2 \cdot 1}x^5,$$
$$= 1 + 5x + 10x^2 + 10x^3 + 5x^4 + x^5.$$

Exercise 2.1

(i) Expand $(1 + x)^7$

(ii) Write out the three lowest order terms in the expansion of $(1 + x)^{12}$

Example 2.2 Expand $(2 + x)^5$.

To use the definition from equation (2.1) we need to put this in the form $(1 + \ldots)^n$. We therefore extract the factor 2 to give

$(2 + x)^5 = 2^5(1 + x/2)^5.$

We can now use the binomial expansion to write

$(2 + x)^5 = 2^5 (1 + 5(x/2) + 10(x/2)^2 + 10(x/2)^3 + 5(x/2)^4 + (x/2)^5)$

$= 32 + 80x + 80x^2 + 40x^3 + 10x^4 + x^5.$

Exercise 2.2 Expand

(i) $(3 + 2x)^4$

(ii) $(1 + x^2)^3$

(iii) $(a + x)^n$

Factorials

For n a positive integer, we define $n!$ (pronounced "n factorial") as the product

$$n! = n \times (n - 1) \times (n - 2) \times \cdots \times 2 \times 1. \qquad (2.2)$$

For $n = 0$ we define $0! = 1$. This definition has no deep meaning; $0!$ is simply defined this way for convenience.

Example 2.3 Find $5!$.

$5! = 5 \times 4 \times 3 \times 2 \times 1 = 120.$

Factorials are common in mathematical physics; for example, they are heavily used in statistical mechanics. Returning to equation (2.1), we can rewrite the denominator of each coefficient: $r \cdot (r - 1) \cdots 3 \cdot 2 \cdot 1 = r!$. Can we find a more compact way to write the numerator?

Example 2.4 Verify that

$$n(n-1)\ldots(n-r+1) = \frac{n!}{(n-r)!}$$

From equation (2.2) we have

$$n! = n(n-1)\cdots 3\cdot 2\cdot 1,$$

and

$$(n-r)! = (n-r)(n-r-1)\cdots 3\cdot 2\cdot 1.$$

Taking the ratio of these two we have

$$\frac{n!}{(n-r)!} = \frac{n(n-1)\ldots 3\cdot 2\cdot 1}{(n-r)(n-r-1)\ldots 3\cdot 2\cdot 1}.$$

We can now cancel terms from the right end of the numerator and denominator. The terms in the denominator cancel with the terms $(n-r)\cdot(n-r-1)\cdots 3\cdot 2\cdot 1$ in the numerator, leaving $(n-r+1)$ and larger terms on the top,

$$\frac{n!}{(n-r)!} = n(n-1)\ldots(n-r+1).$$

We now use factorials to write the binomial expansion (Equation 2.1) in a more compact way,

$$(1+x)^n = \sum_{r=0}^{n} \frac{n!}{r!(n-r)!} x^r. \tag{2.3}$$

Here the "big sigma" symbol Σ indicates that we sum over all terms with an index r running from 0 to n (i.e. we put $r = 0, r = 1, \ldots, r = n$ and add the results).

Exercise 2.3 Using the definition from equation (2.3) verify that $(1 + x)^3 = 1 + 3x + 3x^2 + x^3$.

Exercise 2.4 Evaluate $\dfrac{5!}{r!(5-r)!}$ for $r = 0, 1, 3, 4, 5$. Hence write out equation (2.3) for $n = 5$ and compare with Example 2.1.

The coefficients in front of the x^r terms, i.e. $n!/r!(n-r)!$, are the *binomial coefficients*. Sometimes $n!/r!(n-r)!$ is considered as a function of n and r and is called "binomial n, r" or "n choose r." There are also other notations that are commonly used for this expression; for example, $\binom{n}{r}$ or ^nC_r mean the same thing. So you might see the binomial series written as

$$(1+x)^n = \sum_{r=0}^{n} {}^nC_r x^r, \quad \text{or} \quad (1+x)^n = \sum_{r=0}^{n} \binom{n}{r} x^r.$$

Binomial Series

What happens if n is not a positive integer? We first consider the case $n = -1$. Writing out the binomial series, equation (2.1), with $n = -1$ gives the infinite series:

$$(1+x)^{-1} = 1 - x + \frac{(-1)(-2)}{2}x^2 + \frac{(-1)(-2)(-3)}{3\cdot2}x^3 + \ldots \quad (2.4)$$

$$= 1 - x + x^2 - x^3 + \ldots \quad (2.5)$$

This is correct provided $|x| < 1$ because, then, the right hand side is just a geometric series (section A.7) with ratio $(-x)$. If $|x| < 1$, then the left hand side is the correct expression for the sum of an infinite geometric series. If $|x| > 1$ the RHS is the sum (or difference) of ever larger terms which grows arbitrarily large in magnitude and so cannot equal the LHS, which is a finite number.

In general, the binomial expansion of $(1 + x)^n$ is valid when n is not a positive integer, and $|x| < 1$, but will be an infinite series of terms, not the finite series of equation (2.1).

$$(1+x)^n = 1 + nx + \frac{n(n-1)}{2\cdot1}x^2 + \frac{n(n-1)(n-2)}{3\cdot2\cdot1}x^3 + \ldots \quad (2.6)$$

Example 2.5 Write down the first three terms of the binomial expansion of $(1 + 2x)^{1/2}$ and state the range of x for which the expansion is valid.

Using equation (2.6) we have

$$(1+2x)^{1/2} = 1 + \frac{1}{2}(2x) + \frac{\frac{1}{2}\left(\frac{1}{2}-1\right)}{2!}(2x)^2 + \ldots \quad (2.7)$$

$$= 1 + x - \frac{1}{2}x^2 + \ldots. \tag{2.8}$$

The expansion is valid for $|2x| < 1$, i.e. for $-1/2 < x < 1/2$.

Exercise 2.5 Write down the binomial expansions of the following, stating the range of values of x for which each is valid:

(i) $(1-x)^{-1}$

(ii) $(1+3x)^{-1/2}$

Exercise 2.6 Use your answer from Exercise 2.5 (i) to obtain the expansion

$$\frac{a}{b-x} = \frac{a}{b} + \frac{ax}{b^2} + \frac{ax^2}{b^3} + \ldots,$$

valid for $|x| < |b|$. Use this expansion to show that a radio frequency of 198 (= 200 − 2) kHz corresponds to a wavelength of approximately 1515m. (Wavelength=wavespeed/frequency, and the speed of radio waves is approximately 3×10^8 m s^{-1}.)

Exercise 2.7 By considering the binomial series expansion of $(1 + \delta)^{1/3}$ and then putting $\delta = x - 1$, show that

$$x^{1/3} \approx \frac{5}{9} + \frac{5}{9}x - \frac{1}{9}x^2$$

near $x = 1$. Verify that the value obtained when $x = 1000/729$ differs from the true value of $x^{1/3}$ by less than 0.25%.

Exercise 2.8 If δ is small compared with $|x|$ and 1, find the expansion (in powers of δ) of

$$\frac{(3x+\delta)}{(1+\delta^2)(x-3\delta)}$$

to first order and to second order in δ.

Use of the binomial expansion to approximate rational expressions containing small quantities is an important tool that is commonly used in scientific work.

2.2. MACLAURIN SERIES

The binomial expansion is a special case of a more general result that enables us to represent *any* well-behaved function as a power series.

If $f(x)$ is a well-behaved function, the Maclaurin series of $f(x)$ gives an expansion for $f(x)$ as a power series valid for small x,

$$f(x) = f(0) + xf'(0) + \frac{1}{2!}x^2 f''(0) + \ldots + \frac{1}{p!}x^p f^{(p)}(0) + \ldots . \qquad (2.9)$$

Here, $f'(0)$ stands for $f'(x) = df/dx$ evaluated at $x = 0$; $f''(0)$ stands for $f''(x) = d^2f/dx^2$ evaluated at $x = 0$; $f^{(p)}(0)$ stands for $f^{(p)}(x) = d^p f/dx^p$ evaluated at $x = 0$. The series can be written very compactly as

$$f(x) = \sum_{p=0}^{\infty} \frac{x^p}{p!} f^{(p)}(0). \qquad (2.10)$$

For sufficiently small x the sum of all the terms after the nth is negligibly small compared with the sum of the first n terms. For such x we can approximate a function by the first few terms of its Maclaurin series.

Exercise 2.9 For the function $f(x) = (1 + x)^{1/2}$, evaluate $f(0)$, $f'(0)$, $f''(0)$, $f\left(\dfrac{5}{4}\right), f'\left(\dfrac{5}{4}\right)$ and $f''\left(\dfrac{5}{4}\right)$.

Example 2.6 Find the first three terms of the Maclaurin series of $(1 + 2x)^{1/2}$.

We have

$$f(x) = (1+2x)^{1/2} \Rightarrow f(0) = 1,$$

$$f'(x) = \frac{1}{2}(1+2x)^{-1/2} \cdot 2 \Rightarrow f'(0) = 1,$$

$$f''(x) = -\frac{1}{2}(1+2x)^{-3/2} \cdot 2 \Rightarrow f''(0) = -1.$$

Inserting these into equation (2.9) we get

$$\left(1+2x\right)^{1/2} = 1 + x - \frac{x^2}{2} + \ldots.$$

Notice that this result agrees with equation (2.8) where we expanded $(1 + 2x)^{1/2}$ using the binomial expansion. Any function has only one series expansion about zero. The Maclaurin series is just a general way of finding this.

Example 2.7 Find the Maclaurin series for $\sin(x)$. Write down the general term.

We have
$$f(x) = \sin(x) \Rightarrow f(0) = 0,$$
$$f'(x) = \cos(x) \Rightarrow f'(0) = 1,$$
$$f''(x) = -\sin(x) \Rightarrow f''(0) = 0,$$
$$f'''(x) = -\cos(x) \Rightarrow f'''(0) = -1.$$

Therefore, for the Maclaurin series (Equation 2.9) we get

$$\sin(x) = 0 + x + 0 - \frac{x^3}{3!} + \ldots = x - \frac{x^3}{6} + \ldots. \quad (2.11)$$

The general term is simply the expression for an arbitrary power of x in the expansion. From equation (2.11) we can see that the sequence of terms in the expansion will look like $x - x^3/3! + x^5/5! - x^7/7! + \ldots$

Each successive term in the series changes sign, and only odd powers of x appear in the expansion. If p is an integer that labels each term in the series (see Equation 2.10), then we can make each successive term change sign by including a factor $(-1)^p$, and $2p + 1$ will count all the odd numbers. This means that the general term is $(-1)^p x^{2p+1}/(2p + 1)!$ where p is an integer. Hence

$$\sin(x) = x - \frac{x^3}{6} + \ldots + \frac{(-1)^p x^{2p+1}}{(2p+1)!} + \ldots \quad (2.12)$$

$$= \sum_{p=0}^{\infty} \frac{(-1)^p x^{2p+1}}{(2p+1)!}. \quad (2.13)$$

Exercise 2.10 Show that the next non-zero term of equation (2.11) is $x^5/120$.

When we only require the first few terms of the expansion of a function of a function it is often easier to combine Maclaurin series than to start afresh. In Example 2.8 we find the expansion of a function, $f(x)$, where this is the case. If we were to try to work out the Maclaurin series of this function directly we would, of course, obtain exactly the same answer, but it would take much more work.

Example 2.8 Find the Maclaurin series for $(1 + \sin(x))^{-1/2}$ up to terms of order x^2.

Begin by writing out the inner Maclaurin series for $\sin(x)$ keeping only relevant terms.

$$(1 + \sin(x))^{-1/2} = [1 + (x - x^3/6 + \ldots)]^{-1/2}.$$

Expand the outer series as far as necessary using the binomial form for $(1 + x)^{-1/2}$

$$= 1 - \left(x - x^3/6 + \ldots\right)/2 + \frac{3}{8}\left(x - x^3/6 + \ldots\right)^2 + \ldots$$

$$= 1 - x/2 + 3x^2/8 + \ldots.$$

Exercise 2.11 Obtain the following expansions as Maclaurin series and in (i)–(iii) write down the general term

(i) $(1 + x)^{-1} = 1 - x + x^2 - \ldots,$

(ii) $\cos(x) = 1 - x^2/2 + x^4/24 + \ldots,$

(iii) $\ln(1 + x^2) = x^2 - x^4/2 + x^6/3 + \ldots,$

(iv) $(1 + \ln(\cos(x)))^{-1} = 1 + x^2/2 + \ldots.$

In Exercise 2.11 the series (i) is correct only for $-1 < x < 1$, whereas the cosine series (ii) (and the sine series also) are valid for all values of x (though not very useful for numerical computation if x is large). The series (iii) is valid for $-1 \le x \le 1$. It is beyond the scope of this course to explain how this can be worked out.

2.3. TAYLOR SERIES

Maclaurin series are valid near $x = 0$, and enable us to expand a function as a series for x near to 0. Suppose we want an approximation for a function near some other point. For example, we know that $\sin(\pi/2) = 1$ but what is $\sin(\pi/2 - \delta)$ approximately equal to for small δ? If we want an expansion near a point $x = x_0$ we use the Taylor series

$$f(x) = f(x_0) + (x - x_0)f'(x_0) + \frac{1}{2!}(x - x_0)^2 f''(x_0) + \ldots + \frac{1}{p!}(x - x_0)^p f^{(p)}(x_0) + \ldots$$

(2.14)

Here $f'(x_0)$ stands for $f'(x)$ evaluated at $x = x_0$; $f''(x_0)$ stands for $f''(x)$ evaluated at $x = x_0$ etc. Note that the Taylor series gives us $f(x)$ as a power series in $(x - x_0)$. The Maclaurin series is a special case of the Taylor series with $x_0 = 0$. By putting $x = x_0 + \delta$ we obtain an equivalent form of the Taylor series

$$f(x_0 + \delta) = f(x_0) + \delta f'(x_0) + \frac{1}{2!}\delta^2 f''(x_0) + \ldots + \frac{1}{p!}\delta^p f^{(p)}(x_0) + \ldots \quad (2.15)$$

which resembles the Maclaurin series with x_0 replacing 0, and δ replacing x. The series can be written compactly as

$$f(x_0 + \delta) = \sum_{p=0}^{\infty} \frac{\delta^p}{p!} f^{(p)}(x_0).$$

Example 2.9 Find the Taylor series of the function $x^{1/3}$ about the point $x_0 = 1$ up to terms of order $(x - 1)^2$.

We have

$$f(x) = x^{1/3} \text{ and } x_0 = 1 \Rightarrow f(x_0) = 1,$$
$$f'(x) = \frac{1}{3}x^{-2/3} \Rightarrow f'(1) = \frac{1}{3},$$
$$f''(x) = -\frac{2}{3}\cdot\frac{1}{3}x^{-5/3} \Rightarrow f''(1) = -\frac{2}{9}.$$

Therefore, for the Taylor series (Equation 2.14) we get

$$x^{1/3} = 1 + \frac{1}{3}(x - 1) - \frac{2}{9}(x - 1)^2 + \ldots$$

This is just the binomial expansion of $(1 + (x - 1))^{1/3}$.

Exercise 2.12 Show that the Taylor series expansion of $\sin(\theta)$ about $\theta = \pi/2$ is

$$\sin\left(\pi/2+\delta\right) = 1 - \frac{\delta^2}{2} + \dots,$$

and hence that $\sin\left(85^\circ\right) \approx 1 - \frac{1}{2}\left(\pi/36\right)^2$. Verify that the error is less than 0.00025%.

Taylor's Theorem: Loosely speaking, Taylor's theorem says that we can approximate a smooth function (which is infinitely differentiable) at a given point, as an nth order polynomial function of the displacement from that point (a Taylor series), and that the remainder goes to zero (the approximation converges on the true function) as n increases to infinity.

2.4.   EQUILIBRIUM POINTS

One of the major applications of the Taylor series is in the study of physical systems near points of equilibrium. An equilibrium point of a system characterized by a potential energy $V(x)$ is a point at which $V'(x) = 0$. An equilibrium point is stable if it gives a minimum of $V(x)$ and unstable if it gives a maximum.

Example 2.10 Suppose the potential energy of a certain system with one degree of freedom is given as

$$V(x) = x^4 - 8x^2.$$

Sketch a graph of $V(x)$, find and classify the equilibrium points, and obtain an approximate expression for $V(x)$ near to the point of equilibrium where $x > 0$.

At equilibrium, by definition $V'(x) = 0$. We have $V(x) = x^4 - 8x^2$, so $V'(x) = 4x^3 - 16x = 0$. This has solutions $x = 0$ and $x = \pm 2$ and these are the points of equilibrium. Now $V''(x) = 12x^2 - 16$.

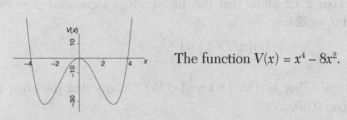

The function $V(x) = x^4 - 8x^2$.

- At $x = 0$, $V''(0) = -16 < 0$, so this is a maximum, hence unstable.

- At $x = +2$, $V''(2) = 32 > 0$, so this is a minimum, hence stable.

- At $x = -2$, $V''(-2) = 32 > 0$, so this is another minimum, hence stable.

Near $x = +2$ the Taylor series for $V(x)$ is

$$V(x) = V(2) + (x-2)V'(2) + \frac{(x-2)^2}{2}V''(2) + \ldots$$

$$= -16 + (x-2) \cdot 0 + \frac{(x-2)^2}{2} 32 + \ldots$$

$$\approx -16 + 16(x-2)^2.$$

Note how we find all the stationary (equilibrium) points and then consider them one by one. Some may be maxima, hence unstable equilibria; some may be minima, hence stable equilibria.

Exercise 2.13 Let the potential $V(x)$ be given by

$$V(x) = \frac{x^4}{4} + \frac{x^3}{3} - x^2 + \frac{8}{3}.$$

Find and classify the equilibrium points of this potential. By considering the shape of the graph, show that $x = -2$ gives a global minimum (this means $V(x) \geq V(-2)$ for every x, not just near $x = -2$) and find an approximate expression for $V(x)$ near this point.

Harmonic Behavior Close to An Equilibrium Point

Near to equilibrium *every* physical system behaves in an analogous way to a mass on a spring.

For example, for any potential energy $V(x)$, we can show that the corresponding force near to a minimum, $F = dV/dx$, is proportional to δ, the distance away from the minimum, i.e. $F = -k\delta$, the same as for a mass on a spring.

Close to a minimum, the Taylor series of $V(x)$ cannot contain any linear terms, because $dV/dx = 0$ at the minimum. From the general Taylor series,

$$V(x_0 + \delta) = \sum_{p=0}^{\infty} \frac{\delta^p}{p!} f^{(p)}(x_0),$$

$$\approx V(x_0) + \frac{V''(x_0)}{2}\delta^2,$$

where we have neglected higher order terms because these will be small if the distance from the minimum, δ, is small. Then, the potential

$$V(\delta) = V(x_0) + \frac{1}{2}k\delta^2,$$

where $V(x_0)$ and $k = V''(x_0)$ are constants. For *any* physical system close to a minimum $V(\delta) \propto \delta^2$ and the force $F = -dV/dx$ is proportional to the distance, $F = -k\delta$.

This means that, close to a minimum in the potential energy, *every* physical system behaves in an analogous way to a harmonic oscillator.

Example 2.11 The Leonard-Jones potential,

$$V(x) = V_0\left[\left(\frac{x_0}{x}\right)^{12} - 2\left(\frac{x_0}{x}\right)^6\right],$$

describes the potential energy of two atoms separated by a distance, x. This has a minimum value of $-V_0$ when $x = x_0$. Show that, near to $x = x_0$, the restoring force felt by the atoms is proportional to the distance, δ, from the minimum.

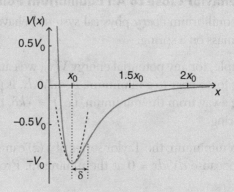

The Leonard-Jones potential (blue line above) has a minimum when the atomic separation is $x = x_0$. If atoms are too close together $(x < x_0)$, or too far apart $(x > x_0)$, their potential energy is large $(V(x) > -V_0)$.

Close to $x = x_0$ we can expand $V(x)$ as a Taylor series,

$$V(x_0) = -V_0$$
$$V'(x) = -12V_0x_0^{12}x^{-13} + 12V_0x_0^6x^{-7} \Rightarrow V'(x_0) = 0$$
$$V''(x) = 156V_0x_0^{12}x^{-14} - 84V_0x_0^6x^{-8} \Rightarrow V''(x_0) = 72V_0x_0^{-2}.$$

Then

$$V(x_0 + \delta) = -V_0 + \frac{1}{2}\left(\frac{72V_0}{x_0^2}\right)\delta^2,$$

and the potential near to the minimum varies quadratically (dashed curve above).

The restoring force is $F = -dV/dx = -dV/d\delta$, so

$$F = -\left(\frac{72V_0}{x_0^2}\right)\delta = -k\delta.$$

Thus, the force is proportional to the distance from the minimum and, close to their equilibrium separation, a pair of atoms will behave like a mass on a spring.

2.5. DEFINITION OF e^x (THE EXPONENTIAL FUNCTION)

The number e is defined by the formula

$$e = 1 + \frac{1}{1!} + \frac{1}{2!} + \frac{1}{3!} + \frac{1}{4!} + \ldots$$
$$= 1 + 1 + 0.5 + 0.166\ldots + 0.04166\ldots + \ldots \tag{2.16}$$

It can be calculated to any number of decimal places by adding together a sufficient number of terms of this infinite series. In fact

$$e = 2.718281828459045235360287 4. \ldots$$

The exponential function is defined as e^x (i.e. the number e raised to the power x). Often we write $\exp(x)$ for e^x. Figure 2.1 illustrates the fact that e^x is an increasing function of x and $e^x \to +\infty$ as $x \to +\infty$, while $e^{-x} \to 0$ as $x \to +\infty$.

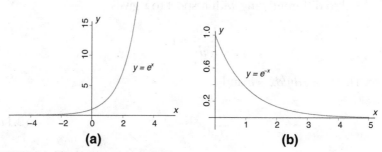

FIGURE 2.1: The functions (a) e^x and (b) e^{-x}.

2.6. THE INVERSE FUNCTION OF e^x

Using logarithms to the base e (see section A.3) we have

(i) $e^{\ln(x)} = x$ (by the definition of $\ln(x)$ as the power to which e must be raised to obtain x)

(ii) $\ln(e^x) = x \ln(e) = x$ (by taking logs) since $\ln(e) = 1$. Therefore the function $x \to \ln(x)$ is the inverse function to $x \to e^x$ (because the application of one mapping followed by the other brings us back to the starting value – see Section 2.10).

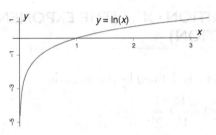

FIGURE 2.2: Graph of ln(x). Note that log(x) (for any base) is only defined for x > 0.

2.7. DERIVATIVE OF THE EXPONENTIAL

If we assume that the properties of $\ln(x)$ are given we can obtain the derivative of e^x by implicit differentiation (see Section 1.5). Let $y = e^x$, then

$$\ln(y) = x$$

and so differentiating with respect to x gives

$$\frac{1}{y}\frac{dy}{dx} = 1.$$

Therefore $dy/dx = y$ and

$$\frac{d}{dx}\{e^x\} = e^x. \tag{2.17}$$

In words, the rate of change of e^x is e^x. This is a fundamental result that distinguishes the number e; it is not true if we replace e by any other number. (See Exercise 2.17.)

Example 2.12 Find the Maclaurin series for e^x.

Let $f(x) = e^x$. We begin by computing $f'(x), f''(x)$ etc. as needed for the Maclaurin series.

$$f(x) = e^x \Rightarrow f(0) = e^0 = 1,$$
$$f'(x) = e^x \Rightarrow f'(0) = e^0 = 1,$$
$$f''(x) = e^x \Rightarrow f''(0) = e^0 = 1.$$

And so the Maclaurin series is

$$f(x) = f(0) + xf'(0) + \frac{x^2}{2!}f''(0) + \ldots$$

Therefore

$$e^x = 1 + x + \frac{x^2}{2!} + \frac{x^3}{3!} + \ldots + \frac{x^n}{n!} + \ldots \qquad (2.18)$$

Equation (2.18) may also be taken as the definition of the function e^x.

Exercise 2.14 Find $e^{3/2}$ to two decimal places by summing sufficiently many terms of the Maclaurin series. Why would it be impractical to calculate e^{10} or e^{-10} by the same method?

Exercise 2.15 From the Maclaurin series for e^x verify that $\frac{d}{dx}\{e^x\} = e^x$.

Example 2.13 Find $\frac{d}{dx}\{e^{5x}\}$.

We use the chain rule. Let $y = 5x$, then

$$\frac{de^y}{dx} = \frac{d}{dy}\{e^y\} \times \frac{dy}{dx} = e^y \cdot 5 = 5e^{5x}.$$

Exercise 2.16 Use the chain rule to find dy/dx if

(i) $y = e^{ax}$,

(ii) $y = e^{x^2}$,

(iii) $y = e^{f(x)}$.

Exercise 2.17 By putting $y = a^x$, taking logs and then differentiating show that

$$\frac{d}{dx}\{a^x\} = a^x \ln(a).$$

So in general

$$\frac{d}{dx}\{a^x\} \propto a^x$$

with equality only if $a = e$.

Example 2.14 Calculate the Taylor expansion of $\exp(-x^2)$ about $x = 1$ to first and second order.

$$f(x) = e^{-x^2} \text{ and } x_0 = 1 \Rightarrow f(x_0) = e^{-1},$$
$$f'(x) = -2xe^{-x^2} \Rightarrow f'(1) = -2e^{-1},$$
$$f''(x) = -2e^{x^2} + 4x^2 e^{-x^2} \Rightarrow f''(1) = 2e^{-1}.$$

Therefore, to first order, the Taylor series for $\exp(-x^2)$ is

$$\exp(-x^2) = e^{-1} - 2(x-1)e^{-1} = e^{-1}(3 - 2x).$$

To second order, the Taylor series expansion is

$$\exp(-x^2) = e^{-1}(1 - 2(x-1) + (x-1)^2) = e^{-1}(4 - 4x + x^2).$$

A sketch of the function together with its Taylor expansion to various orders is shown in Figure 2.3.

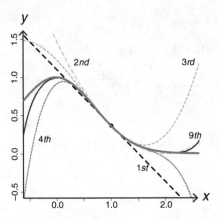

FIGURE 2.3: The function $\exp(-x^2)$ (solid blue line) together with its Taylor expansion to various orders (see Example 2.14). The black dashed line shows the first order approximation, and the long dashed grey line shows the Taylor expansion to second order; 3rd, 4th and 9th order approximations are also shown. Close to $x = 1$ the expansion is very good with just a few terms. For example, the first order (linear) approximation is accurate to within 2% for $0.82 < x < 1.12$. Further from $x = 1$ we need to use a higher order expansion to give a good approximation to the function.

Exercise 2.18 Calculate the Taylor series expansion of $\exp(-x)$ to second order about the point $x = 1$.

2.8. INTEGRATION OF EXPONENTIALS

We now know how to find the derivatives of exponential functions. The next step is to use this knowledge to find the integrals of exponential functions.

Since

$$e^x = \frac{d}{dx}\{e^x\},$$

integration of both sides gives

$$\int e^x dx = e^x + \text{constant}.$$

Example 2.15 Find $I = \int e^{ax}\, dx$ where a is a constant.
Put $u = ax$. Then $du = adx$ and

$$I = \frac{1}{a}\int e^u du = \frac{1}{a}e^{ax} + C,$$

where C is a constant.

Exercise 2.19 Find $\int xe^{-ax^2}$ (Hint: put $u = -ax^2$.)

Note that the indefinite integral $\int e^{-x^2} dx$ (without the additional factor of x in the integrand) cannot be expressed in terms of elementary functions. Do not waste time trying! This is known as the *Gaussian integral* and is very important in probability theory and quantum mechanics.

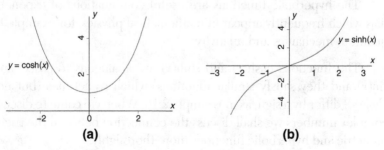

FIGURE 2.4: Graphs of the hyperbolic functions (a) cosh(x) and (b) sinh(x).

2.9. HYPERBOLIC FUNCTIONS

We define new functions, called *hyperbolic functions*, as combinations of e^x and e^{-x}:

$$\cosh(x) = \frac{1}{2}\left(e^x + e^{-x}\right),$$

and

$$\sinh(x) = \frac{1}{2}\left(e^x - e^{-x}\right),$$

where sinh is pronounced either "shine" or "sinsh."

Example 2.16 Show that $\cosh^2(x) - \sinh^2(x) = 1$.

We have

$$\cosh^2(x) = \left[\frac{1}{2}\left(e^x + e^{-x}\right)\right]^2 = \frac{1}{4}\left(e^{2x} + 2 + e^{-2x}\right),$$

$$\sinh^2(x) = \left[\frac{1}{2}\left(e^x - e^{-x}\right)\right]^2 = \frac{1}{4}\left(e^{2x} - 2 + e^{-2x}\right).$$

Hence

$$\cosh^2(x) - \sinh^2(x) = \frac{1}{4}(2) - \frac{1}{4}(-2) = 1.$$

(This result should be memorized and can be used without proof.)

The hyperbolic functions are useful combinations of exponentials which frequently appear in mathematical physics, for example in quantum mechanics and relativity.

The functions $\cosh(x)$ and $\sinh(x)$ are analogous to $\cos(x)$ and $\sin(x)$ and they satisfy similar identities, which sometimes (but not always!) differ by a sign, as in Example 2.16. When we come to discuss complex numbers we shall discuss the connection between the trigonometric and hyperbolic functions more thoroughly.

We complete the analogy with trigonometric functions by defining

$$\tanh(x) = \frac{\sinh(x)}{\cosh(x)},$$

$$\coth(x) = \frac{1}{\tanh(x)},$$

$$\operatorname{cosech}(x) = \frac{1}{\sinh(x)},$$

$$\operatorname{sech}(x) = \frac{1}{\cosh(x)}.$$

$\tanh(x)$ is often pronounced as "tansh," cosech(x) as "coshek" and sech(x) as "shek."

Exercise 2.20 Show that $1 - \tanh^2(x) = \operatorname{sech}^2(x)$ and that $\coth^2(x) - 1 = \operatorname{cosech}^2(x)$.

Example 2.17 Show that $\dfrac{d}{dx}\cosh(x) = \sinh(x)$.

$$\frac{d}{dx}\{\cosh(x)\} = \frac{d}{dx}\left\{\frac{e^x + e^{-x}}{2}\right\} = \frac{e^x - e^{-x}}{2} = \sinh(x).$$

Exercise 2.21 Show that $\dfrac{d}{dx}\{\sinh(x)\} = \cosh(x)$.

Exercise 2.22 Show that sinh(x) has the same sign as x for all x and hence that cosh(x) has a global minimum at $x = 0$.

2.10. INVERSE FUNCTIONS

Example 2.18 If $y = x^2$ then $x = y^{1/2}$ is the inverse function.

However, because we usually write an independent variable as x, we usually say that $y = x^{1/2}$ is the inverse function to $y = x^2$. (Alternatively, $x \to x^{1/2}$ is the function inverse to $x \to x^2$.)

Example 2.19 If $y = \sin(x)$ then $x = \sin^{-1}(y)$; so $y = \sin^{-1}(x)$ is the inverse function to $y = \sin(x)$. The notation $\arcsin(x)$ is also used for $\sin^{-1}(x)$.

Note again that $\sin^{-1}(x)$ is not the same as $(\sin(x))^{-1}$.

If $y = f(x)$ then $x = f^{-1}(y)$ is the function inverse to f that associates (one or more) values of x with any given y (in the range of f).

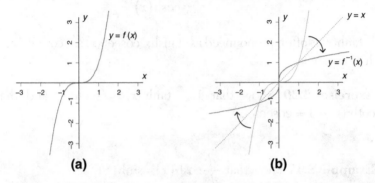

FIGURE 2.5: (a) A graph of a function, $y = f(x)$. (b) The graph of the inverse function $y = f^{-1}(x)$ is obtained by reflection in the line $y = x$.

TABLE 2.1: Table of some standard functions and their inverse functions. The third and fourth rows give restrictions on the x and y values for which the inverse function $y = f^{-1}(x)$ is defined.

Function	x^2	e^x	$\sin(x)$	$\cos(x)$	$\tan(x)$
Inverse	$x^{1/2}$	$\ln(x)$	$\sin^{-1}(x)$	$\cos^{-1}(x)$	$\tan^{-1}(x)$
Range of x	$x \geq 0$	$x > 0$	$-1 \leq x \leq 1$	$-1 \leq x \leq 1$	$-\infty < x < \infty$
Range of y	$y \geq 0$	(none)	$-\pi/2 \leq y \leq \pi/2$	$0 \leq y \leq \pi$	$-\pi/2 < y < \pi/2$

In both the examples the inverse function is many-valued: for $y = x^2$, both $y = +x^{1/2}$ and $y = -x^{1/2}$ are inverse functions (for $x \geq 0$). Often the non-negative square root is intended if we write, say, $\sqrt{2}$. In example 2.19, for $y = \sin^{-1}(x)$ (where x is between -1 and 1), the value in the range $-\pi/2 \leq y \leq \pi/2$ is often intended, but any of the functions $\sin^{-1}(x) + 2n\pi$ or $(2n + 1)\pi - \sin^{-1}(x)$, with n an integer, is also an inverse to $y = \sin(x)$. The other inverse trigonometric functions $\cos^{-1}(x)$ and $\tan^{-1}(x)$ are similarly many-valued. See Table 2.1.

Exercise 2.23 Find all distinct real solutions of the equation $\tan(3\theta) = 1$ in the range $0 < \theta < \pi$.

2.11. INVERSE HYPERBOLIC FUNCTIONS

Since the sinh and cosh functions are combinations of exponentials, their inverses involve the logarithm.

Example 2.20 Show that $\sinh^{-1}(x) = \ln\left(x + \sqrt{x^2 + 1}\right)$.

We need to turn this into a problem about the sinh function whose properties we know. So to begin we invert $y = \sinh^{-1}(x)$. Now we can use the definition of sinh,

$$x = \sinh(y) = (e^y - e^{-y})/2.$$

We try to turn this into an equation we can solve for y. To begin, multiply through by $2e^y$ and rearrange to give

$$e^{2y} - 2xe^y - 1 = 0.$$

This is a quadratic in e^y. Let $u = e^y$ (which means we are seeking $y = \ln(u)$); then,

$$u^2 - 2xu - 1 = 0.$$

This is a simple quadratic equation that has solutions,

$$u = x \pm \sqrt{x^2 + 1}.$$

We choose the + sign because otherwise $u < 0$ and $\ln(u)$ does not exist. Then, finally,

$$y = \ln\left(x + \sqrt{x^2 + 1}\right).$$

Exercise 2.24 Show that $\cosh^{-1}(x) = \ln\left(x \pm \sqrt{x^2 - 1}\right)$ and hence, or otherwise, show that

$$\frac{d}{dx}\{\cosh^{-1}(x)\} = \pm\frac{1}{\sqrt{x^2 - 1}}.$$

Explain the ± signs, with reference to a graph.

Revision Notes

After completing this chapter you should be able to

- Compute $n!$
- Write down the general coefficient in the binomial expansion and appreciate the use of this expansion
- Compute Maclaurin and Taylor series for combinations of elementary functions
- Define the exponential function e^x and know its properties
- Define the hyperbolic functions sinh and cosh and derive their properties
- Find the inverses of simple trigonometric and hyperbolic functions

2.12. EXERCISES

1. (a) Use Taylor series to show that
$$\cos\left(61^\circ\right) \approx \frac{1}{2} - \frac{\pi\sqrt{3}}{360}.$$

   (b) By regarding $\sqrt{10}$ as $(10.24 + \delta)^{1/2}$ where $\delta = -0.24$, and using the fact that $(10.24)^{1/2} = 3.2$, calculate the approximate value 3.1625 for $\sqrt{10}$. (The correct value is 3.1622778.)

   (c) By a similar method, show that $(2.2)^{1=3} \approx 1.30059$.

2. Show that if an angle A is measured *in degrees* as $90 - \delta$ where δ is small, then
$$\sin\left(A\right) \approx 1 - \frac{\pi^2\delta^2}{64800}.$$

3. Show that $\sin\left(\dfrac{\pi}{4}+\delta\right)=\dfrac{1}{\sqrt{2}}\left\{1+\delta-\dfrac{\delta^2}{2}-\dfrac{\delta^3}{6}+\ldots\right\}$.

4. By combining known series, find the Taylor series of $\left(\dfrac{1+x}{1-x}\right)$ about $x=0$, and use it to show that $\ln(1.25)$ is approximately equal to 488/2187.

5. Write down the first three terms in the binomial expansion of $\sqrt{4+x}$. Use this truncated series to estimate $\sqrt{4.5}$ and calculate the percentage error in your result.

6. Find the first four non-zero terms in the Taylor series of $(\ln x)^2$ about $x=1$.

7. Find the first non-vanishing term of the Taylor series for $y=\sin^4(x)$ near $x=0$.

8. Use a Taylor series to show that $\sin(32°)$ is approximately equal to $\dfrac{1}{2}+\dfrac{2\pi}{360}\sqrt{3}$.

9. Write down the series for $\sin(\theta)$ in terms of θ and deduce the approximate formula

$$\theta \approx \sin(\theta)+\dfrac{1}{6}\theta^3.$$

   By considering the case $\theta=\pi/6$ obtain the estimate $\pi=3\dfrac{1}{8}$.

10. By equating coefficients of x^n on both sides of

$$(1+x)^n(1+x)^n=(1+x)^{2n}$$

   deduce the identity

$$\sum_{r=0}^{n}\binom{n}{r}^2=\binom{2n}{n}.$$

11. Sketch the graph of $y=1+\sin(x)$ for $-\pi/2 \le x \le \pi/2$ and hence that of its inverse function. For what values of the variable is the latter defined?

12. By writing 10^x in the form e^y evaluate the indefinite integral $\int 10^x dx$.

13. From first principles show that $1 - \tanh^2(x) = \operatorname{sech}^2(x)$ and deduce that

$$\coth^2(x) - 1 = \operatorname{cosech}^2(x).$$

What is the value of $\cosh(x)$ when $3\cosh(x) = 5\sinh(x)$?

14. Write down the Maclaurin series for $\sin(x)$ and obtain the series for $\cos(x)$ by differentiation.

Verify that when the square

$$\left(x - \frac{x^3}{3!} + \frac{x^5}{5!} - \ldots\right) \times \left(x - \frac{x^3}{3!} + \frac{x^5}{5!} - \ldots\right)$$

is formed and terms of the same degree collected together, the coefficients of x^2 and x^4 agree with those obtained by using the series for $\cos(2x)$ in the formula

$$\sin^2(x) = \frac{1}{2}\left(1 - \cos(2x)\right).$$

15. Sketch the graph of $f(x) = \dfrac{\sinh(x)}{\cosh(x)}$. Has $f(x)$ any maxima or minima?

16. Define the functions $\cosh(x)$ and $\sinh(x)$, and deduce that each is the derivative of the other and that $\cosh(0) = 1$, $\sinh(0) = 0$. Hence obtain the Maclaurin series for $\cosh(x)$ as

$$1 + \frac{x^2}{2!} + \frac{x^4}{4!}$$

and show that

$$e^{0.1} + e^{-0.1} \approx 2.010008333\ldots$$

2.13. PROBLEMS

1. From the formula $\tanh^{-1}(x) = \frac{1}{2}\ln(\frac{1+x}{1-x})$ deduce that

$$\frac{d}{dx}(\tanh^{-1}(x)) = \frac{1}{1-x^2}$$ and hence show (with the help of an integration by parts) that $\int \tanh^{-1}(x)\,dx$ can be written as

$$x\tanh^{-1}(x) + \frac{1}{2}\ln(1-x^2) + C.$$

2. (a) Show that the function $\tanh^{-1}(x)$ may be written as
   $\ln\sqrt{1+x} - \ln\sqrt{1-x}$ and that

   $$\frac{d}{dx}\tanh^{-1}(x) = (1-x^2)^{-1} \quad \text{and} \quad \frac{d}{dx}\tanh(x) = 1 - \tanh^2(x).$$

   (b) Sketch the curves $y = \tanh(x)$, $y = \tanh^{-1}(x)$,
   $y = 1 - \tanh^{-2}(x)$ and $y = (1 - x^2)^{-1}$, indicating any axis
   crossings and the asymptotic behavior for each.

   (c) Show that the curves $y = \tanh(x)$ and $y = 1 - \tanh^2(x)$

   cross at the point $\left(\left(\ln(2+\sqrt{5})\right)/2, (\sqrt{5}-1)/2\right)$ and by

   examining the limit $\displaystyle\lim_{x\to 1}\frac{\tanh^{-1}(x)}{(1-x^2)^{-1}}$ or otherwise, that the

   curve $y = \tanh^{-1}(x)$ never becomes as large as
   $y = (1 - x^2)^{-1}$.

3. Show that $\sinh^{-1}(x) = \ln(x + \sqrt{x^2+1})$ and hence that

   $$\frac{d}{dx}\sinh^{-1}(x) = \frac{1}{\sqrt{x^2+1}}.$$

   Use an integration by parts to show that

   $$\int \sinh^{-1}(x)\,dx = x\sinh(x) - \sqrt{x^2+1} + C.$$

4. Verify that the following three methods for expanding
   $f(t) = e^{3t+t^2}$ in a power series

   $$a_0 + a_1 t + a_2 t^2 + \ldots$$

   all yield the same values for a_0, a_1 and a_2.

   (i)   Substitute $3t+t^2$ for x in the series for e^x and collect powers of t.

   (ii) Multiply together the power series for e^{3t} and e^{t^2}.

   (iii) Obtain the Taylor coefficients $f(0)$, $f'(0)$,

   $\dfrac{1}{2} f''(0)$ using

   $$f'(t) = (3 + 2t)f(t), \quad \text{and} \quad f''(t) = (3 + 2t)f'(t) + 2f(t).$$

3

FUNCTIONS, LIMITS, AND SERIES

In physics we will often describe physical processes or quantities in terms of mathematical functions. Understanding the behavior of functions is therefore an important part of understanding the physics. In this chapter we will look at curve sketching, at the symmetry of functions, at approximations to functions and, importantly, at the asymptotic behavior of a function as some variable tends to a limit. We will also introduce the ideas of convergent and divergent series.

3.1. CURVE SKETCHING: QUADRATICS

It is often very useful to visualize a function to get an overview of its behavior. This is particularly the case when a function $f(x)$ is used to describe a physical quantity. One simple way to visualize $f(x)$ is to sketch a graph of $f(x)$ against x.

To sketch a quadratic function we need to find its maximum or minimum, where it crosses the axes and how it goes to infinity.

Example 3.1 Find the stationary point and zeros of the quadratic $y = x^2 - 4x + 3$ and sketch its graph.

We could find the stationary points by setting $dy/dx = 0$, and find the zeros from the formula $x = \left[-b \pm \sqrt{(b^2 - 4ac)}\right] / 2a$.

Alternatively we can use the following procedure which begins by "completing the square":

$$y = x^2 - 4x + 3 = (x - 2)^2 - 1.$$

This has its smallest value (of -1) at $x = 2$ (since the smallest value of the squared term is 0). So $x = 2$ gives a minimum. Next determine the axis crossings – first the x-axis:

$(x - 2)^2 - 1 = 0,$

$\Rightarrow (x - 2)^2 = 1,$

$\Rightarrow x - 2 = \pm 1.$

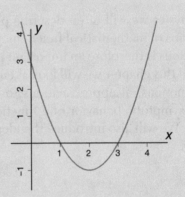

Hence $x = 3$ or 1. The graph crosses the y axis when $x = 0$, which gives $y = 3$. Finally, $y \to +\infty$ (where "$\to$" means "tends to") for both large positive and large negative values of x. The graph is therefore as shown in the figure.

Exercise 3.1 By following the indicated procedure, sketch the graphs of the following, labeling the maximum or minimum and axes crossings.

   (i) $y = -4x^2 + 12x - 9$

   (ii) $y = 4x^2 - 12x + 10$

3.2. CURVE SKETCHING: GENERAL POLYNOMIALS

A *polynomial function* of degree n is a function of the form

$$y = a_n x^n + a_{n-1}x^{n-1} + \cdots + a_3 x^3 + a_2 x^2 + a_1 x + a_0,$$

where $a_n, a_{n-1}, \ldots, a_1, a_0$ are constant *coefficients*. An nth degree polynomial has at most n roots (values of x where $y = 0$). Finding these can be difficult for $n = 3, 4$, and there is no general method for finding the roots of higher degree polynomials.

Nevertheless, in sketching graphs of any function we usually look for:

- behavior at $+\infty$ and $-\infty$ (this gives us the *asymptotes*),

- crossing of axes ($y = 0, x = 0$),

- maxima and minima ($dy/dx = 0$; $d^2y/dx^2 < 0$ or > 0),

- points of inflection ($d^2y/dx^2 = 0$).

Example 3.2 Sketch the graph of the cubic polynomial $y = 4x^3 - 3x$.

First, the asymptotes: when $|x|$ is large $4x^3 >> 3x$ so we can neglect the $3x$ in comparison with $4x^3$. Then, for large $|x|$, y varies like $4x^3$. This means as $x \to +\infty$, y also tends to $+\infty$, and as $x \to -\infty$, y also tends to $-\infty$.

Now for the x-axis crossing. We see $y = 0$ when $x = 0$ or $x = \pm\sqrt{3/2}$.

Next to find the maxima and minima:

$$dy/dx = 12x^2 - 3 = 0$$

when $x = \pm 1/2$.

To join smoothly onto the curve for $x \to \pm\infty$, $x = -1/2$ must give a maximum and $x = 1/2$ a minimum. This can also be determined from the sign of d^2y/dx^2.

Also,

$$d^2y/dx^2 = 24x = 0$$

when $x = 0$, hence when $y = 0$. So the origin is a point of inflection. Approaching the origin from $x < 0$ the curve is getting steeper; beyond the origin it becomes less steep, i.e. the slope gets less negative.

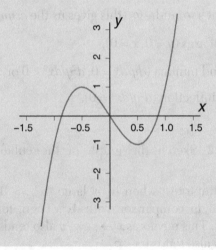

Exercise 3.2 Sketch the graphs of the cubic polynomials

   (i) $y = x^3 - 3x^2$,

   (ii) $y = x^3 - 3x^2 + 2$.

Note that a quadratic has at most two zeros, while a cubic has at most three and at least one zero. In general, a polynomial of degree n (i.e. having highest order term x^n) has at most n zeros; if n is odd it has at least one zero (because a polynomial of odd degree must range between $-\infty$ and $+\infty$ and hence must cross zero).

3.3. CURVE SKETCHING: RATIONAL FUNCTIONS

A *rational function* is a polynomial divided by another polynomial, i.e. a fraction with polynomials as numerator and denominator. See Section A.6.

Example 3.3 Sketch the graph of $y = 1/x$.

First, the behavior at infinity. For large $|x|$, y tends to zero. For small positive x, $y \to +\infty$, and for small negative x, $y \to -\infty$. Also, $y \neq 0$ for any finite x.

Next, stationary points.

$$dy/dx = -1/x^2 \neq 0,$$

for any finite x. And

$$d^2y/dx^2 = 2/x^3 \neq 0,$$

meaning there are no points of inflection. So the graph is as shown below.

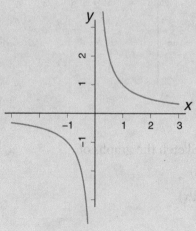

Example 3.4 Sketch the graph of $y = x/(x-2)$.

Asymptotes: since we want to look at large x, write y in a form that involves $1/x$ for large x (which we can then neglect).

$$y = \frac{1}{1-2/x} = \left(1-\frac{2}{x}\right)^{-1}$$

$$y = 1 + \frac{2}{x} + \frac{4}{x^2} + \ldots$$

(using the binomial series; see section 2.1). So $y \to 1$ as $|x| \to \infty$. If x is near to 2 then $|y|$ will get very large. But the sign of y depends on whether $x > 2$ or < 2. For $x - 2$ small and positive, $y \to +\infty$; for $x - 2$ small and negative, $y \to -\infty$.

We can see that $y = 0$ for $x = 0$, and $dy/dx \neq 0$ for any finite x. We can also find that $d^2y/dx^2 \neq 0$. So the graph is as shown below.

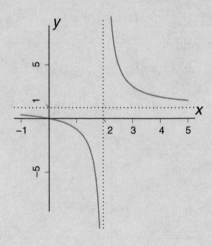

Exercise 3.3 Sketch the graphs of

(i) $y = (x-3)^{-1}$,

(ii) $y = x + (2/x)$.

3.4. GRAPHICAL SOLUTION OF INEQUALITIES

A good starting point when solving an inequality is to sketch the form of the curves involved. As we shall show below, the graphical method is foolproof while, in contrast, there are some pitfalls to algebraic methods.

Example 3.5 What is the region determined by the condition $y > 0, y < 3 - x, y < x$?

First we draw the lines $y = 0, y = 3 - x, y = x$ and then shade the disallowed regions.

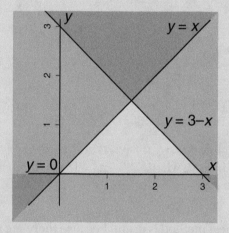

It is then easy to see that the permitted region is the interior of the triangle with vertices at the points $(0, 0)$, $\left(\dfrac{3}{2}, \dfrac{3}{2}\right)$, $(3, 0)$.

Exercise 3.4 What is the region determined by the conditions $y < 4, y > -x, y > 2x + 1$?

Example 3.6 For what range(s) of x is $1/(3-x) < 1/(x-1)$?

First, we let

$$y_1 = 1/(3-x),$$

$$y_2 = 1/(x-1),$$

then sketch the graphs of y_1 and y_2.

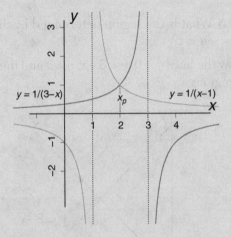

From the graph $y_1 < y_2$ for $x > 3$ and between x_p and 1. The crossover point is x_p, where $1/(3-x_p) = 1/(x_p - 1)$, i.e. $x_p - 1 = 3 - x_p$ so $x_p = 2$. So $1/(3-x) < 1/(x-1)$ for $1 < x < 2$ and $x > 3$.

The graphical method is useful because there are some common pitfalls associated with trying to solve an inequality algebraically.

If we were to try to solve Example 3.6 algebraically it would be tempting to proceed via the following incorrect procedure. We require $1/(3 - x) < 1/(x - 1)$, hence, multiplying through by $(x - 1)(3 - x)$, we need $(x - 1) < (3 - x)$ or $2x < 4$, and so $x < 2$.

This is wrong because, for $x < 1$, multiplication by $(x - 1)$ multiplies the inequality by a negative number. This reverses the sign of the inequality, a fact that has been ignored. This can be remedied by squaring the multiplier (i.e. multiply through by $(x-1)^2(3-x)^2$), but we recommend that you stick to the graphical method.

Exercise 3.5 Determine where

$$\frac{x}{x-6} < \frac{1}{x-4}.$$

3.5. THE SYMMETRY OF FUNCTIONS

Symmetry is a very important idea in theoretical physics. For example, the symmetries of the laws of physics determine the elementary particles found in nature. Here we will introduce the idea of *even* and *odd* functions. As we shall see, we can often use the symmetry of a function to make mathematical operations, like integration, much easier.

Even and Odd Functions

If $f(-x) = f(x)$, the function, $f(x)$, is *even*. This means that the graph of $f(x)$ is symmetric about the y-axis; $f(x) = x^2$ and $f(x) = \cos(x)$ are examples of even functions.

If $g(-x) = -g(x)$, the function, $g(x)$, is *odd*. This means that $g(x)$ is antisymmetric about the y-axis; $g(x) = x^3$ and $g(x) = \sin(x)$ are examples of odd functions. Notice that the graph of an odd function must pass through the origin: $g(0) = 0$.

A function may be neither even nor odd. For example, both $h(x) = \exp(x)$ and $h(x) = x + \cos(x)$ are functions that are neither even nor odd.

It is always possible to write a function as a sum of even and odd functions,

$$h(x) = f(x) + g(x),$$

where

$$f(x) = \frac{1}{2}(h(x) + h(-x)) \quad \text{is even,}$$

and

$$g(x) = \frac{1}{2}(h(x) - h(-x)) \quad \text{is odd.}$$

▋ **Exercise 3.6** Write e^x as the sum of an even and an odd function.

Products of Even and Edd Functions

The product of two even functions is always even. So, because $f_1(x) = x^2$ and $f_2(x) = \cos(x)$ are both even, then $h(x) = f_1(x)f_2(x) = x^2 \cos(x)$ is even.

The product of two odd functions is always even. So, because $g_1(x) = x$ and $g_2(x) = \sin(x)$ are both odd, then $h(x) = g_1(x)g_2(x) = x \sin(x)$ is even.

The product of an even function and an odd function is always odd. So, because $f(x) = x^2$ is even and $g(x) = \sin(x)$ is odd, then $h(x) = f(x)g(x) = x^2 \sin(x)$ is odd.

Integrals of Even and Odd Functions

We can make use of symmetry to simplify the integrals of even and odd functions between limits that are symmetric about the origin. The integral of any odd function $g(x)$ between $-a$ and a is simply zero. We can see this graphically in Figure 3.1: the area under the

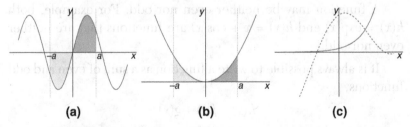

(a)   **(b)**   **(c)**

FIGURE 3.1: (a) An example of an odd function, $f(x) = \sin 2x$. (b) An example of an even function, $f(x) = x^2$. (c) Two functions, $h_1(x) = \exp(x)$ (solid line) and $h_2(x) = x + \cos x$ (dashed line) that are neither odd nor even.

curve between $-a$ and 0 is equal and opposite to the area under the curve between 0 and a. For any odd function,

$$\int_{-a}^{a} g(x)\,dx = 0.$$

For any even function we have

$$\int_{-a}^{0} f(x)\,dx = \int_{0}^{a} f(x)\,dx.$$

Again, this is illustrated graphically in Figure 3.1.

In general,

$$\int_{-a}^{a} h(x)\,dx = \begin{cases} 0 & \text{if} \quad h(x) \text{ is odd} \\ 2\int_{0}^{a} h(x)\,dx & \text{if} \quad h(x) \text{ is even.} \end{cases} \tag{3.1}$$

Exercise 3.7 Examine the following integrals. Decide whether the integrand is even or odd and use this to evaluate the definite integral.

(i) $\displaystyle\int_{-1}^{1} \sinh(x)\,dx$,

(ii) $\displaystyle\int_{-1}^{1} \cosh(x)\,dx$.

3.6. LIMITS

In Section 3.1 we looked at the asymptotic limits of some simple functions and introduced the notation "$x \to a$" which is read as "x tends to a." Here we will show how to find the limits of a function in some more complicated cases.

Limits by Substitution

Consider the function $y = x^2$. As x tends to 4, for example, then y gets and remains arbitrarily close to 16, the value of y when $x = 4$.

We write that as $x \to 4$, $y \to 16$. Equivalently we say that 16 is the limit of y as x tends to 4 and write

$$\lim_{x \to 4} y = 16 \quad \text{or} \quad \lim_{x \to 4} x^2 = 16.$$

Exercise 3.8 What is $\lim_{x \to \pi/2} \sin(x)$?

Example 3.7 If $y = 1/x$, what is $\lim_{x \to +\infty} y$?

As $x \to +\infty$, $y \to 0$ because as x increases y becomes and remains arbitrarily close to 0.

Example 3.8 Find the limit of $y = e^{2+(1/x)}$ as $x \to +\infty$.

We have $2 + 1/x \to 2$ as $x \to +\infty$ since $1/x \to 0$ from Example 3.7. So $y \to e^2$ as $x \to +\infty$.

Example 3.9 Find the limit of $y = 1/x$ as $x \to 0$.

Here we need to be careful. As x tends to zero through positive values (which we write as $x \to 0^+$), y increases arbitrarily through positive values so $\lim_{x \to 0^+} y = +\infty$. But as x tends to zero through negative values (which we write as $x \to 0^-$), y gets increasingly large and negative: so $\lim_{x \to 0^-} y = -\infty$.

Exercise 3.9 Find the limits of

(i) $y = (x - 3)^{-1}$ as $x \to 3$ (consider the two cases $x \to 3^+$ and $x \to 3^-$)

(ii) $y = x^p$ as $x \to +\infty$ (consider the cases $p > 0$, $p = 0$, $p < 0$ separately)

(iii) $y = x/(x - 3)$ as $x \to +\infty$.

3.7. INDETERMINATE FORMS

In many cases we cannot evaluate limits by simple substitution as we did previously. For example, consider the following limits:

$$(a)\lim_{x\to 0}\frac{\sin(x)}{x}, \quad (b)\lim_{x\to 0^+}x\ln(x), \quad (c)\lim_{x\to +\infty}x^{1/x}, \quad (d)\lim_{x\to +\infty}\left(1+\frac{1}{x}\right)^x.$$

Direct substitution gives the *meaningless* results 0/0, 0 × (−∞), ∞⁰, 1^∞, respectively. It is hopeless to try to proceed further like this. In particular 0/0 is not 1, but meaningless. And 0 × ∞ is not 0, or ∞, or any other number, but meaningless.

Such expressions may nevertheless approach definite limits (finite or infinite) – for example, 1, 0, 1, *e*, respectively, in the above four cases. But we must use a different method to evaluate these *indeterminate* limits. One method that sometimes works is expansion in series. Often limits can be related to those already known. Alternatively, a method that usually works is l'Hôpital's rule.

Example 3.10 Find $\lim_{x\to 0}\dfrac{\sin(x)}{x}$.

Substituting $x = 0$ we get sin(0)/0 which is indeterminate. To work out the limit we expand sin(x) as a series

$$\frac{\sin(x)}{x}=\frac{x-\dfrac{x^3}{6}+\dots}{x}=1-\frac{x^2}{6}+\dots$$

Putting $x = 0$ in the resulting series leads to a determinate result.

So

$$\lim_{x\to 0}\frac{\sin(x)}{x}=1.$$

Example 3.11 Find $\lim\limits_{x \to +\infty} \dfrac{e^x}{x^n}$ where n is a positive integer.

This is of the form ∞/∞, so indeterminate. To work out the limit, we expand e^x as a series.

$$\frac{e^x}{x^n} = \frac{1 + x + \dfrac{x^2}{2!} + \ldots + \dfrac{x^{n-1}}{(n-1)!} + \dfrac{x^n}{n!} + \dfrac{x^{n+1}}{(n+1)!} + \ldots}{x^n}$$

$$= x^{-n} + x^{1-n} + \frac{x^{2-n}}{2!} + \ldots + \frac{x^{-1}}{(n-1)!} + \frac{1}{n!} + \frac{x}{(n+1)!} + \ldots$$

Now take the limit as $x \to +\infty$

$$\frac{e^x}{x^n} \to 0 + 0 + 0 + \ldots + 0 + \frac{1}{n!} + \infty + \ldots$$

The negative powers of x become arbitrarily small, but the terms involving positive powers of x become arbitrarily large. Therefore the sum tends to

$$\lim_{x \to +\infty} \frac{e^x}{x^n} = +\infty.$$

This is an important result which should be remembered (and can generally be quoted without proof): as $x \to +\infty$, e^x grows faster than *any* power of x. As a corollary to this result (exercise 3.10) we have that, as $x \to +\infty$, e^{-x} decreases faster than any power of x.

Exercise 3.10 Show that $\lim_{x \to +\infty} x^n e^{-x} = 0$. (Hint: this is $(e^x/x^n)^{-1}$.)

Example 3.12 Find $\lim\limits_{x \to 0^+} x \ln(x)$.

This is indeterminate of the form $0 \times (-\infty)$. Since we cannot expand $\ln(x)$ as a series about 0 we try to relate this limit to one we know. This we can do by substituting $x = e^{-u}$

$$x \ln(x) = -e^{-u} u.$$

Then $u \to +\infty$ as $x \to 0^+$. Therefore

$$\lim_{x\to 0+} x\ln(x) = -\lim_{u\to+\infty} ue^{-u}$$
$$= 0.$$

And we used the result of Exercise 3.10 in the last line.

Example 3.13 Find $\lim_{x\to+\infty} x^{1/x}$.

This has the indeterminate form ∞^0. Since we cannot expand as a series we relate this to a limit we already know. This we can do by taking logs.

$$L = \lim_{x\to+\infty} x^{1/x}$$

$$\ln(L) = \lim_{x\to+\infty} \frac{\ln(x)}{x}.$$

To relate this to example 3.12 let $u = 1/x$. So $u \to 0^+$ as $x \to +\infty$.

$$= -\lim_{u\to 0^+} u\ln(u)$$
$$= 0.$$

Now we see that $\ln(L) = 0$ which implies $L = 1$.

3.8.   L'HÔPITAL'S RULE

An alternative method for finding limits that usually works is l'Hôpital's rule:

$$\lim_{x\to x_0} \frac{f(x)}{g(x)} = \lim_{x\to x_0} \frac{f'(x)}{g'(x)} \tag{3.2}$$

This is only valid when the limit (on the left) is indeterminate, and both $f(x_0)$ and $g(x_0)$ are zero, or both are $\pm\infty$. Notice that to apply the rule we differentiate both $f(x)$ and $g(x)$ and then divide $f'(x)$ by $g'(x)$. This is not the derivative of (f/g). Also, if the limit on the right is also indeterminate, you may apply the rule again.

Example 3.14 Find $\lim_{x \to 0^+} x \ln(x)$ using l'Hôpital's rule.

To apply l'Hôpital's rule we have to state the problem in the form $\lim_{x \to 0^+} (f/g)$:

$$\lim_{x \to 0^+} \frac{\ln(x)}{1/x} = \frac{-\infty}{\infty} \text{ which is indeterminate}$$

Using l'Hôpital's rule with $f(x) = \ln(x)$ and $g(x) = 1/x$ we get

$$= \lim_{x \to 0^+} \frac{(\ln(x))'}{(1/x)'} = \lim_{x \to 0^+} \frac{1/x}{-1/x^2}$$

$$= \lim_{x \to 0^+} (-x) = 0.$$

Exercise 3.11 Find

(i) $\lim_{x \to 0} \dfrac{1 - \cos(x)}{x^2}$,

(ii) $\lim_{x \to +\infty} \left(1 + \dfrac{a}{x}\right)^x$ $(a = \text{constant})$,

(iii) $\lim_{x \to +\infty} \dfrac{\sin(x) - x}{x^3}$.

3.9. LIMITS OF INTEGRALS

Integrals with infinite limits: By definition

$$\int_a^\infty f(x)\,dx = \lim_{b \to \infty} \int_a^b f(x)\,dx.$$

An integral that gives a finite answer is said to *converge*; when the integral is not convergent it is said to *diverge*.

Exercise 3.12 For which values of the real number a does the integral $\int_1^\infty x^a \, dx$ converge?

Integrals with infinite integrands: If $f(x)$ becomes infinite at $x = a$ we define

$$\int_a^b f(x)\,dx = \lim_{l \to a^+} \int_l^b f(x)\,dx.$$

Example 3.15 Evaluate $\int_0^1 x^{-1/3} \, dx$.

We have

$$\int_0^1 x^{-1/3} \, dx = \lim_{l \to 0^+} \int_l^1 x^{-1/3} dx$$

$$= \lim_{l \to 0^+} \left[\frac{3}{2} x^{2/3} \right]_l^1$$

$$= \lim_{l \to 0^+} \left(\frac{3}{2} - \frac{3}{2} l^{2/3} \right)$$

$$= \frac{3}{2}.$$

Exercise 3.13 For what values of a does the integral $\int_0^1 x^a \, dx$ converge?

Exercise 3.14 By use of integration by parts evaluate $I = \int_0^\infty xe^{-x} \, dx$.

Exercise 3.15 Let

$$\Gamma(n) = \int_0^\infty t^{n-1} e^{-t}\, dt, \quad \text{for} \quad n = 1, 2, 3, \ldots$$

Use integration by parts to show that $\Gamma(n) = (n - 1)\, \Gamma(n - 1)$ for $n > 1$ and hence deduce that $\Gamma(n) = (n - 1)!$.

$\Gamma(x)$, with n replaced by a real variable x, is called the *gamma function* and satisfies the same reduction formula (but the integral cannot be evaluated in an elementary manner except when x is an integer).

3.10. APPROXIMATION OF FUNCTIONS

Consider the function $\cosh(x) = \dfrac{1}{2}\left(e^x + e^{-x}\right)$. As $x \to +\infty$, e^{-x} becomes negligible compared to e^x. Thus $\cosh(x)$ behaves like e^x for large x. Neglecting parts of complicated expressions under certain conditions can often simplify calculations enormously, but we need a precise rule for what we are allowed to do. To emphasize that a rule is involved mathematicians use a special terminology: we say a function $f(x)$ is *asymptotically equivalent* to a function $g(x)$ as $x \to x_0$, and write $f \sim g$ as $x \to x_0$, if f and g behave similarly in the following sense:

Definition: $f(x) \sim g(x)$ as $x \to x_0$ if $\dfrac{f(x)}{g(x)} \to 1$ as $x \to x_0$.

"$f \sim g$" is read "f is asymptotically equivalent to g," but scientists often say "f is approximately equal to g" while using the same $\sim$ notation and meaning the same thing. This definition encapsulates the idea that the difference between the two functions, when x is near x_0, is vanishingly small, relative to the size of the functions themselves.

Note that this is not the same as taking a limit. For example, as $x \to 0$ we have $\sin(x) \to 0$ but (as the next example shows) $\sin(x) \sim x$. To say this again: "$\to$" and "$\sim$" are not the same because the former gives a number and the latter a function.

Example 3.16 Show that $\sin(x) \sim x$ as $x \to 0$.

First we look at the definition. We need to verify that

$$\frac{\sin(x)}{x} \to 1 \text{ as } x \to 0.$$

This will be true if

$$\lim_{x \to 0} \frac{\sin(x)}{x} \to 1 \text{ as } x \to 0.$$

This was proved earlier in Example 3.10.

More generally we may write down formulae such as $\sin(x) \sim x - x^3/6$ as $x \to 0$. What is intended is that $x - x^3/6$ is a more accurate approximation for $\sin(x)$ near $x = 0$ than x is. More precisely $\sin(x) - x \sim -x^3/6$ as $x \to 0$.

Exercise 3.16 Verify that $\sin(x) - x \sim -x^3/6$.

Exercise 3.17 Use the definition of asymptotic equivalence to show that

  (i) $(e^x - 1)^{-1} \sim 1/x$ as $x \to 0$,

  (ii) $(e^x + 1)^{-1} \sim e^{-x}$ as $x \to +\infty$,

  (iii) $(e^x + 1)^{-1} - 1 \sim -e^x$ as $x \to -\infty$,

  (iv) $(\cosh(x)/\sinh(x)) - 1 \sim 2e^{-2x}$ as $x \to +\infty$.

An approximation frequently useful in applications of mathematics is Stirling's formula

$$n! \sim \left(\frac{n}{e}\right)^n \sqrt{2\pi n}, \tag{3.3}$$

from which follow, with increasing degrees of accuracy,

  (i) $\ln(n!) \sim n \ln(n)$,

  (ii) $\ln(n!) \sim n \ln(n) - n$,

(iii) $\ln(n!) \sim n\ln(n) - n + \frac{1}{2}\ln(n),$

(iv) $\ln(n!) \sim n\ln(n) - n + \frac{1}{2}\ln(n) + \frac{1}{2}\ln 2\pi.$

The second one is usually sufficient in physics and should be memorized. Note that you cannot get back to equation (3.3) by exponentiating (ii), although you can do so from the more accurate form (iv). In general asymptotic equivalences do not remain valid when exponentiated.

Exercise 3.18 Use trial and error on a pocket calculator to determine how large n must be for formula (ii) to give an answer accurate to within 10%. For this value of n, how accurate an estimate does this imply for $n!$ itself, and how accurate an estimate does (iv) give for $n!$?

If we are not given the asymptotic form we proceed by trying to identify the leading term, as in the following example.

Example 3.17 Find the asymptotic form for $\cosh(x)$ as $x \to +\infty$.

Identify the dominant term

$$\cosh(x) = (e^x + e^{-x})/2 \sim e^x/2 \text{ as } x \to +\infty,$$

since $e^{-x} \to 0$. Check by using the definition

$$\frac{\cosh(x)}{e^x/2} = \frac{(e^x + x^{-x})}{e^x} = 1 + e^{-2x} \to 1 \text{ as } x \to \infty.$$

Exercise 3.19 Find the asymptotic form for $\sinh^2(x)$ as $x \to +\infty$.

3.11. SEQUENCES AND SERIES

A *sequence* is simply an ordered set of numbers, such as {2, 4, 6, 8}, which may be finite or infinite. In this example, 6 is the third element of the sequence. A *series* is an unevaluated sum of a sequence, in this case 2 + 4 + 6 + 8. The sum of this series is 20. (See section A.7 for a reminder.) We can find the limit of an infinite sequence of numbers from the limit of a "corresponding" function.

Example 3.18 Find the limit of the sequence $1/n$ as $n \to +\infty$ through positive integer values.

The function $1/x$ is equal to the sequence for $x = 1, 2, \ldots$. So the limit of the sequence is the same as $\lim_{x \to +\infty}(1/x) = 0$. This can also be seen directly because $1/n$ becomes and remains arbitrarily close to 0 as n increases through positive integers. The next example is less obvious.

Example 3.19 Find the limit of the sequence $1, 2^{1/2}, 3^{1/3}, \ldots$

The general term is $n^{1/n}$. The function $x^{1/x}$ is equal to the sequence for $x = 1, 2, \ldots$. So the limit of the sequence is the same as $\lim_{x \to +\infty} x^{1/x} = 1$ from Example 3.13.

Exercise 3.20 Find the limits as $n \to +\infty$ of

(i) $\dfrac{n(n-1)}{n^2 - 1}$

(ii) $\left(1 + \dfrac{a}{n}\right)^n$ (Hint: use the result of exercise 3.11(ii).)

3.12. INFINITE SERIES

An important application of limits is to series (sums of sequences). Given a sequence of numbers, such as $\{a_1, a_2, a_3, \ldots, a_n, \ldots\}$, the new sequence defined by

$$S_1 = a_1$$
$$S_2 = a_1 + a_2$$
$$S_3 = a_1 + a_2 + a_3$$
$$\vdots$$
$$S_n = \sum_{i=1}^{n} a_i$$
$$\vdots$$

is the sequence of *partial sums*. If the sequence of partial sums converges to some limit L, then we say the series converges with sum L. If the series of partial sums does not converge, we say the series is divergent.

Example 3.20 What is the sum of the series $S = 1 + 1/2 + 1/4 + 1/8 + \ldots$?

Imagine trying to construct the sum in an obvious way: we start with 1; then add 1/2 to get $1 + 1/2 = 3/2$; then add 1/4 to get $1 + \dfrac{1}{2} + \dfrac{1}{4} = 7/4$ and so on. The sequence $S_1 = 1$, $S_2 = 3/2$, $S_3 = 7/4 \ldots$ has a limit which is the sum of the *infinite series*. To obtain this limit we construct the general term of the sequence S_n. This is the sum of the first n terms

$$S_n = 1 + \frac{1}{2} + \frac{1}{4} + \ldots + \frac{1}{2^{n-1}} = \frac{1 - \dfrac{1}{2^n}}{1 - \dfrac{1}{2}}.$$

The expression on the right comes from the formula for the sum of a geometric progression with ratio 1/2 (see Section A.7). Now to find the limit of the sequence

$$S = \lim_{n \to \infty} S_n = \lim_{n \to \infty} \frac{1 - \dfrac{1}{2^n}}{1 - \dfrac{1}{2}} = 2,$$

(since $1/2^n \to 0$ as $n \to \infty$). Thus

$$1 + 1/2 + 1/4 + \ldots = 2.$$

There are very few series that can be summed in this way because we cannot usually find a simple *closed form*[1] for the nth partial sum, S_n, from which to obtain the limit. (Note that this is not the same thing as a formula for the nth term of the series which is often easy!) Instead, we might try to get an approximation to the sum by summing a sufficiently large finite number of terms. For example, an approximate value for the sum of the series $1 + 1/2 + 1/4 + 1/8 + \ldots$ obtained by summing the first four terms is $S \approx S_4 = 15/8 = 1.875$ (which is in fact in error by 6.25 %). If you add enough terms on your calculator you will end up with the value of $2.000. \ldots$ This is *not* a proof that the sum of the series is 2. You have run out of decimal places on your calculator and you have no proof that the sum of the infinite number of small terms you have not yet considered is less than the accuracy of your calculator.

Here is another example where your calculator would manifestly be useless. Consider the series $S = 1 + 1/2 + 1/3 + 1/4 + \ldots$. The first 4 terms give $S_4 = 2.08$. Is this a good approximation? The next 4 terms give $S_8 = 2.7145. \ldots$ What about 12 terms? Or 100? In fact, the sum creeps up and up very slowly. (The sum of the first 100 million terms is less than 20.) We can guess that even though each additional term is getting smaller there are so many of them that the sum is always growing toward infinity. This is true, but you cannot prove it on your calculator. (This will only show that the sum is beyond the maximum number your calculator will store, not that it is beyond any number any calculator can store!) The following is a proper proof that the series does not have a finite sum.

---

[1] A "closed form" expression is one involving a finite number of mathematical operations on the standard functions. This definition is intentionally a little vague as the set of allowed functions may vary depending on the context.

The essence of the proof is to show that the given series is greater than a series which is known not to have a finite sum. There is no rule for how to do this – just educated guesswork. Here we use the trick of bracketing terms together. With a suitable choice of how to bracket the terms we have

$$S = 1 + 1/2 + 1/3 + 1/4 + 1/5 + 1/6 + 1/7 + 1/8 + \dots$$
$$= 1 + 1/2 + (1/3 + 1/4) + (1/5 + 1/6 + 1/7 + 1/8) + \dots$$
$$\geq 1 + 1/2 + (1/4 + 1/4) + (1/8 + 1/8 + 1/8 + 1/8) + \dots,$$

where the inequality arises because each term in the series has been replaced by something equal or smaller. But for the new series we can form the partial sums:

$$S > 1 + \frac{1}{2} + \frac{1}{2} + \frac{1}{2} + \dots,$$

which clearly goes on getting bigger and bigger, i.e. the series *diverges* to $+\infty$.

Thus, before we try to approximate a series by summing a number of terms, we need to know whether the series does have a finite sum (in which case we say the series *converges*) or whether it diverges. Courses on pure mathematics and mathematical reference books give a large number of methods for doing this, based essentially on comparison with known series. (The large number of methods has arisen because often a particular test will be inconclusive; one goes through them in order until one finds a test that works.) We give an example of comparison in which the bracketing technique given above is used to establish convergence.

Example 3.21 Show that $S = 1 + 1/2^2 + 1/3^2 + 1/4^2 + \dots$ has a finite sum.

We are given some help in the statement of the question that the series is in fact convergent. We therefore want to bracket the terms in such a way that they are smaller than some known convergent series. There is no rule for doing this – you just have to spot the possibilities. (There is usually more than one.)

$$S = 1 + (1/2^2 + 1/3^2) + (1/4^2 + 1/5^2 + 1/6^2 + 1/7^2) + \ldots$$
$$< 1 + (1/2^2 + 1/2^2) + (1/4^2 + 1/4^2 + 1/4^2 + 1/4^2) + \ldots$$
$$= 1 + 2/2^2 + 4/4^2 + \ldots$$
$$= 1 + 1/2 + 1/4 + \ldots = 2.$$

Therefore the series has a finite sum. In fact $S = \pi^2/6$ as can be shown by the use of Fourier series.

The same method works for the series $1 + 1/2^k + 1/3^k + \ldots$ for any constant $k > 1$. Henceforth you may generally assume that $\sum_{n=1}^{\infty} 1/n^k$ is convergent for $k > 1$ and divergent for $k \leq 1$. Convergence or divergence of other series can often be settled by comparison with one of these.

Example 3.22 Determine whether the following two series are convergent or divergent:

(i) $\quad S = \dfrac{2}{1 \times 1} + \dfrac{3}{2 \times 2} + \dfrac{4}{3 \times 3} + \ldots$

(ii) $\quad S = \dfrac{3^2 - 2^2}{1^3} + \dfrac{4^2 - 2^2}{2^3} + \dfrac{5^2 - 4^2}{3^3} + \ldots$

(i) The terms in the numerator are growing linearly, those in the denominator quadratically, so the series looks something like $1/n$. So try to prove it diverges by comparing with $1/n$.

$$S = \dfrac{2}{1 \times 1} + \dfrac{3}{2 \times 2} + \dfrac{4}{3 \times 3} + \ldots$$
$$> \dfrac{1}{1 \times 1} + \dfrac{2}{2 \times 2} + \dfrac{3}{3 \times 3} + \ldots$$
$$= \dfrac{1}{1} + \dfrac{1}{2} + \dfrac{1}{3} \ldots$$

The given series is therefore greater term by term than a known divergent series and hence divergent.

(*ii*) The numerator is screaming out to be factorized as a square minus a square. So we guess it grows linearly, whereas the denominator is cubed. So overall we expect this to look like $1/n^2$, hence to converge. However, it's difficult to see what is happening term by term, so we write out the general term.

$$a_n = \frac{(n+2)^2 - (n+1)^2}{n^3}$$

$$= \frac{2n+3}{n^3}$$

$$= \frac{2}{n^2} + \frac{3}{n^3}.$$

S is the sum of two series each of which is convergent (from the discussion following Example 3.21, and since a constant multiple of a convergent series is convergent)

$$S = \sum_n a_n = 2\sum \frac{1}{n^2} + 3\sum \frac{1}{n^3}.$$

Hence it is convergent.

Direct comparison of the above kind works only for series of positive terms. A series containing both infinitely many positive and infinitely many negative terms will certainly be convergent if the positive and negative parts are separately convergent. Other series such as $1 - 1/2 + 1/3 - 1/4 + \ldots$, or more exotic examples such as $\left(\frac{3}{2}\right)^2 - \left(\frac{2}{1}\right)^1 + \left(\frac{5}{4}\right)^4 - \left(\frac{4}{3}\right)^3 + \left(\frac{6}{5}\right)^5 + \ldots$, in which both positive and negative parts are divergent, cannot be tested in this way. (Although they do converge to $\ln 2$ and $0.4456\ldots$, respectively.)

Exercise 3.21 Show that the series

$$S = \frac{1}{1 \times 2} + \frac{1}{2 \times 3} + \frac{1}{3 \times 4} + \dots$$

has a finite sum, but the series

$$S = \frac{1}{1} + \frac{1}{\sqrt{2}} + \frac{1}{\sqrt{3}} + \dots$$

is divergent. Does the series

$$S = 1 - \frac{1}{2^2} + \frac{1}{3^2} - \frac{1}{4^2} + \dots$$

converge? (Hint: consider the positive and negative parts separately.)

Revision Notes

After completing this chapter you should be able to

- Sketch graphs of *polynomials* and *rational functions*
- Solve *inequalities* graphically
- Write down conditions for a function to be *continuous*
- Identify indeterminate limits and use *l'Hôpital's rule* or series methods to find them
- Define what is meant by *convergence* of a series
- Define *asymptotic approximations* of functions and verify given examples
- Calculate integrals over an infinite range as limits

3.13. EXERCISES

1. Sketch the graph of $y = x^3 - 13x^2 + 40x$, indicating its intersections with the axes and any stationary points and their nature.

2. Show that the derivative of $g(x) = x^4 - 4x^3 + 4x^2$ can be written as $4x(x - 1)(x - 2)$, and hence sketch the graph of $y = g(x)$, indicating the positions of any maxima or minima.

3. State why it is obvious that $x > 3(x - 2)/(x - 4)$ for all large positive values of x. Determine the precise ranges of x for which the inequality is true by rewriting it in the form $(x - A)(x - B)/(x - 4) > 0$ and noting where the left side changes sign, or otherwise.

4. Sketch the region consisting of the points in the x, y plane for which the inequalities $|x| \leq 1$ and $x \leq y \leq 1$ are simultaneously satisfied, describe its shape and determine the vertices.

5. Verify that $\dfrac{1}{\sqrt{n} + \sqrt{n+1}} = \sqrt{n+1} - \sqrt{n}$, and hence write down a formula for the sum of the first n terms of the series

$$\frac{1}{\sqrt{1} + \sqrt{2}} + \frac{1}{\sqrt{2} + \sqrt{3}} + \frac{1}{\sqrt{3} + \sqrt{4}} + \ldots$$ and deduce that this series is divergent. How many terms are needed to give a partial sum greater than 10?

6. Use l'Hôpital's rule or the series for $\sin(x)$ to evaluate the limit of $\dfrac{\sin(x)}{x}$ as $x \to 0$.

7. Write down a short formula for the sum of the first N terms of the infinite series

$$\sum_{n=1}^{\infty} \left[n \sin\left(\frac{\pi}{2n}\right) - (n+1)\sin\left(\frac{\pi}{2(n+1)}\right) \right]$$

and hence show that this series is convergent, with

sum $1 - \dfrac{\pi}{2}$.

8. Show that $\lim\limits_{x \to 0} \dfrac{1 + \cos(2x) - 2\cos(x)}{\sin^2(x)} = -1$ using l'Hôpital's rule.

9. Use l'Hôpitals rule to find the limit of $\dfrac{e^x - e^{-x} - 2x}{x - \sin(x)}$ as $x \to 0$.

10. Show that
$$\lim_{x \to \pi/4} \frac{\sin(x) - \cos(x)}{16x^2 - \pi^2} = \frac{\sqrt{2}}{8\pi}.$$

11. Prove that the series $\displaystyle\sum_{n=1}^{\infty} \left[\frac{1}{2n-1} \cdot \frac{1}{2n+1} \right]$ is convergent.

12. Evaluate the sum $\displaystyle\sum_{n=1}^{\infty} 2^{-n}$. Hence show that the series $\displaystyle\sum_{n=1}^{\infty} 2^{-n^2}$ converges.

13. Evaluate the limit of $(e^x - 1)/x$ as $x \to 0$. Deduce the asymptotic form of $e^x - 1$ as $x \to 0$. Show that $(e^{a/x} - 1)^{-1} \sim x/a$ as $x \to \infty$.

3.14. PROBLEMS

1. The function
$$f(x) = \frac{x^3 + 4x^2 + Ax + 1}{x^2 + 2x} - \frac{x^3 + x^2 - 4}{x^2 - 4}$$

tends to a finite limit L as $x \to -2$. Show that $A = 9/2$, find L, and also discuss the behavior of $f(x)$ as $x \to 0^+$, $x \to 0^-$, $x \to 1$ and $x \to \pm\infty$.

2. Discuss the behavior of

$$\frac{x^3 + x^2 - 4x - 2}{x^2 - 3x + 2} - \frac{x^3 + 3x^2 + 4x}{x^2 - 1}$$

as $x \to \pm 1$, $x \to 2$ and $x \to \infty$.

3. A function $f(z)$ is defined by the formula

$$f(z) = \frac{3z^4 - 9z^3 + 10z - 2}{z^3 - 3z^2 + 2z}. \tag{3.4}$$

(a) Obtain the expression (polynomial plus partial fractions)

$$f(z) = 3z - \frac{1}{z} - \frac{2}{z-1} - \frac{3}{z-2} \tag{3.5}$$

showing the method used. (Do not merely verify that (3.4) reduces to (3.5).)

(b) Assuming that for $|z| < 1$ (and only then)

$$\frac{1}{1-z} = 1 + z + z^2 + ...;$$

obtain the terms a_{-1}, a_0, a_1 in the expression, valid for $0 < |z| < 1$,

$$f(z) = \sum_{n=-1}^{\infty} a_n z^n.$$

(c) Show that if $0 < a < b < 1$ then

$$\int_a^b f(z)\,dz = \frac{3}{2}\left(b^2 - a^2\right) - \ln\left(\frac{b}{a}\right) - 2\ln\left(\frac{1-b}{1-a}\right) - 3\ln\left(\frac{1 - \frac{1}{2}b}{1 - \frac{1}{2}a}\right),$$

and state to what extent the result would differ if $0 < a < 1 < b$ or $1 < a < b < 2$.

4. From the fact that $(\cos(x))' = -\sin(x)$ obtain the derivative of $\sec(x)$ as $\sec(x)\tan(x)$, and also write down the derivative of $\tan(x)$. Derive expressions (in terms of $\sec(x)$

and tan(x)) for the second, third, and fourth derivatives of sec(x). Hence show that all non-zero coefficients in the Maclaurin series for sec(x) are positive, and that this series begins

$$1+\frac{x^2}{2}+\frac{5x^4}{24}+....\ .$$

Deduce a formula of the form

$$\cos(x) + \sec(x) - 2 \sim Ax^p \text{ as } x \to 0.$$

where A and p are constants. Show also that the left side can be written as

$$\frac{4\sin^4\left(\frac{x}{2}\right)}{\cos(x)}$$

and use this to verify your asymptotic formula.

5. It is known that (for $x \neq -1$)

$$\frac{3x-2}{(x+1)(x^2+4)}=\frac{A}{x+1}+\frac{Bx+C}{x^2+4} \qquad (3.6)$$

where A, B, C are constants. Obtain the values of A, B, C by the following method.

(i) Multiply both sides by $x+1$; show that substitution of $x = 1$ in the resulting identity gives $A = 1$.

(ii) If $f(x)$ denotes the right hand side of (3.6) show that $xf(x) \to A + B$ as $x \to +\infty$, and hence that $A + B = 0$.

(iii) Finally, show that C can now be found by putting $x = 0$ in (3.6).

6. From the identity

$$\sum_{n=1}^{\infty}\frac{x^n}{n(n+1)}=\sum_{n=1}^{\infty}(\tfrac{1}{n}-\tfrac{1}{n+1})x^n=\sum_{n=1}^{\infty}\frac{x^n}{n}-\frac{1}{x}\sum_{n=1}^{\infty}\frac{x^{n+1}}{n+1}$$

(valid for $-1 \le x < 1$), deduce the formula

$$1+(\frac{1}{x}-1)\ln(1-x)$$

for the sum of the series on the left. (You may quote the Maclaurin series for $\ln(1-x)$.) What is the limit of this expression as $x \to 1$ from below?

7. Write down the expansion of $\cos(\theta)$ in powers of θ (valid for all real θ).

   (a) By considering the identity $\cos(\theta)\sec(\theta) = 1$, show that the first four coefficients in the expansion

   $$\sec(\theta) = a_0 + a_2\theta^2 + a_4\theta^4 + a_6\theta^6 + \ldots$$

   are given by $a_0 = 1$, $a_2 = 1/2$, $a_4 = 5/24$, $a_6 = 61/720$.

   (b) Hence write down the value of $\dfrac{d^{2k}}{d\theta^{2k}}(\sec(\theta))$ at $\theta = 0$, for $k = 0,1,2,3$.

   (c) Calculate the limit of $(\sec(\theta) - 1)/\theta^2$ as $\theta \to 0$ both from the series and also with the help of l'Hôpital's rule.

8. Write down the power series for e^x and deduce the series for $\cosh(x)$. Hence show that

   $$2\cosh(\sqrt{x}) - 1 < e^x \quad \text{for} \quad x > 0,$$

   and

   $$2\cosh(\sqrt{x}) - 2 \sim e^x - 1 \quad \text{for} \quad x \to 0^+.$$

   Show also that

   $$\int_0^X \cosh(\sqrt{x})\,dx < \frac{1}{2}(e^x + X - 1) \quad \text{for} \quad X > 0.$$

   Are these two quantities asymptotically equal as $X \to 0^+$?

9. (i) Write down the expansion of $1/(1-x)$ in an infinite series of powers of x, and by integration obtain the expansion

   $$-\ln(1-x) = x + \frac{x^2}{2} + \frac{x^3}{3} + \frac{x^4}{4} + \ldots \tag{3.7}$$

(ii) Derive the inequalities

$$x+\frac{x^2}{2}+<-\ln(1-x)<x+\frac{x^2}{2}+\frac{x^3}{3(1-x)}.$$

By choosing a suitable x, show that the value of ln(10/9) differs from 21/200 by less than $\frac{1}{3}\%$.

(iii) Verify that the values for the limit

$$\lim_{x\to 0}(\frac{-\ln(1-x)-x}{x})$$

obtained from (3.7) and from l'Hôpital's rule are equal.

10. (i) It is known that if $\sum a_n$ is a convergent series of positive numbers and $b_n \sim ka_n$ as $n \to \infty$ (where k is a constant) then $\sum b_n$ is also convergent. Given that $\sum_{n=1}^{\infty} n^{-p}$ (where p is a constant) is convergent if and only if $p > 1$, examine for convergence the series

(a) $\sum \dfrac{1}{n(n+1)}$

(b) $\sum \dfrac{1}{(\sqrt{n}+\sqrt{n+1})}$

(ii) Find the first four terms in the Maclaurin expansion of $(1-x)^{-1/2}$, and show that the coefficient a_n of x_n can be written as

$$\frac{1\times 3\times 5\times...\times(2n-1)}{2\times 4\times 6\times...\times 2n}$$

and also show (by multiplying numerator and denominator by $2 \times 4 \times 6 \times \ldots \times 2n$) that this is equal to $(2n)!/2^{2n}(n!)^2$.

(iii) With the help of Stirling's formula deduce that

$a_n \sim 1/\sqrt{(\pi n)}$ as $n \to \infty$ and hence the series $\sum_{n=0}^{\infty} a_n x^n$

is divergent when $x = 1$.

(iv) This series is, however, convergent when $x = -1$; what should be the value S of its sum? Show that one would expect over $N = 3000$ terms to be needed, before the partial sum $\sum_{n=0}^{N-1}(-1)^n a_n$ together with all subsequent partial sums, is within 0.01 of S.

11. (a) Show that $\int(x-1)e^{-x}dx = xe^x + C$ and also find $\int(x-1)e^{-x}dx$.

(b) The function $f(x)$ is defined by

$$f(x) = \begin{cases} (x-1)e^{-x} & \text{for} \quad x > 2, \\ A(x+1)e^x & \text{for} \quad x < 2 \end{cases}$$

and is continuous at $x = 2$. Show that $A = \frac{1}{3}e^{-4}$.

(c) Evaluate, for $X > 2$, the integral $\int_2^X f(x)dx$. What is the *meaning* of $\int_2^{\infty} f(x)dx$ and what is its value?

(d) Also find the values of $\int_Y^2 f(x)dx$ $(Y < 2)$ and $\int_{-\infty}^2 f(x)dx$ and deduce that $\int_{-\infty}^{\infty} f(x)dx = \frac{8}{3}e^{-2}$.

12. (a) Find the values of the function $(\sin(x))/x$ and its first two derivatives at $x = 0$. Hence or otherwise write down the Maclaurin series for $(\sin(x))/x$ up to terms in x^2.

(b) What is the ratio of the values of x which make the truncated Maclaurin series in part (a) and the original function $(\sin(x))/x$ zero? (Use the smallest possible absolute value of x in the latter case.)

4

VECTORS

The physical world (as we perceive it) exists in three spatial dimensions. Physical quantities such as velocities – which have both magnitude and direction – need three components for their specification. We call these *vectors*. (In fact, even more than this, a chemical system, or an economic model may have, say, 20 components; such a system can often be represented conveniently by a vector in the appropriate number of dimensions (20 in this example): the algebra is the same.) You are going to learn to do algebra with vectors (add and multiply them) in a geometrically meaningful way.

The first part of the chapter deals with the algebra and the second with the geometrical picture. You can think of this in two ways: either that the algebra enables you to do complicated geometrical things rather mechanically (by following the algebraic rules), or that the geometry enables you to get a picture of the algebra. In either

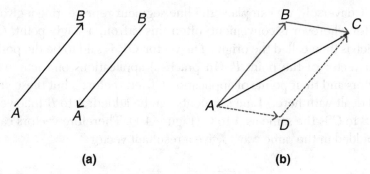

(a)   **(b)**

FIGURE 4.1: (a) The vectors $\overrightarrow{AB}$ and $\overrightarrow{BA}$. (b) The parallelogram law of vector addition.

case you have to learn to switch smoothly between the two methods, and this requires practice. Your reward will be command of a mathematical tool that pervades physics.

4.1. BASIC PROPERTIES

Definition: A vector is an object with a magnitude and a direction.

Therefore any vector can be *represented* by a line segment with the specified length (or magnitude) and direction. Note that this picture provides a representation of any physical quantity, for example a force, that can be described by a magnitude and direction in space. It does not say that a vector is an arrow, only that they behave geometrically in the same way. (Bodies are accelerated by forces, not by arrows on a page.) This is important: many physical quantities are not represented by spatial arrows (velocities and forces in relativity theory, for example).

A line segment with end-points A and B can be written $\overrightarrow{AB}$. The line segment $\overrightarrow{BA}$ is a vector of the same magnitude but opposite direction (indicated by the direction of the arrowhead in Figure 4.1).

Note that vectors as defined here ("free vectors") do not have a specific point of application. So $\overrightarrow{AB} = -\overrightarrow{BA}$ in Figure 4.1 even though the line segments do not lie on one another; indeed all line segments of the same magnitude and direction represent the same vector.

Conversely, we can place the line segment representing a given vector wherever is convenient; often this is from a single point, O, which is then called the origin. The vector $\overrightarrow{OP}$ is said to be the position vector of the point P. (In practical applications one may use vectors and their points of application ("fixed vectors") but these are not dealt with here.) Line segments can be added: A to B followed by B to C is the same as A to C (Figure 4.1). Therefore vectors can be added in the same way to give a resultant vector:

$$\overrightarrow{AB} + \overrightarrow{BC} = \overrightarrow{AC}. \tag{4.1}$$

The sum is in the same plane as the summands. Equation (4.1) is usually referred to as the *parallelogram law*. Figure 4.1 shows also that vectors can be added in any order: we have also

$$\overrightarrow{AB} + \overrightarrow{BC} = \overrightarrow{AD} + \overrightarrow{DC} = \overrightarrow{AC}.$$

Line segments can be multiplied by real numbers to give further line segments; for example in Figure 4.2, A to C is one and a half times A to B. Therefore vectors can be multiplied in the same way:

$$\overrightarrow{AC} = 1.5\overrightarrow{AB}. \tag{4.2}$$

Here $\overrightarrow{AC}$ is the vector in the same direction as $\overrightarrow{AB}$, 1.5 times as long. (In this context real numbers are also called *scalars*, so in equation (4.2) we say that the vector is multiplied by the scalar 1.5.) We can do both multiplication by scalars and addition, so, if λ and μ are scalars, expressions like

$$\lambda\overrightarrow{AB} + \mu\overrightarrow{CD}$$

are vectors. Such expressions are referred to as *linear combinations* of $\overrightarrow{AB}$ and $\overrightarrow{CD}$.

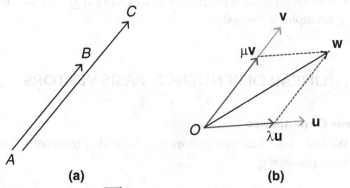

(a)   **(b)**

FIGURE 4.2: (a) $\overrightarrow{AC} = 1.5\,\overrightarrow{AB}$. (b) Linear dependence, with $\mathbf{w} = \lambda\mathbf{u} + \mu\mathbf{v}$.

Exercise 4.1 If $\overrightarrow{AB}$ is 2 units due north and $\overrightarrow{AC}$ is 1 unit due east, what is $\overrightarrow{AB} + 2\overrightarrow{AC}$?

Exercise 4.2 A person swims across a river with a speed relative to the water of 4 ft s^{-1} at right angles to the current which is flowing at 3 ft s^{-1}. Draw a figure to show the swimmer's resultant motion and check that it accords qualitatively with your intuition.

Exercise 4.3 Rain falls vertically at 10 m s^{-1}; draw a vector diagram to show how it appears to a person running at 5 m s^{-1}.

Exercise 4.4 A, B, C, D are four points in space; and the segments AB, BC, CD, DA have mid-points P, Q, R, S, respectively. Show that

(i) $2\overrightarrow{AP} = \overrightarrow{AB}$

(ii) $2\overrightarrow{PQ} = \overrightarrow{AC}$

(iii) PQRS is a parallelogram.

Usually, instead of writing $\overrightarrow{AB}$, say, for the line segment representing a vector, we give a vector a name, **v** say. The bold font indicates that this is the name of a vector; in hand written work we use $\underline{v}$ or $\vec{v}$ or $\utilde{v}$.

- The *magnitude* of a vector **v** is written as $|\mathbf{v}|$ or v.

- A linear combination of two vectors **u** and **v** would be written $\lambda\mathbf{u} + \mu\mathbf{v}$.

- The *zero vector* is printed as **0** (and written $\underline{0}$), so that, for example, $\mathbf{u} - \mathbf{u} = \mathbf{0}$.

4.2. LINEAR DEPENDENCE; BASIS VECTORS

Linear Dependence

We have seen that two vectors may be added to make a third in the same plane; e.g.

$$\mathbf{w} = \lambda\mathbf{u} + \mu\mathbf{v}. \tag{4.3}$$

If a relation of the form (4.3) holds we say that **u**, **v**, and **w** are *linearly dependent*.

Conversely, if **u**, **v**, and **w** are vectors lying in a plane, with **u** and **v** not parallel, then we can always find λ and μ such that equation (4.3) holds by completing the parallelogram of vectors as in Figure 4.2(b). Thus, in the plane any two (non-parallel) vectors may be given independently but any third vector must be expressible in terms of these two as in equation (4.3). Thus, any three (non-parallel) vectors in the plane are linearly dependent and any one can be expressed in terms of the other two.

A set of vectors is said to be *linearly independent* if no vector in the set can be decomposed into a linear sum of the other vectors. If we rearrange equation (4.3) we get the criterion for linear dependence

$$\lambda \mathbf{u} + \mu \mathbf{v} + \nu \mathbf{w} = \mathbf{0}. \tag{4.4}$$

If this can only be satisfied when the coefficients are all zero (i.e. $\lambda = \mu = \nu = 0$) then the vectors **u**, **v**, **w** are linearly independent. There is a close connection between the maximum number of linearly independent vectors and the dimension of a given space: the dimension of a "vector space" is the largest number of linearly independent vectors we can choose.

In three-dimensional space we can specify three non-coplanar vectors arbitrarily (say **u**, **v**, **w**) and construct any other vector (say **x**) as a linear combination of these:

$$\mathbf{x} = \lambda \mathbf{u} + \mu \mathbf{v} + \nu \mathbf{w}.$$

We say that **x**, **u**, **v**, and **w** are linearly dependent. It follows that any four (or more) vectors in three-dimensional space are linearly dependent and any one can be expressed in terms of the other three.

Basis Vectors

We can use these observations to provide a useful way of specifying vectors. We make a convenient choice of the maximum number of independent vectors (two non-parallel vectors in two dimensions, three non-coplanar vectors in three dimensions). These we call our *basis vectors*. Any other vector in that space can then expressed as a linear combination of these basis vectors.

It is often most convenient to choose the basis vectors to be of unit length and mutually perpendicular. Two such vectors in the plane are usually called **i** and **j**. In three-dimensional space, three such vectors are labelled **i**, **j**, **k** (see Figure 4.3). These are said to form an *orthonormal basis*: the *ortho-* indicates they are mutually orthogonal, the *normal* means they are unit vectors (normalized to have unit magnitude, i.e. of length 1) and *basis* means that they are linearly independent of each other, and every other vector is linearly dependent on them (can be made from a linear combination of the basis vector).

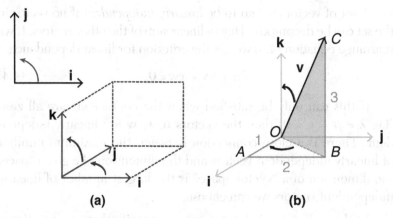

FIGURE 4.3: (a) Basis vectors in 2D and 3D. (b) Diagram for example 4.1.

Unless stated otherwise, **i**, **j**, **k** are taken to be oriented to form a right–handed set; i.e. a rotation from **i** to **j** is clockwise when viewed in the direction of **k**. Sometimes the fact that these are vectors of unit length ("unit vectors") is symbolized by a carat symbol (a "hat")– e.g. $\hat{\mathbf{i}}, \hat{\mathbf{j}}, \hat{\mathbf{k}}$. However, we shall take it that the notation **i**, **j**, **k** itself signifies unit vectors.

Example 4.1 Find the magnitude and direction of the vector **v** = **i** + 2**j** + 3**k**.

The vector **v** is obtained by starting at the origin O and moving 1 unit along the **i** direction (call this point A), then 2 along the **j** direction (call this point B), and 3 along the **k** direction (call this point C). The length of **v** (line $\overrightarrow{OC}$) can be calculated using

Pythagoras's theorem on the triangles OAB (to give the length of $\overrightarrow{OB}$) and OBC (to give the length of $\overrightarrow{OC}$).

The length of the hypotenuse of OAB is $\left|\overrightarrow{OB}\right| = \sqrt{1^2 + 2^2} = \sqrt{5}$. The length of the hypotenuse of OBC is $\left|\overrightarrow{OC}\right| = |\mathbf{v}| = \sqrt{1^2 + 2^2 + 3^2} = \sqrt{14}$ units.

By considering the triangle OBC we see the angle between $\mathbf{v}$ and $\mathbf{k}$ is $\cos^{-1}\left(3/\sqrt{14}\right)$; by considering the triangle OAB we see the angle between the plane containing $\mathbf{v}$ and $\mathbf{k}$ and the vector $\mathbf{i}$ is $\cos^{-1}\left(1/\sqrt{5}\right)$.

Exercise 4.5 Draw a figure showing the magnitude and direction of the vector $\mathbf{v} = \mathbf{i} + \mathbf{j} + \sqrt{2}\mathbf{k}$.

In general, the magnitude of a vector $\mathbf{v} = v_1\mathbf{i} + v_2\mathbf{j} + v_3\mathbf{k}$ is

$$|\mathbf{v}| = \left(v_1^2 + v_2^2 + v_3^2\right)^{1/2}.$$

You should remember this formula.

Example 4.2 If $\mathbf{a} = \mathbf{i} - 3\mathbf{j} - 6\mathbf{k}$ and $\mathbf{b} = 7\mathbf{i} + 2\mathbf{k}$, find $3\mathbf{a} + 2\mathbf{b}$.

To get $3\mathbf{a}$ we multiply the coefficients in front of each of the basis vectors by 3; similarly to get $2\mathbf{b}$. Then we add the coefficients for each of the basis vectors.

$$3\mathbf{a} + 2\mathbf{b} = 3(\mathbf{i} - 3\mathbf{j} - 6\mathbf{k}) + 2(7\mathbf{i} + 2\mathbf{k})$$
$$= 3\mathbf{i} - 9\mathbf{j} - 18\mathbf{k} + 14\mathbf{i} + 4\mathbf{k}$$
$$= 17\mathbf{i} - 9\mathbf{j} - 14\mathbf{k}.$$

Exercise 4.6 If $\mathbf{u} = 3\mathbf{i} + \mathbf{j}$ and $\mathbf{v} = \mathbf{i} - 2\mathbf{j} + 0.5\mathbf{k}$ find expressions, in terms of the basis vectors $\mathbf{i}, \mathbf{j}, \mathbf{k}$, for

(i) $\mathbf{u} + \mathbf{v}$,

(ii) $2\mathbf{v} - \mathbf{u}$.

Example 4.3 Find the unit vector in the direction of the vector
v = **i** + 2**j** + 3**k**.

To obtain the unit vector we must divide each component
by the magnitude of **v**. We know from example 4.1 that
$|\mathbf{v}| = \sqrt{1^2 + 2^2 + 3^3} = \sqrt{14}$, so the unit vector is

$$\hat{\mathbf{v}} = \frac{\mathbf{v}}{|\mathbf{v}|} = \frac{1}{\sqrt{14}}(\mathbf{i} + 2\mathbf{j} + 3\mathbf{k}).$$

The introduction of a set of basis vectors provides the crucial link between the geometrical picture of a vector (an object
with magnitude and direction) and the algebra of vectors (objects
formed from linear combinations of the basis vectors). By passing
between the two approaches we can formulate a problem in geometric terms, solve it using algebra and then return to a geometric
picture, if that is appropriate. We shall see many examples of this
in what follows: it is one of the reasons why vectors are so useful in
physical science.

4.3. THE SCALAR (DOT) PRODUCT

We have seen how to add vectors both geometrically and in
terms of a basis, and we have seen how to multiply a vector by a
scalar but so far we have given no meaning to multiplying vectors.
We are free to choose a definition for the product of two vectors,
but if the resulting construct is to have any use it must correspond
to some geometrical operation on directed line segments. It turns
out that there are two such operations. The first of these is called
the scalar or "dot" product and is defined here. Later we shall
meet another product called the vector product. These are the
only two meaningful ways of multiplying vectors. (Actually there
is another, called the tensor product, but that will not concern
us here.)

Definition: Let θ be the angle between the two vectors **a** and **b**. The *scalar product* (or "dot product") of two vectors **a** and **b**, written **a** · **b**, is defined to be the number $|\mathbf{a}||\mathbf{b}| \cos(\theta)$:

$$\mathbf{a} \cdot \mathbf{b} = |\mathbf{a}||\mathbf{b}| \cos(\theta) \qquad (4.5)$$

Note that the definition implies the important result

$$\mathbf{a} \cdot \mathbf{a} = |\mathbf{a}|^2.$$

Notice that the scalar product takes two vectors and produces one scalar (a number). Also notice, from the definition (Equation 4.5), that the scalar product is *commutative*:

$$\mathbf{a} \cdot \mathbf{b} = \mathbf{b} \cdot \mathbf{a} \qquad (4.6)$$

(the order of the two vectors in the scalar product does not affect the result).

If $\hat{\mathbf{n}}$ is a vector of unit length making an angle θ with a vector **v**, then

$$\hat{\mathbf{n}} \cdot \mathbf{v} = |\hat{\mathbf{n}}||\mathbf{v}|\cos(\theta) = |\mathbf{v}|\cos(\theta)$$

is the (orthogonal) projection of **v** on $\hat{\mathbf{n}}$ (see Figure 4.4). In Section 4.9 we shall find an alternative expression for the projection of one vector on another that does not explicitly involve the angle θ.

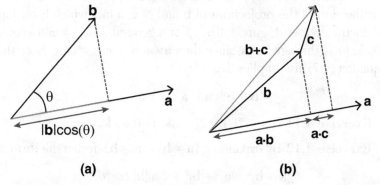

(a) **(b)**

FIGURE 4.4: (a) Projection of **b** and **a**. (b) The sum of projections onto vector **a** (which can be assumed to be a unit vector).

Example 4.4 If **a** is a vector of magnitude 3, and **b** a vector of magnitude 1/2, and if the angle between **a** and **b** is 60°, what is **a** · **b**?

$$\mathbf{a} \cdot \mathbf{b} = |\mathbf{a}||\mathbf{b}|\cos\theta = 3 \times \frac{1}{2} \times \cos(60°) = 3 \times \frac{1}{2} \times \frac{1}{2} = \frac{3}{4}.$$

Exercise 4.7 If **a** · **b** = 0 and $|\mathbf{a}| \neq 0$, $|\mathbf{b}| \neq 0$, show that **a** and **b** are orthogonal (i.e. perpendicular to one another).

Exercise 4.8

(i) What is **a** · **b** if **a** and **b** are parallel? When is **a** · **b** = $-|\mathbf{a}||\mathbf{b}|$?

(ii) Show that **i** · **i** = **j** · **j** = **k** · **k** = 1 and that **i** · **j** = **j** · **k** = **k** · **i** = 0.

Exercise 4.9 If $\mathbf{a} \cdot \mathbf{b} = 1/\sqrt{2}$ and $|\mathbf{a}| = |\mathbf{b}| = 1$, what is the angle between **a** and **b**?

Exercise 4.10 Show that the scalar product may be taken in either order; i.e. prove equation (4.6).

We shall need the distributive law for the scalar product

$$\mathbf{a} \cdot (\mathbf{b} + \mathbf{c}) = \mathbf{a} \cdot \mathbf{b} + \mathbf{a} \cdot \mathbf{c}. \tag{4.7}$$

To establish equation (4.7) assume first that **a** is a unit vector. Then equation (4.7) says that the orthogonal projection of **b** + **c** on **a** is the sum of the projections of **b** and of **c**, a fact which is obvious geometrically (see Figure 4.4(b)). For a general **a** (not a unit vector) divide (4.7) through by $|\mathbf{a}|$ since then $\mathbf{a}/|\mathbf{a}|$ is a unit vector. Note that equation (4.7) also implies that

$$(\mathbf{b} + \mathbf{c}) \cdot \mathbf{a} = \mathbf{a} \cdot \mathbf{b} + \mathbf{a} \cdot \mathbf{c}.$$

Exercise 4.11 If **a** = 7**i** + 0.2**j** + 3**k**, find **a** · **k**.

Exercise 4.12 By expanding (**a** + **b**) · (**a** + **b**) derive the formula

$$|\mathbf{a} + \mathbf{b}|^2 = |\mathbf{a}|^2 + |\mathbf{b}|^2 + 2|\mathbf{a}||\mathbf{b}|\cos(\theta),$$

where θ is the angle between **a** and **b**. Draw a figure to demonstrate that this is the usual cosine rule of trigonometry. Also deduce that $|\mathbf{a} + \mathbf{b}| \leq |\mathbf{a}| + |\mathbf{b}|$ (which is called the "triangle inequality").

Here we summarize the most important algebraic properties of the scalar product

$$\mathbf{a} \cdot \mathbf{a} = |\mathbf{a}|^2$$
$$\mathbf{a} \cdot \mathbf{b} = \mathbf{b} \cdot \mathbf{a} \qquad \text{(commutative)}$$
$$\mathbf{a} \cdot (\mathbf{b} + \mathbf{c}) = \mathbf{a} \cdot \mathbf{b} + \mathbf{a} \cdot \mathbf{c} \qquad \text{(distributive)}$$

4.4. THE VECTOR (CROSS) PRODUCT

Our second way of multiplying vectors is the vector or "cross" product.

Definition: Let θ be the angle between the two vectors $\mathbf{a}$ and $\mathbf{b}$. The vector product (also called the "cross product") of $\mathbf{a}$ and $\mathbf{b}$, written $\mathbf{a} \times \mathbf{b}$ (or sometimes $\mathbf{a} \wedge \mathbf{b}$) and read as "a cross b", is defined to be the vector of magnitude $|\mathbf{a}||\mathbf{b}|\sin(\theta)$ in a direction $\hat{\mathbf{n}}$ perpendicular to both $\mathbf{a}$ and $\mathbf{b}$ such that $\mathbf{a}$, $\mathbf{b}$ and $\hat{\mathbf{n}}$ form a right-handed set. Thus rotation from $\mathbf{a}$ to $\mathbf{b}$ when viewed in the direction of $\hat{\mathbf{n}}$ is clockwise (see Figure 4.5) and

$$\mathbf{a} \times \mathbf{b} = |\mathbf{a}||\mathbf{b}|\sin(\theta)\hat{\mathbf{n}}. \qquad (4.8)$$

Notice the $\hat{\mathbf{n}}$ in the definition: the vector product gives a vector; the scalar product gives a scalar!

Example 4.5 Show that if $\mathbf{a}$ is parallel (or anti-parallel) to $\mathbf{b}$, then $\mathbf{a} \times \mathbf{b} = \mathbf{0}$ and conversely.

$|\mathbf{a} \times \mathbf{b}| = |\mathbf{a}||\mathbf{b}|\sin(\theta) = 0$ if and only if $\sin(\theta) = 0$ or $\mathbf{a} = \mathbf{0}$ or $\mathbf{b} = \mathbf{0}$. So, $|\mathbf{a} \times \mathbf{b}| = \mathbf{0}$ if and only if $\theta = 0$ (or π), i.e. if and only if $\mathbf{a}$ and $\mathbf{b}$ are parallel (or anti-parallel). If $\mathbf{a} = \mathbf{0}$ or $\mathbf{b} = \mathbf{0}$ it is regarded as parallel to every vector.

■ **Exercise 4.13** What is $\mathbf{a} \times \mathbf{b}$ if $\mathbf{a}$ and $\mathbf{b}$ are perpendicular?

■ **Exercise 4.14** Show that $\mathbf{i} \times \mathbf{j} = \mathbf{k}$. What are $\mathbf{j} \times \mathbf{k}$ and $\mathbf{k} \times \mathbf{i}$?

■ **Exercise 4.15** Show that $|\mathbf{a} \times \mathbf{b}|$ is the area of the parallelogram formed from $\mathbf{a}$ and $\mathbf{b}$ in Figure 4.5(b).

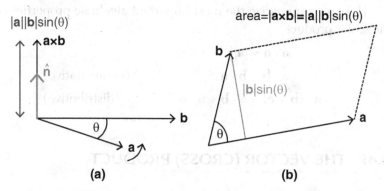

FIGURE 4.5: (a) The vector product $\mathbf{a} \times \mathbf{b}$. (b) The vector product $\mathbf{a} \times \mathbf{b}$ interpreted as a vector with magnitude equal to the area of the parallelogram formed by $\mathbf{a}$ and $\mathbf{b}$.

▌ **Exercise 4.16** Show that $\mathbf{a} \times \mathbf{b} = -(\mathbf{b} \times \mathbf{a})$.

Unlike the scalar product, the order in which we take the vector product does matter. We need the distributive law for vector multiplication:

$$\mathbf{a} \times (\mathbf{b} + \mathbf{c}) = \mathbf{a} \times \mathbf{b} + \mathbf{a} \times \mathbf{c}.$$

We won't prove this relation here (the proof is quite tedious) but you can look it up if you are interested in the details.

▌ **Exercise 4.17** If $\mathbf{v} = 3\mathbf{i} + 2\mathbf{j} - \mathbf{k}$, show that $\mathbf{v} \times \mathbf{k} = 2\mathbf{i} - 3\mathbf{j}$. What is $\mathbf{k} \times \mathbf{v}$?

Here we summarize the most important algebraic properties of the vector product:

$$\mathbf{a} \times \mathbf{a} = 0$$

$$\mathbf{a} \times \mathbf{b} = -(\mathbf{b} \times \mathbf{a}) \qquad \text{(anti-commutative)}$$

$$\mathbf{a} \times (\mathbf{b} + \mathbf{c}) = (\mathbf{a} \times \mathbf{b}) + (\mathbf{a} \times \mathbf{c}) \qquad \text{(distributive)}$$

Note that

$$\mathbf{a} \times (\mathbf{b} \times \mathbf{c}) \neq (\mathbf{a} \times \mathbf{b}) \times \mathbf{c} \qquad \text{(not associative)}$$

in general.

4.5. MULTIPLE PRODUCTS

From three vectors **a**, **b**, **c** we can form the scalar (i.e. number) (**a**×**b**) · **c**, by taking the scalar product of **c** with the vector product of **a** and **b**. This is called the scalar triple product of **a**, **b**, and **c**

a × **b** is the vector with magnitude equal to the area of the base of the parallelepiped formed from **a**, **b**, and **c** (Exercise 4.15), in the direction perpendicular to the plane of **a** and **b** (Figure 4.6). Therefore

$$(\mathbf{a} \times \mathbf{b}) \cdot \mathbf{c} = (\text{area of base}) \times (\text{perpendicular height})$$
$$= \text{volume } V \text{ of the parallelepiped.}$$

This assumes that **a**, **b**, **c** are a right-handed set, by which we mean that **a** × **b** makes an acute angle with **c**; otherwise (**a** × **b**) · **c** = −**V** (the sign is reversed).

> **Exercise 4.18** By considering the volume represented in each case show that
>
> $$(\mathbf{a} \times \mathbf{b}) \cdot \mathbf{c} = (\mathbf{c} \times \mathbf{a}) \cdot \mathbf{b} = (\mathbf{b} \times \mathbf{c}) \cdot \mathbf{a}. \qquad (4.9)$$
>
> Deduce that (**a** × **b**) · **c** = **a** · (**b** × **c**).

The scalar triple product is usually written as [**a**, **b**, **c**]. From exercise 4.18, [**a**, **b**, **c**] may be taken to mean (**a** × **b**) · **c** or **a** · (**b** × **c**) (or any of the permutations of Exercise 4.18); i.e. we can interchange the dot and cross, as well as permute the vectors cyclically. If we interchange the vectors in the vector product then this will change the sign of the result. Thus, for example (**b** × **a**) · **c** = − (**a** × **b**) · **c**.

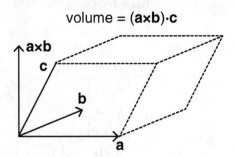

volume = (a×b)·c

FIGURE 4.6: The scalar triple product and the volume of a parallelepiped.

In terms of the notation [a, b, c] for the scalar triple product the above properties can be written

[a, b, c] = [c, a, b] = [b, c, a] = – [a, c, b] = –[c, b, a] = – [b, a, c].

Exercise 4.19 Show that [a, b, c] =0 if and only if a, b, c lie in a plane.

Exercise 4.20 Show that

$$A = \frac{b \times c}{[a, b, c]}$$

satisfies $A \cdot a = 1$, $A \cdot b = 0$, $A \cdot c = 0$ (if $[a, b, c] \neq 0$).

Exercise 4.21 Let a, b, c be three non-coplanar vectors and suppose that $v = v_1 a + v_2 b + v_3 c$. By taking the scalar product of v with the vector $A = \dfrac{1}{[a, b, c]} b \times c$ show that

$$v_1 = \frac{[v, b, c]}{[a, b, c]},$$

and find similar expressions for v_2 and v_3.

The vector A in Exercise 4.20 is called the *reciprocal vector* of a. A set of three vectors a, b, c has a reciprocal set of vectors A, B, C as follows:

$$A = \frac{b \times c}{[a, b, c]},$$

$$B = \frac{c \times a}{[a, b, c]},$$

$$C = \frac{a \times b}{[a, b, c]}. \tag{4.10}$$

These exist if [**a**, **b**, **c**] ≠ 0, which means that **a**, **b**, **c** are not coplanar (Exercise 4.19). Reciprocal vectors are particularly useful in the study of crystal structure.

From three vectors **a**, **b**, **c** we can also form the vector (**a** × **b**) × **c**, called the *vector triple product*. We shall later establish a formula, equation (4.22), for this product in terms of the operations we have already met, so no special symbol is needed.

Exercise 4.22 In the plane of which two of the vectors **a**, **b**, **c** does (**a** × **b**) × **c** lie?

4.6. EQUATION OF A LINE

In this section and the next we shall treat some geometrical properties of lines and planes directly in terms of vectors. Only after we have demonstrated how this can be done do we introduce coordinates and components which enable us (if we wish) to do these calculations algebraically.

Let **r** be the position vector of a general point P on the straight line l in the direction of the unit vector $\hat{\mathbf{t}}$ passing through the point A having position vector **a** (Figure 4.7), and let $\overrightarrow{AP}$ have length λ. Then

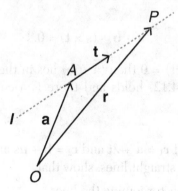

FIGURE 4.7: Equation of a line: $\mathbf{r} = \mathbf{a} + \lambda\hat{\mathbf{t}}$.

$$\mathbf{r} = \mathbf{a} + \lambda\hat{\mathbf{t}}. \tag{4.11}$$

As λ varies between $-\infty$ and $+\infty$ the points P (reached by the position vector $\mathbf{r}$) describe the whole of the line l. Thus equation (4.11) is the equation of the line l.

Example 4.6 Let $\mathbf{r}_1 = \mathbf{a} + \lambda\mathbf{t}$, $\mathbf{r}_2 = \mathbf{b} + \mu\mathbf{s}$ be the equations of two non-parallel straight lines as λ and μ vary (with $|\mathbf{s}| = |\mathbf{t}| = 1$). Show that the two lines intersect if and only if $(\mathbf{a} - \mathbf{b}) \cdot (\mathbf{s} \times \mathbf{t}) = 0$.

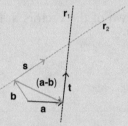

Since the lines are not parallel they intersect if, and only if, $(\mathbf{a} - \mathbf{b})$, $\mathbf{s}$ and $\mathbf{t}$ lie in a plane (see figure). Three vectors lie in a plane if their scalar triple product vanishes (exercise 4.19). That is, $(\mathbf{a} - \mathbf{b}) \cdot (\mathbf{s} \times \mathbf{t}) = 0$.

(**Alternative solution**) If the lines intersect $\mathbf{r}_1$ and $\mathbf{r}_2$ have a common value at a point of intersection. So, the two lines intersect where

$$\mathbf{a} + \lambda\mathbf{t} = \mathbf{b} + \mu\mathbf{s}, \tag{4.12}$$

for some λ and μ, i.e. if

$$\mathbf{a} - \mathbf{b} = \mu\mathbf{s} - \lambda\mathbf{t}.$$

Since $\mathbf{s} \cdot (\mathbf{s} \times \mathbf{t}) = 0$ and $\mathbf{t} \cdot (\mathbf{s} \times \mathbf{t}) = 0$, we take the scalar product of both sides with $\mathbf{s} \times \mathbf{t}$ to get rid of the unknowns λ and μ. Hence

$$(\mathbf{a} - \mathbf{b}) \cdot (\mathbf{s} \times \mathbf{t}) = 0.$$

If $(\mathbf{a} - \mathbf{b}) \cdot (\mathbf{s} \times \mathbf{t}) = 0$ then $(\mathbf{a} - \mathbf{b})$ lies in the plane of $\mathbf{s}$ and $\mathbf{t}$, so equation (4.12) holds and there is a common point of intersection.

Exercise 4.23 If $\mathbf{r}_1 = \mathbf{a} + \lambda\mathbf{t}$ and $\mathbf{r}_2 = \mathbf{b} + \mu\mathbf{s}$ are the equations of two non-parallel straight lines, show that

  (i) $\mathbf{a} - \mathbf{b}$ is a vector joining the lines,

  (ii) $\dfrac{\mathbf{s} \times \mathbf{t}}{|\mathbf{s} \times \mathbf{t}|}$ is a unit vector perpendicular to both lines, and hence

> (iii) the shortest distance p between the two lines satisfies
> $|[\mathbf{a}, \mathbf{s}, \mathbf{t}] - [\mathbf{b}, \mathbf{s}, \mathbf{t}]| = p \, |\mathbf{s} \times \mathbf{t}|$.
>
> (Hint: the shortest distance is the length of the projection of any vector joining the lines on the direction perpendicular to both lines.)

4.7. EQUATION OF A PLANE

Let $\mathbf{a}$ be the position vector of a point A on the plane Σ which contains the unit vectors $\mathbf{s}$ and $\mathbf{t}$. Let $\mathbf{r}$ be the position vector of the point R obtained by moving from A a distance λ in the direction $\mathbf{s}$ and μ in the direction $\mathbf{t}$ (Figure 4.8); then

$$\mathbf{r} = \mathbf{a} + \lambda\mathbf{s} + \mu\mathbf{t}. \tag{4.13}$$

As λ and μ vary, R describes the whole plane Σ. Thus equation (4.13) is the equation of Σ.

Example 4.7 Find a vector normal to the plane Σ given by equation (4.13).

$\mathbf{s} \times \mathbf{t}$ is normal to $\mathbf{s}$ and to $\mathbf{t}$ by the properties of the vector product. You can see that $\mathbf{s} \times \mathbf{t}$ is normal to any vector in the plane by drawing $\mathbf{s} \times \mathbf{t}$ in the diagram. Formally, any vector in the plane can be written as $\mathbf{r} - \mathbf{a} = \lambda\mathbf{s} + \mu\mathbf{t}$ for some λ and μ and hence $(\mathbf{r} - \mathbf{a}) \cdot \mathbf{s} \times \mathbf{t} = 0$ (see Example 4.6). So, the vector $\mathbf{s} \times \mathbf{t}$ is normal to both $\mathbf{s}$ and $\mathbf{t}$ and hence it must also be normal to Σ.

Example 4.8 Show that as $\mathbf{r}$ ranges over points in Σ (equation 4.13)

$$\mathbf{r} \cdot (\mathbf{s} \times \mathbf{t}) = \text{constant}.$$

Since the equation we are aiming at does not involve λ or μ we try to eliminate these terms. This can always be done by taking the scalar product of the equation with a vector perpendicular to both, as in Example 4.6. Let

$$\mathbf{r} = \mathbf{a} + \lambda\mathbf{s} + \mu\mathbf{t},$$

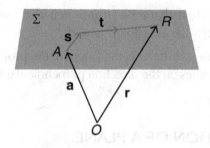

FIGURE 4.8: Equation of a plane: **r = a + λs + μt.**

so,

$$\mathbf{r} \cdot (\mathbf{s} \times \mathbf{t}) = \mathbf{a} \cdot (\mathbf{s} \times \mathbf{t}) + \lambda \mathbf{s} \cdot (\mathbf{s} \times \mathbf{t}) + \mu \mathbf{t} \cdot (\mathbf{s} \times \mathbf{t})$$

$$= \mathbf{a} \cdot (\mathbf{s} \times \mathbf{t}) + 0 + 0$$

$$= \text{constant.}$$

We let $\hat{\mathbf{n}} = \dfrac{\mathbf{s} \times \mathbf{t}}{|\mathbf{s} \times \mathbf{t}|}$; then it follows from Example 4.8 that we can write, for any **r** in the plane Σ,

$$\mathbf{r} \cdot \hat{\mathbf{n}} = p, \qquad (4.14)$$

where $p = \mathbf{a} \cdot \hat{\mathbf{n}} = \text{constant}$ (by Example 4.8). Since equation (4.14) holds if and only if **r** is in Σ, this is also (an alternative form of) the equation of the plane Σ. In applications we use whichever of equation (4.13) and equation (4.14) is more convenient.

Example 4.9 Show that $p = |\mathbf{a} \cdot \hat{\mathbf{n}}|$ is the shortest distance to the plane Σ from the origin O.

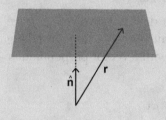

The shortest distance occurs along the direction from the origin perpendicular to the plane – i.e. in the direction of the normal $\hat{\mathbf{n}}$. This distance is the projection of **r** (any vector to the plane) on $\hat{\mathbf{n}}$. The distance is $\mathbf{r} \cdot \hat{\mathbf{n}} = p$.

Exercise 4.24 Show that two distinct planes with unit normals $\hat{\mathbf{n}}_1$ and $\hat{\mathbf{n}}_2$ intersect if and only if $\hat{\mathbf{n}}_1 \times \hat{\mathbf{n}}_2 \neq 0$.

4.8. COMPONENTS OF VECTORS

We turn now to the algebraic approach to vectors. We have seen that any vector **a** (in three dimensions) may be represented by introducing unit basis vectors **i**, **j**, **k**:

$$\mathbf{a} = a_1\mathbf{i} + a_2\mathbf{j} + a_3\mathbf{k}. \qquad (4.15)$$

Thus, we can refer to **a** in terms of its components a_1, a_2, a_3 without writing out equation (4.15) in full. We say that **a** is represented by the row vector (a_1, a_2, a_3) in this basis. Often one writes this as

$$\mathbf{a} = (a_1, a_2, a_3),$$

with the meaning of equation (4.15). Once we are given a basis, any vector in three dimensions can be represented by an (ordered) triplet of numbers like this. (A vector in two dimensions obviously requires only two numbers and two basis vectors.) The same vector would be represented by a different triplet of numbers by someone using a

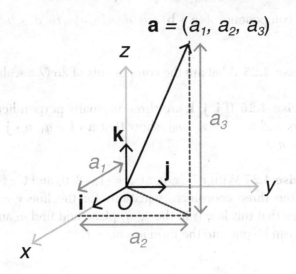

FIGURE 4.9: Vector components.

different basis. Thus, a triplet of numbers is not a vector and a vector is not another name for triplet of numbers – a vector is a triplet of numbers in a specified basis.(In more sophisticated mathematical treatments, a vector is the set of triplets by which it is represented in all bases.)

If axes are labeled such that $\mathbf{i}$, $\mathbf{j}$, $\mathbf{k}$ are unit vectors along the x, y and z axes, respectively, from some origin O, then a_1, a_2, a_3 are, respectively, the x-component of $\mathbf{a}$, the y-component and the z-component. So sometimes one writes (a_x, a_y, a_z) for $\mathbf{a}$.

Thus, $\mathbf{a}$ is represented by the point with coordinates (a_1, a_2, a_3) relative to O. When dealing with the position vector $\mathbf{r}$ of a point relative to O, one usually writes (x, y, z) for the components of $\mathbf{r}$ (instead of (r_1, r_2, r_3) or (r_x, r_y, r_z)).

Example 4.10 What are the components of $\mathbf{a} + \mathbf{b}$ in terms of the components of $\mathbf{a}$ and $\mathbf{b}$?

Write each vector as a linear combination of the basis vectors (equation 4.15) and rearrange

$$\mathbf{a} + \mathbf{b} = (a_1\mathbf{i} + a_2\mathbf{j} + a_3\mathbf{k}) + (b_1\mathbf{i} + b_2\mathbf{j} + b_3\mathbf{k})$$
$$= (a_1 + b_1)\mathbf{i} + (a_2 + b_2)\mathbf{j} + (a_3 + b_3)\mathbf{k},$$

so the components of $(\mathbf{a} + \mathbf{b})$ are $(a_1 + b_1, a_2 + b_2, a_3 + b_3)$.

Exercise 4.25 What are the components of $\lambda\mathbf{a}$ (λ a scalar)?

Exercise 4.26 If $\mathbf{i}$, $\mathbf{j}$, $\mathbf{k}$ are three mutually perpendicular unit vectors and $\mathbf{a} = (a_1, a_2, a_3)$, show that $\mathbf{a} \cdot \mathbf{i} = a_1$, $\mathbf{a} \cdot \mathbf{j} = a_2$ and $\mathbf{a} \cdot \mathbf{k} = a_3$.

Exercise 4.27 With $\mathbf{r} = (x, y, z)$, $\mathbf{a} = (1, -2, 0)$ and $\mathbf{t} = (-1, 3, 0)$, write the three coordinate equations for the line $\mathbf{r} = \mathbf{a} + \lambda\mathbf{t}$. Deduce that this line is in the (x, y) plane and find m and c such that it can be put into the form $y = mx + c$.

4.9. SCALAR (DOT) PRODUCT (COMPONENT FORM)

In terms of components we have the important expression for the scalar product of vectors **a** and **b**:

$$\mathbf{a} \cdot \mathbf{b} = a_1 b_1 + a_2 b_2 + a_3 b_3. \tag{4.16}$$

This follows because

$$\mathbf{a} \cdot \mathbf{b} = (a_1 \mathbf{i} + a_2 \mathbf{j} + a_3 \mathbf{k}) \cdot (b_1 \mathbf{i} + b_2 \mathbf{j} + b_3 \mathbf{k}) \tag{4.17}$$
$$= a_1 b_1 + a_2 b_2 + a_3 b_3. \tag{4.18}$$

Note that all the cross terms (e.g. $a_1 \mathbf{i} \cdot b_2 \mathbf{j}$ etc.) are zero since $\mathbf{i} \cdot \mathbf{j} = 0$ etc. Only the terms $(a_1 \mathbf{i} \cdot b_1 \mathbf{i})$, $(a_2 \mathbf{j} \cdot b_2 \mathbf{j})$, $(a_3 \mathbf{k} \cdot b_3 \mathbf{k})$ remain, and $\mathbf{i} \cdot \mathbf{i} = 1$ etc.

Thus we can work out $\mathbf{a} \cdot \mathbf{b}$ in either of two ways, whichever is most convenient:

$$\mathbf{a} \cdot \mathbf{b} = |\mathbf{a}||\mathbf{b}| \cos(\theta)$$
$$= a_1 b_1 + a_2 b_2 + a_3 b_3. \tag{4.19}$$

Note particularly the following important application of equation (4.19) to finding the angle between two vectors given in component form.

Example 4.11 Find (in radians as a number between 0 and π) the angle between the vector $\mathbf{a} = \left(-1, \sqrt{3}, 1\right)$ and $\mathbf{b} = (2, 0, -1)$.

Since we are given **a** and **b** in component form we can use the components to work out the scalar product.

$$a_1 b_1 + a_2 b_2 + a_3 b_3 = -1 \times 2 + \sqrt{3} \times 0 + 1 \times -1 = -3.$$

Also

$$|\mathbf{a}| = (\mathbf{a} \cdot \mathbf{a})^{1/2} = \left(-1 \cdot -1 + \sqrt{3} \cdot \sqrt{3} + 1 \cdot 1\right)^{1/2} = \sqrt{5},$$
$$|\mathbf{b}| = (\mathbf{b} \cdot \mathbf{b})^{1/2} = \left(2 \cdot 2 + 0 \cdot 0 + -1 \cdot -1\right)^{1/2} = \sqrt{5}.$$

But we also have expression (4.19) for the scalar product involving the angle we are trying to find. Hence

$$\cos(\theta) = \frac{\mathbf{a} \cdot \mathbf{b}}{|\mathbf{a}||\mathbf{b}|} = \frac{-3}{\sqrt{5}\sqrt{5}} = -\frac{3}{5},$$

Take the inverse of cos in the correct quadrant to find the angle θ:

$$\theta = \cos^{-1}(-3/5) = \pi - \cos^{-1}(3/5) = 2.214 \ldots \text{ radians}$$

Exercise 4.28 Find in radians the angle between the vectors $\mathbf{a} = (-2, -2, 1)$ and $\mathbf{b} = (2, 0, -1)$.

Exercise 4.29 Find the projection of the vector $\mathbf{a} = (3, 5, 1)$ on the direction $\hat{\mathbf{n}} = \frac{1}{\sqrt{3}}(1,1,1)$.

Exercise 4.30 Find a vector $\mathbf{N}$ and a scalar P such that the equation $x - y + 2z = 3$ can be written in the form $\mathbf{r} \cdot \mathbf{N} = P$ (where $\mathbf{r} = (x, y, z)$ as usual). Hence write the equation in the form $\mathbf{r} \cdot \hat{\mathbf{n}} = p$ where $\hat{\mathbf{n}}$ is a unit vector. Deduce that the equation represents a plane and obtain the shortest distance from the origin to the plane.

For readers familiar with matrix algebra, we can write the scalar product of two vectors as a row vector multiplied by a column vector

$$\mathbf{a} \cdot \mathbf{b} = \overline{(a_1, a_2, a_3)} \begin{pmatrix} b_1 \\ b_2 \\ b_3 \end{pmatrix} \downarrow = a_1 b_1 + a_2 b_2 + a_3 b_3. \quad (4.20)$$

4.10. VECTOR (CROSS) PRODUCT (COMPONENT FORM)

In terms of the components of two vectors $\mathbf{a}$ and $\mathbf{b}$ their vector product is

$$\mathbf{a} \times \mathbf{b} = (a_2 b_3 - a_3 b_2,\ a_3 b_1 - a_1 b_3,\ a_1 b_2 - a_2 b_1). \quad (4.21)$$

Proof:

$\mathbf{a} \times \mathbf{b} = (a_1\mathbf{i} + a_2\mathbf{j} + a_3\mathbf{k}) \times (b_1\mathbf{i} + b_2\mathbf{j} + b_3\mathbf{k})$

$\qquad = a_1\mathbf{i} \times (b_2\mathbf{j} + b_3\mathbf{k}) + a_2\mathbf{j} \times (b_1\mathbf{i} + b_3\mathbf{k}) + a_3\mathbf{k} \times (b_1\mathbf{i} + b_2\mathbf{j}),$

expanding out using the distributive law and remembering that $\mathbf{i} \times \mathbf{i} = \mathbf{j} \times \mathbf{j} = \mathbf{k} \times \mathbf{k} = \mathbf{0}$. Then making use of $\mathbf{i} \times \mathbf{j} = -\mathbf{j} \times \mathbf{i} = \mathbf{k}$ etc, we have that

$$= a_1b_2\mathbf{k} - a_1b_3\mathbf{j} - a_2b_1\mathbf{k} + a_2b_3\mathbf{i} + a_3b_1\mathbf{j} - a_3b_2\mathbf{i}$$
$$= (a_2b_3 - a_3b_2)\mathbf{i} + (a_3b_1 - a_1b_3)\mathbf{j} + (a_1b_2 - a_2b_1)\mathbf{k}.$$

We can use either equation (4.21) or equation (4.8) to evaluate a vector product – depending on the information available. It is important to know *both* expressions and to be able to spot which is appropriate.

Exercise 4.31 If $\mathbf{a} = (-2, -2, 1)$ and $\mathbf{b} = (2, 0, -1)$, find $\mathbf{a} \times \mathbf{b}$. Show by explicit calculation that

(i) $\mathbf{a} \times \mathbf{b}$ is orthogonal to $\mathbf{a}$ and $\mathbf{b}$,

(ii) $(\mathbf{a} \times \mathbf{b}) \times \mathbf{a}$ lies in the same plane as $\mathbf{a}$ and $\mathbf{b}$. (Express it as a linear combination of these two vectors.)

Exercise 4.32 Find the unit normal to the plane containing the vectors $\mathbf{s} = (1, 1, 2)$ and $\mathbf{t} = (-1, -3, 0)$ and hence write the equation of the plane containing these vectors which passes through the point $(0, 1, -1)$ in the form $ax + by + cz = d$.

For readers familiar with matrix algebra, we can write the vector product of two vectors as a determinant as follows:

$$\mathbf{a} \times \mathbf{b} = \begin{vmatrix} \mathbf{i} & \mathbf{j} & \mathbf{k} \\ a_1 & a_2 & a_3 \\ b_1 & b_2 & b_3 \end{vmatrix}$$
$$= (a_2b_3 - a_3b_2)\mathbf{i} + (a_3b_1 - a_1b_3)\mathbf{j} + (a_1b_2 - a_2b_1)\mathbf{k}.$$

4.11. SCALAR TRIPLE PRODUCT

Now we have expressions for both scalar and vector products in component form it is possible to deduce the component form of the triple products.

Example 4.12 Find an expression for the scalar triple product $[\mathbf{a}, \mathbf{b}, \mathbf{c}] = \mathbf{a} \cdot (\mathbf{b} \times \mathbf{c})$ in terms of the components of the three vectors $\mathbf{a}$, $\mathbf{b}$ and $\mathbf{c}$.

We find $\mathbf{b} \times \mathbf{c}$ from the components of $\mathbf{b}$ and $\mathbf{c}$ using formula (4.21).

$$\mathbf{a} \cdot (\mathbf{b} \times \mathbf{c}) = (a_1, a_2, a_3) \cdot (b_2c_3 - b_3c_2, b_3c_1 - b_1c_3, b_1c_2 - b_2c_1).$$

Now find the scalar product of the result with $\mathbf{a}$ using equation (4.16):

$$\mathbf{a} \cdot (\mathbf{b} \times \mathbf{c}) = a_1b_2c_3 - a_1b_3c_2 + a_2b_3c_1 - a_2b_1c_3 + a_3b_1c_2 - a_3b_2c_1.$$

Exercise 4.33 If $\mathbf{a} = (a, 0, 0)$, $\mathbf{b} = (0, b, 0)$ and $\mathbf{c} = (0, 0, c)$ compute $(\mathbf{a} \times \mathbf{b}) \cdot \mathbf{c}$ and explain the result geometrically.

For readers familiar with matrix algebra, we can write the scalar triple product as a determinant

$$\mathbf{a} \cdot (\mathbf{b} \times \mathbf{c}) = \begin{vmatrix} a_1 & a_2 & a_3 \\ b_1 & b_2 & b_3 \\ c_1 & c_2 & c_3 \end{vmatrix}$$
$$= a_1b_2c_3 - a_1b_3c_2 + a_2b_3c_1 - a_2b_1c_3 + a_3b_1c_2 - a_3b_2c_1.$$

4.12. VECTOR TRIPLE PRODUCT

The following important result should be memorized (but not the proof):

$$\mathbf{a} \times (\mathbf{b} \times \mathbf{c}) = \mathbf{b}(\mathbf{a} \cdot \mathbf{c}) - \mathbf{c}(\mathbf{a} \cdot \mathbf{b}). \qquad (4.22)$$

This is sometimes known as the "bac minus cab" rule, the name chosen to help recall the order of the vectors on the right side of the equation.

Proof: Since no axes have been specified we may choose them to simplify the problem. Choose the $\mathbf{i}$-axis perpendicular to $\mathbf{b}$ and $\mathbf{c}$, so that $b_1 = c_1 = 0$. Then

$$\mathbf{a} \times (\mathbf{b} \times \mathbf{c}) = (a_1, a_2, a_3) \times (b_2 c_3 - b_3 c_2, 0, 0)$$
$$= (0, a_3(b_2 c_3 - b_3 c_2), -a_2(b_2 b_3 - b_3 c_2)),$$

while

$$\mathbf{b}(\mathbf{a} \cdot \mathbf{c}) - \mathbf{c}(\mathbf{a} \cdot \mathbf{b}) = (0, b_2, b_3)(a_2 c_2 + a_3 b_3) - (0, c_2, c_3)(a_2 b_2 + a_3 b_3)$$
$$= (0, a_3 c_3 b_2 - a_3 b_3 c_2, a_2 c_2 b_3 - a_2 b_2 c_3)$$

which is the same.

4.13. VECTOR IDENTITIES

We can use equation (4.22) together with the properties of the scalar triple product (section 4.5) to simplify various complicated products of vectors.

Example 4.13 Show that if $\mathbf{v} = \mathbf{w} \times \mathbf{r}$ then $|\mathbf{v}|^2 = |\mathbf{w}|^2 |\mathbf{r}|^2 - (\mathbf{w} \cdot \mathbf{r})^2$.

From the definition of $|\mathbf{v}|$:

$$|\mathbf{v}|^2 = \mathbf{v} \cdot \mathbf{v}.$$

The expression involves a double vector product so we try to put it in a form where we can use the expansion (4.22). Using $\mathbf{v} = \mathbf{w} \times \mathbf{r}$

$$\mathbf{v} \cdot \mathbf{v} = (\mathbf{w} \times \mathbf{r}) \cdot \mathbf{v}.$$

Interchanging scalar and vector products (dot and cross)

$$= \mathbf{w} \cdot (\mathbf{r} \times \mathbf{v}) = \mathbf{w} \cdot (\mathbf{r} \times (\mathbf{w} \times \mathbf{r})).$$

We now have a triple vector product so use equation (4.22)

$$= \mathbf{w} \cdot ((\mathbf{r} \cdot \mathbf{r})\mathbf{w} - (\mathbf{r} \cdot \mathbf{w})\mathbf{r}).$$

Expanding out the parenthetical by the distributive law

$$= (\mathbf{r} \cdot \mathbf{r})(\mathbf{w} \cdot \mathbf{w}) - (\mathbf{r} \cdot \mathbf{w})(\mathbf{w} \cdot \mathbf{r}).$$

Finally using the definition of "| |"

$$= |\mathbf{w}|^2 |\mathbf{r}|^2 - (\mathbf{w} \cdot \mathbf{r})^2,$$

as required.

▌**Exercise 4.34** Show that $(\mathbf{a} \times \mathbf{b}) \cdot (\mathbf{a} \times \mathbf{c}) = (\mathbf{b} \cdot \mathbf{c})(\mathbf{a} \cdot \mathbf{a}) - (\mathbf{a} \cdot \mathbf{c})(\mathbf{b} \cdot \mathbf{a})$.

▌**Exercise 4.35** Show that

$$(\mathbf{a} \times \mathbf{b}) \times (\mathbf{c} \times \mathbf{d}) = [\mathbf{a}, \mathbf{b}, \mathbf{d}]\mathbf{c} - [\mathbf{a}, \mathbf{b}, \mathbf{c}]\mathbf{d}.$$

Find a similar expression for $(\mathbf{a} \times \mathbf{b}) \times (\mathbf{c} \times \mathbf{d})$ as a linear combination of $\mathbf{a}$ and $\mathbf{b}$, and hence deduce that

$$[\mathbf{a}, \mathbf{b}, \mathbf{c}]\mathbf{d} = [\mathbf{d}, \mathbf{b}, \mathbf{c}]\mathbf{a} + [\mathbf{a}, \mathbf{d}, \mathbf{c}]\mathbf{b} + [\mathbf{a}, \mathbf{b}, \mathbf{d}]\mathbf{c}.$$

(Compare Exercise 4.21)

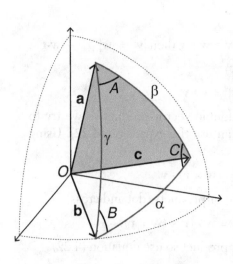

Let $\mathbf{a}$, $\mathbf{b}$, $\mathbf{c}$ be the (unit) vectors from the center O to points on a unit sphere centered on O. A spherical triangle is made from arcs of great circles joining the points at the ends of $\mathbf{a}$, $\mathbf{b}$ and $\mathbf{c}$. Let a, α, β and γ denote the sides of the triangle; because the sphere is of unit radius the length α is the same as the angle between $\mathbf{b}$ and $\mathbf{c}$ etc. Since $\mathbf{a} \times \mathbf{b}$ is normal to the plane containing $\mathbf{a}$ and $\mathbf{b}$, $\mathbf{a} \times \mathbf{c}$ is normal to the plane containing $\mathbf{a}$ and $\mathbf{c}$, and A is the angle between these planes, we have

$$(\mathbf{a} \times \mathbf{b}) \cdot (\mathbf{a} \times \mathbf{c}) = |\mathbf{a} \times \mathbf{b}||\mathbf{a} \times \mathbf{c}| \cos(A) = \sin(\gamma) \sin(\beta) \cos(A). \quad (4.23)$$

▌**Exercise 4.36** Deduce from equation (4.23) and Exercise 4.34 that

$$\cos(a) = \cos(\beta) \cos(\gamma) + \sin(\beta) \sin(\gamma) \cos(A).$$

This is the *cosine formula* of spherical trigonometry.

Example 4.14 If **a**, **b** and $\lambda > 0$ are given, solve for **x** the equation

$$\mathbf{x} \times \mathbf{a} + \lambda \mathbf{x} = \mathbf{b}. \tag{4.24}$$

Take scalar and vector products with the constant vectors in the problem and attempt to simplify the result. Here the possible independent constant vectors we might want to use are **a**, **b** and **a** × **b**. Taking the scalar product with **a** on both sides gives

$$0 + \lambda \mathbf{a} \cdot \mathbf{x} = \mathbf{a} \cdot \mathbf{b}. \tag{4.25}$$

Therefore

$$\mathbf{a} \cdot \mathbf{x} = \lambda^{-1} (\mathbf{a} \cdot \mathbf{b}). \tag{4.26}$$

Now taking the vector product of **a** with both sides gives another equation:

$$\mathbf{a} \times (\mathbf{x} \times \mathbf{a}) + \lambda \mathbf{a} \times \mathbf{x} = \mathbf{a} \times \mathbf{b}. \tag{4.27}$$

And note that, by equation (4.24),

$$\mathbf{a} \times \mathbf{x} = \lambda \mathbf{x} - \mathbf{b}. \tag{4.28}$$

Now, expand the vector triple product equation (4.27) using equation (4.22). Then substitute for **a** · **x** using equation (4.26), and substitute for **a** × **x** using equation (4.28). The result should be an equation that is linear in **x** and can be solved with a little algebra. See Exercise 4.37. The solution is

$$X = \frac{1}{|\mathbf{a}|^2 + \lambda^2} \left(\lambda \mathbf{b} + \lambda^{-1} (\mathbf{a} \cdot \mathbf{b}) \mathbf{a} + \mathbf{a} \times \mathbf{b} \right).$$

Example 4.15 An alternative method for solving

$$\mathbf{x} \times \mathbf{a} + \lambda \mathbf{x} = \mathbf{b}.$$

Express the unknown **x** as a linear combination of the constant vectors of the problem with coefficients to be determined. Let

$$\mathbf{x} = a\mathbf{a} + \beta\mathbf{b} + \gamma\mathbf{a} \times \mathbf{b}.$$

Substitute in (4.24)

$$(a\mathbf{a} + \beta\mathbf{b} + \gamma\mathbf{a} \times \mathbf{b}) \times \mathbf{a} + (a\mathbf{a} + \beta\mathbf{b} + \gamma\mathbf{a} \times \mathbf{b})\,\lambda = \mathbf{b}. \qquad (4.29)$$

Multiplying out and comparing coefficients of $\mathbf{a}$, $\mathbf{b}$ and $\mathbf{a} \times \mathbf{b}$ will yield three equations for the three unknowns a, β and γ.

Exercise 4.37

(i) Complete the missing algebra steps in example 4.14.

(ii) Complete the missing algebra steps in example 4.15 to obtain the same solution.

Exercise 4.38 Given vectors $\mathbf{a}$ and $\mathbf{b}$ such that $\mathbf{a} \cdot \mathbf{b} = 0$, solve the equation $\mathbf{x} \times \mathbf{a} = \mathbf{b}$ subject to the condition $\mathbf{x} \cdot \mathbf{a} = k$ (a given constant).

4.14. EXAMPLES FROM PHYSICAL SCIENCE

Work is done by a force when the force acts on a body that moves, and the force has a component along the line of motion.

Example 4.16 A particle acted on by a force $\mathbf{F} = 4\mathbf{i} - 2\mathbf{j}$ N is given a displacement $\mathbf{l} = 2\mathbf{i} + 8\mathbf{j} + 2\mathbf{k}$ m. How much work is done by the force?

The work done is the component of force in the direction of motion times the distance. Since we are given the component form of the force and the displacement we use the component form of the scalar product. Thus

$$\begin{aligned} \text{Work done} = \mathbf{F} \cdot \mathbf{l} &= F_x l_x + F_y l_y + F_z l_z \\ &= 4 \times 2 - 2 \times 8 + 0 \times 2 \\ &= -8 \text{ J}. \end{aligned}$$

Exercise 4.39 How much work is done by a force $\mathbf{F} = 8\mathbf{i} + 2\mathbf{k}$ N acting on a particle which is displaced through a distance $\mathbf{l} = -2\mathbf{i} + 4\mathbf{j} + 10\mathbf{k}$ m?

Exercise 4.40 A particle displaced through $\mathbf{l} = 0.5\mathbf{i} + 0.2\mathbf{j}$ m is acted on by two forces, $\mathbf{F}_1 = 3\mathbf{i} + 3\mathbf{j}$ N and $\mathbf{F}_2 = 4\mathbf{i} + 3\mathbf{k}$ N.

(i) What is the work done by $\mathbf{F}_1$ and $\mathbf{F}_2$?

(ii) What is the net force $\mathbf{F}_1 + \mathbf{F}_2$?

The power supplied by a force is equal to the rate at which the force does work. This can be written as $\mathbf{P} = \mathbf{F} \cdot \mathbf{v}$, where $\mathbf{v}$ is the velocity of a particle acted on by a force $\mathbf{F}$.

Exercise 4.41 What is the power input of a force $2\mathbf{i} + 3\mathbf{j} + 2\mathbf{k}$ N that acts on a particle moving with velocity $3\mathbf{i} + 2\mathbf{j} - 2\mathbf{k}$ m s^{-1}?

A constant electric field $\mathbf{E}$ passes through a plane surface of area A with unit normal $\hat{\mathbf{n}}$. The electric flux ϕ through A is defined to be $\phi = \mathbf{E} \cdot \hat{\mathbf{n}} A$.

Exercise 4.42 A surface of area 2 cm^2 has a unit normal $\hat{\mathbf{n}} = (2\mathbf{i} + \mathbf{j}) / \sqrt{5}$. The surface is located in a region with a uniform electric field $\mathbf{E} = 4\mathbf{i} + 2\mathbf{j} + 2\mathbf{k}$ V m^{-1}. What is the electric flux through the surface?

The torque τ about the origin exerted by a force $\mathbf{F}$ acting through a point with position vector $\mathbf{r}$ is defined as $\tau = \mathbf{r} \times \mathbf{F}$. Clearly τ is orthogonal to both $\mathbf{r}$ and $\mathbf{F}$.

Exercise 4.43

(i) A force $\mathbf{F} = 150\mathbf{i} + 200\mathbf{j} - 150\mathbf{k}$ N acts on a body through a point at $\mathbf{r} = 2\mathbf{i} - 3\mathbf{j}$ m relative to the origin O. Calculate the torque exerted by this force about O.

(ii) Verify that the torque is orthogonal to the applied force.

The angular momentum $\mathbf{L}$ of a particle with linear momentum $\mathbf{p}$ at a position $\mathbf{r}$ relative to some origin is $\mathbf{L} = \mathbf{r} \times \mathbf{p}$. As with torque, angular momentum is defined relative to an origin. Torque is related to angular momentum: the net torque acting on a body equals the rate of change of angular momentum. If the angular momentum of a body is constant then there is no net torque on the body (and conversely).

> **Exercise 4.44** At some instant a particle has velocity $\mathbf{v} = -3\mathbf{i} + 6\mathbf{j}$ m s^{-1} and is located at a point with position vector $\mathbf{r} = -4\mathbf{i} + 2\mathbf{j}$ m. If the particle has a mass 2.5 kg, what is its angular momentum about the origin?

4.15. EXTENSION: INDEX NOTATION

Many vector equations involve the x, y and z components in equivalent ways. For example, the scalar product of $\mathbf{A}$ and $\mathbf{B}$ is $A_x B_x + A_y B_y + A_z B_z$ where each term in the sum is identical in form but with a different index. If we label the coordinates as x_1, x_2 and x_3 instead of x, y and z we can write the scalar product as

$$\mathbf{A} \cdot \mathbf{B} = \sum_{i=1}^{3} A_i B_i.$$

We can even abbreviate this if we use the convention (called the Einstein summation convention) that a repeated index is to be summed over. With this understanding we simply drop the explicit summation sign and write the expression for the scalar product as

$$\mathbf{A} \cdot \mathbf{B} = A_i B_i.$$

A useful quantity is the *Kronecker delta* defined by

$$\delta_{ij} = \begin{cases} 1 & \text{if} \quad i = j \\ 0 & \text{if} \quad i \neq j. \end{cases}$$

So, for example, $\delta_{11} = 1$ and $\delta_{13} = 0$. A second useful quantity is the permutation symbol defined by

$$\in_{ijk} = \begin{cases} 1 & \text{if } i, j, k = 1,2,3 \text{ or } 2,3,1 \text{ or } 3,1,2 \\ -1 & \text{if } i, j, k = 2,1,3 \text{ or } 1,3,2 \text{ or } 3,2,1 \\ 0 & \text{otherwise.} \end{cases}$$

So, for example, $\in_{132} = -1$ and $\in_{112} = 0$. We can write the cross product using $\in_{ijk}$ and the summation convention in the compact form

$$(\mathbf{A} \times \mathbf{B})_i = \in_{ijk} A_j B_k.$$

So, for example, we deduce that $(\mathbf{A} \times \mathbf{B})_1 = \in_{123}A_2B_3 + \in_{132}A_3B_2$ $= A_2B_3 - A_3B_2$ because all the other potential terms in the sum over j and k vanish.

A useful identity is $\in_{ijk}\in_{ilm} = \delta_{jl}\delta_{km} - \delta_{jm}\delta_{lk}$ from which we can deduce directly the formula for the triple vector product in Section 4.12.

Revision Notes

After completing this chapter you should be able to

- Calculate *linear combinations* of vectors
- Calculate the *magnitude* and *direction* of a vector given in component form
- Calculate the *components* of a vector given as a magnitude and direction
- Calculate the *scalar product* of two vectors from magnitudes and directions or from components
- Calculate the *vector product* of two vectors
- State the properties of the *scalar triple product*
- Give a formula for the *vector triple product*
- Derive *vector identities* for more complex products
- State the *equation of a line* in vector form
- State the *equation of a plane* in two different forms
- Find the *unit normal to a plane* and its distance from the origin
- Compute the *work done* by a force, the *flux* of a constant field through a plane surface, the *torque* exerted by a force about an axis and the *angular momentum* of a particle of a given velocity

4.16. EXERCISES

1. If a = i − j and b = i + j find a unit vector perpendicular to both **b** and **a** × **b**.

2. Find the angle between the vectors (1, −1, −4) and (2, 2, 1).

3. Find the angle between the vectors (3, 2, 4) and (1, 0, 2).

4. Write down the components of a unit vector which is normal to both the vectors (1, −1, 3) and (2, 2, −2).

5. Find the components of a unit vector which is perpendicular to both (1, −1, −1) and (0, 2, 3).

6. Two sides of a triangle are formed by the vectors **i** + 2**j** + **k** and 3**i** + 2**j** − **k**. What is the area of the triangle and the unit normal to the triangle?

7. (i) If vectors **a** and **c** are given such that **a** · **c** = 0, find all solutions to the following equation for the vector **x**:

$$\mathbf{a} \times \mathbf{x} = \mathbf{c}.$$

   (ii) Find an expression for the vector **X** in terms of **A**, **B** and C given that $\mathbf{X} \cdot \mathbf{A} + C = 0$ and $\mathbf{X} \times \mathbf{A} + \mathbf{B} = 0$.

8. If **A** and **B** are orthogonal show that the length of the vector $C\mathbf{A}+\mathbf{A}\times\mathbf{B}$ is $|\mathbf{A}|\sqrt{(|\mathbf{B}|^2 +C^2}$.

9. Find the unit normal to the plane $x + 2y − 2z = 15$. What is the distance of the plane from the origin?

4.17. PROBLEMS

1. Show that the equation of the straight line passing through the points with position vectors **c** and **d** can be written as

$$\mathbf{r} = \mathbf{c} + \lambda \mathbf{t},$$

where $\mathbf{t} = \mathbf{d} - \mathbf{c}$, and λ is a parameter. Show on a diagram which parts of the line correspond to (i) $\lambda < 0$, (ii) $0 \le \lambda \le 1$, (iii) $\lambda > 1$. Show that the equation of the plane that passes through the point with position vector $\mathbf{a}$, and that has normal $\mathbf{n}$, can be written as

$$\mathbf{r} \cdot \mathbf{n} = \mathbf{a} \cdot \mathbf{n}.$$

Show that, if $\mathbf{t} \cdot \mathbf{n} \ne 0$, the straight line intersects the plane in the point with position vector

$$\mathbf{c} + \left\{ \frac{(\mathbf{a} - \mathbf{c}) \cdot \mathbf{n}}{\mathbf{t} \cdot \mathbf{n}} \right\} \mathbf{t}.$$

What does this point become if $\mathbf{a} = \mathbf{d}$?

2. The points A, B, C, D and E have rectangular Cartesian coordinates $(3, 1, 2), (-2, 7, 1), (4, -1, 3), (1, 1, -4), (2, 0, 1)$, respectively. Find the equation of the straight line, L, through D and E. Find an equation for the plane, P, through A, B and C. Find where L intersects P.

3. For $\mathbf{a} = (3, 1, 2)$, $\mathbf{b} = (-2, 7, 1)$, $\mathbf{c} = (4, -1, 3)$ verify the identities

(i) $\mathbf{a} \times (\mathbf{b} \times \mathbf{c}) = (\mathbf{a} \cdot \mathbf{c})\mathbf{b} - (\mathbf{a} \cdot \mathbf{b})\mathbf{c}$,

(ii) $(\mathbf{a} \times \mathbf{b}) \cdot \mathbf{c} = \mathbf{a} \cdot (\mathbf{b} \times \mathbf{c})$.

Now, letting $\mathbf{a}, \mathbf{b}, \mathbf{c}, \mathbf{d}$ be any vectors and assuming that (i) and (ii) are always true, show that

$$(\mathbf{a} \times \mathbf{b}) \cdot (\mathbf{c} \times \mathbf{d}) = (\mathbf{a} \cdot \mathbf{c})(\mathbf{b} \cdot \mathbf{d}) - (\mathbf{b} \cdot \mathbf{c})(\mathbf{a} \cdot \mathbf{d}),$$

and also simplify $(\mathbf{a} \times \mathbf{b}) \times (\mathbf{c} \times \mathbf{d})$.

4. Three points A, B, C have coordinates $(1, 1, 2), (2, -1, -2), (3, 2, 1)$:

(i) Calculate the perimeter of the triangle ABC.

(ii) Find the vectors $\overrightarrow{AB}$ and $\overrightarrow{AC}$.

(iii) Find a vector perpendicular to both $\vec{AB}$ and $\vec{AC}$.

(iv) Find the area of the triangle ABC.

(v) Find the equation of the plane through ABC in the form $ax + by + cz = d$.

(vi) Find the perpendicular distance of this plane from the origin.

5. The plane which has vector equation $\mathbf{k} \cdot \mathbf{r} = 1$ (with $\mathbf{k}$ a unit vector) intersects the sphere which has vector equation $|\mathbf{r}| = 2$. If $\mathbf{a}$ and $\mathbf{b}$ are the position vectors of the two diametrically opposite points on the circle of intersection of the plane and the sphere show that $(\mathbf{a} - \mathbf{b}) = 2\mathbf{a} - 2\mathbf{k}$. Hence find the diameter of the circle.

6. The angle between two non-parallel *unit* vectors $\mathbf{a}$ and $\mathbf{b}$ is θ, and two further vectors $\mathbf{A}$ and $\mathbf{B}$ are defined by

$$\mathbf{A} = \mathbf{a} - \mathbf{b}\cos(\theta) \text{ and } \mathbf{B} = \mathbf{b} - \mathbf{a}\cos(\theta).$$

Show that

$$\mathbf{A} \cdot \mathbf{B} = -\sin^2(\theta)\mathbf{a} \cdot \mathbf{b} \text{ and } \mathbf{A} \times \mathbf{B} = \sin^2(\theta)\mathbf{a} \times \mathbf{b}.$$

An arbitrary vector $\mathbf{d}$ is to be expressed as a linear combination

$$\mathbf{d} = \lambda\mathbf{a} + \mu\mathbf{b} + v\mathbf{a} \times \mathbf{b};$$

show that

$$\lambda\sin^2(\theta) = \mathbf{d} \cdot \mathbf{A}, \mu\sin^2(\theta) = \mathbf{d} \cdot \mathbf{B}, v\sin^2(\theta) = \mathbf{d} \cdot (\mathbf{a} \times \mathbf{b}).$$

A point $\mathbf{d}$ lies on the intersection of the two planes $\mathbf{r} \cdot \mathbf{a} = \alpha$ and $\mathbf{r} \cdot \mathbf{b} = \beta$, and satisfies the condition that $\mathbf{d} \cdot (\mathbf{a} \times \mathbf{b}) = 0$. Show that $\mathbf{d}\sin^2(\theta) = \alpha\mathbf{A} + \beta\mathbf{B}$.

7. (i) Write down the vector equation for the plane through the point $\mathbf{c}$ containing the vectors $\mathbf{a}$ and $\mathbf{b}$.

(ii) If $\mathbf{a} = (1, 1, 4)$, $\mathbf{b} = (2, 3, 1)$ and $\mathbf{c} = (1, 2, 1)$ write down the equation of the plane in the form

$$lx + my + nz = p.$$

(iii) Write down the vector equation of the line through $\mathbf{c}$ in the direction of the normal to the plane, and find the point $\mathbf{d}$ where it cuts the (x, y) plane.

(iv) Find the points $\mathbf{m}_1$, $\mathbf{m}_2$, and $\mathbf{m}_3$ at which this plane cuts the x-, y- and z-axes, respectively.

(v) What is the volume of the tetrahedron with vertices at $\mathbf{d}$, $\mathbf{m}_1$, $\mathbf{m}_2$, and $\mathbf{m}_3$?

8. Find the solution for $\mathbf{x}$ for the following vector equation:

$$\mathbf{a} \times \mathbf{x} + b\mathbf{x} + \mathbf{c} = \mathbf{0}.$$

Show also that $\mathbf{x}$ and $\mathbf{c}$ are equal in length if $\mathbf{a}$ and $\mathbf{c}$ are orthogonal vectors and $a^2 + b^2 = 1$.

9. Find the angle between the vector $\mathbf{a} = (1, 0, 2)$ and the plane $3x - y + z = 1$.

A line is drawn parallel to $\mathbf{a}$ through the origin. Find the co-ordinates of the point where this line cuts the above plane.

10. Find the equation of the line l_1 through the point $(5, 2, 0)$ in the direction of $(2, 1, 2)$ and of the line l_2 through $(-1, 1, -1)$ in the direction of $(4, 0, -1)$. Find the point of intersection A of l_1 and l_2. Hence determine the equation of the plane through A containing l_1 and l_2. What is the distance of this plane from the origin?

11. Let $a = (0, 0, \sqrt{3}/2)$ and $t = (1, 1, 1)\sqrt{3}$. Find the points on the line, l_1, $\mathbf{r} = \mathbf{a} + \lambda\mathbf{t}$ that lie a unit distance from the origin. Find the equation of the line, l_2, joining the origin to one of these points. Find the equation of the plane containing the lines l_1 and l_2.

CHAPTER 5

MATRICES

You should be familiar with the solution of a pair of simultaneous linear equations in two unknowns, by the elimination of one variable. You might think that more equations in more unknowns could be solved by a similar process of successive eliminations. If all you ever want to do is to solve specific equations then this is correct (but not very efficient; imaging trying this for 100 equations in 100 unknowns typical of large engineering projects).

However, if you want to understand, for example, the circumstances under which a system of equations has a solution (since not all of them do), and to apply important tools for analyzing them, you will need to be familiar with the matrix representation of such systems of equations. The theory leads to the use of matrices as representations of transformations (for example a rotation of coordinate axes) and the construction of the "eigenvectors" and "eigenvalues" that turn out to play a crucial role in almost all branches of physical science (wherever small oscillations are important, e.g. in quantum mechanics and in numerical data analysis). These topics (matrices, eigenvectors etc.) are sometimes gathered under the umbrella term *linear algebra*.

Let's begin with what matrices are and some motivation for their introduction. Suppose we want to solve the pair of simultaneous equations

$$3x + 2y = 7$$
$$6x - y = 4$$

We start by (for example) multiplying the second equation by 2:

$$3x + 2y = 7$$
$$12x - 2y = 8$$

Adding the equations and multiplying the first by 5 gives

$$15x + 10y = 35$$
$$15x + 0y = 15$$

and subtraction gives

$$0x + 10y = 20$$
$$15x + 0y = 15$$

from which we can read off $x = 1$ and $y = 2$. Notice, however, that the x and y are acting merely as placeholders: if we keep everything aligned we do not have to write out the x and y in every line. For example, the first two steps of the calculation could be written:

$$\begin{pmatrix} 3 & 2 \\ 12 & -1 \end{pmatrix} = \begin{pmatrix} 7 \\ 4 \end{pmatrix}$$

and

$$\begin{pmatrix} 3 & 2 \\ 12 & -2 \end{pmatrix} = \begin{pmatrix} 7 \\ 8 \end{pmatrix}$$

And so on, provided we have a rule that enables us to put back the x and y. This rule must be such that

$$\begin{pmatrix} 3 & 2 \\ 6 & -1 \end{pmatrix} \begin{pmatrix} x \\ y \end{pmatrix} = \begin{pmatrix} 3x + 27 \\ 6x - y \end{pmatrix}. \tag{5.1}$$

With this rule in place we can manipulate the set of coefficients as an entity, as long as the rules are compatible with the rules for manipulating simultaneous equations. The set of coefficients is then called a matrix. Note that a matrix is not just an array of numbers; it is the array plus the rules for manipulation. Let us now formalize these rules.

5.1. MATRIX REPRESENTATION

Let $\mathbf{a} = (a_1, a_2, a_3)$ and $\mathbf{x} = (x_1, x_2, x_3)$ be vectors. Consider the scalar product $\mathbf{a} \cdot \mathbf{x} = a_1x_1 + a_2x_2 + a_3x_3$ from Section 4.9. We can write this product as

$$\left(a_1, a_2, a_3\right) \begin{pmatrix} x_1 \\ x_2 \\ x_3 \end{pmatrix} = \mathbf{a} \cdot \mathbf{x} = a_1x_1 + a_2x_2 + a_3x_3 \qquad (5.2)$$

where corresponding terms reading across the row and down the column are to be multiplied together ($a_1 \times x_1, a_2 \times x_2, a_3 \times x_3$) and the results added. (The arrows are there for illustration only.) This can be extended thus:

$$\begin{pmatrix} a_1 & a_2 & a_3 \\ b_1 & b_2 & b_3 \\ c_1 & c_2 & c_3 \end{pmatrix} \begin{pmatrix} x_1 \\ x_2 \\ x_3 \end{pmatrix} = \begin{pmatrix} \mathbf{a} \cdot \mathbf{x} \\ \mathbf{b} \cdot \mathbf{x} \\ \mathbf{x} \cdot \mathbf{x} \end{pmatrix} = \begin{pmatrix} a_1x_1 + a_2x_2 + a_3x_3 \\ b_1x_1 + b_2x_2 + b_3x_3 \\ c_1x_1 + c_2x_2 + c_3x_3 \end{pmatrix} \qquad (5.3)$$

where the result of going across each row in turn on the left is placed at the corresponding level on the right. The first array on the left is called a *matrix*. More specifically it is a 3×3 matrix (there are 3 rows running across the page, 3 columns running down the page). A vector is a special case: it can be written as a single column, in which case it is known as a "column vector" or 3×1 matrix, or as a single row, hence known as a "row vector" or 1×3 matrix. Both are used in equation (5.2).

We can use matrices to represent systems of linear equations, as in the following example.

Example 5.1 Represent the system of equations

$$
\begin{array}{rrcl}
\text{(i)} & x + y - z & = & -3 \\
\text{(ii)} & 2x + y - z & = & 0 \\
\text{(iii)} & 4x - 2y - 3z & = & 4
\end{array}
\qquad (5.4)
$$

in matrix form.

Write out the coefficients as a matrix, and the set of unknowns and the constants on the right as column vectors:

$$
\begin{pmatrix} 1 & 1 & -1 \\ 2 & 1 & -1 \\ 4 & -2 & -3 \end{pmatrix}
\begin{pmatrix} x \\ y \\ z \end{pmatrix}
=
\begin{pmatrix} -3 \\ 0 \\ 4 \end{pmatrix}
\qquad (5.5)
$$

Using the rule shown in equation (5.3) on the left side of equation (5.5) gives us (5.4) as required.

However, if we want this representation to be useful, we must establish rules for manipulating the matrix of coefficients that correspond to manipulation of the equations. In particular, writing the system (5.5) in the form

$$
A x = \mathbf{h},
$$

where A stands for the matrix on the left side and $\mathbf{h}$ for the column vector on the right, we should like to "divide through by A" and obtain the matrix solution

$$
x = A^{-1} \mathbf{h}.
\qquad (5.6)
$$

In order to attach a meaning to the matrix A^{-1} we must develop the rules of algebra for matrices.

1. Multiplication by a scalar. We can multiply the system (5.4) by any non-zero number to obtain an equivalent system: e.g. multiplying (5.4) by 2 we get

$$2x + 2y - 2z = -6$$
$$4x + 2y - 2z = 0$$
$$8x - 4y - 6z = 8.$$

In matrix form this can be written as

$$\begin{pmatrix} 2 & 2 & -2 \\ 4 & 2 & -2 \\ 8 & -4 & -6 \end{pmatrix} \cdot \begin{pmatrix} x \\ y \\ z \end{pmatrix} = \begin{pmatrix} -6 \\ 0 \\ 8 \end{pmatrix}.$$

Thus we have shown that

$$2 \times \begin{pmatrix} 1 & 1 & -1 \\ 2 & 1 & -1 \\ 4 & -2 & -3 \end{pmatrix} = \begin{pmatrix} 2 & 2 & -2 \\ 4 & 2 & -2 \\ 8 & -4 & -6 \end{pmatrix},$$

i.e. to multiply a matrix by a scalar we multiply all the entries by the scalar value.

2. Addition (or subtraction). Equation (5.5) can be written (in infinitely many ways) as a sum (or difference) of subsystems: e.g.

$$0x + 0y - 2z + x + y + z = -3$$
$$0x + y + 0z + 2x + 0y - z = 0$$
$$x - 5y - 6z + 3x + 3y + 3z = 4.$$

The choice of coefficients is quite arbitrary provided they sum to those in equation (5.5). In matrix notation this is

$$\left[\begin{pmatrix} 0 & 0 & -2 \\ 0 & 1 & 0 \\ 1 & -5 & -6 \end{pmatrix} + \begin{pmatrix} 1 & 1 & 1 \\ 2 & 0 & -1 \\ 3 & 3 & 3 \end{pmatrix} \right] \begin{pmatrix} x \\ y \\ z \end{pmatrix} = \begin{pmatrix} -3 \\ 0 \\ 4 \end{pmatrix}.$$

Thus to add two matrices, we add the corresponding entries. In order to add or subtract two matrices, they must have the same dimensions (in the above example we add two 3 × 3 matrices to get another 3 × 3 matrix).

Exercise 5.1 Let

$$A = \begin{pmatrix} 3 & 0 & -1 \\ 0 & 1 & 5 \\ 3 & -2 & 3 \end{pmatrix}, \quad B = \begin{pmatrix} 8 & 1 & 1 \\ 5 & 2 & -3 \\ 0 & 1 & -1 \end{pmatrix}, \quad h = \begin{pmatrix} 1 \\ 2 \\ 1 \end{pmatrix}, \quad \text{and} \quad x = \begin{pmatrix} x \\ y \\ z \end{pmatrix}.$$

(i) Write out the system of equations $Ax = h$ explicitly.

(ii) What is the matrix $3A - B$?

Notation: It is sometimes more convenient to use a more systematic notation for the elements of A. If we write

$$A = \begin{pmatrix} a_{11} & a_{12} \\ a_{21} & a_{22} \end{pmatrix} \quad \begin{matrix} \leftarrow & \text{row1} \\ \leftarrow & \text{row2} \end{matrix}$$
$$\begin{matrix} \uparrow & \uparrow \\ \text{col1} & \text{col2} \end{matrix}$$

(the first index labels the row, the second the column) then a_{ij} is the element of A in the ith row and jth column. An $M \times N$ matrix has M rows and N columns. A *square matrix* has the same number of rows as columns, i.e. has dimensions $N \times N$.

Using this notation we can write down the rules above. For multiplication by a scalar $B = \lambda A$ we multiply all the elements of the matrix by the scalar:

$$b_{ij} = \lambda a_{ij},$$

where subscripts i, j run over the dimensions of the matrix. And for the sum of two matrices (of the same dimensions) $C = A + B$, we compute the elements of C as the sum of the corresponding elements of A and B:

$$c_{ij} = a_{ij} + b_{ij}.$$

5.2. SOLUTION OF SYSTEMS OF EQUATIONS

Systems of linear equations can be solved by operations equivalent to those we have just introduced. To solve equations (5.4) we can begin by subtracting $2\times$(i) from (ii), keeping the other equations untouched.

$$
\begin{aligned}
(x + y - z) &+ (+0x + 0y + 0z) &= -3 \\
(2x + y - z) &+ (-2x - 2y + 2z) &= 0 - (-6) = 6. \\
(4x - 2y - 3z) &+ (+0x + 0y + 0z) &= 4
\end{aligned}
$$

In matrix terms this can be written as

$$
\left[\begin{pmatrix} 1 & 1 & -1 \\ 2 & 1 & -1 \\ 4 & -2 & -3 \end{pmatrix} + \begin{pmatrix} 0 & 0 & 0 \\ -2 & -2 & 2 \\ 0 & 0 & 0 \end{pmatrix}\right] \begin{pmatrix} x \\ y \\ z \end{pmatrix} = \begin{pmatrix} -3 \\ 6 \\ 4 \end{pmatrix},
$$

or

$$
\begin{pmatrix} 1 & 1 & -1 \\ 0 & -1 & 1 \\ 4 & -2 & -3 \end{pmatrix} \begin{pmatrix} x \\ y \\ z \end{pmatrix} = \begin{pmatrix} -3 \\ 6 \\ 4 \end{pmatrix}. \tag{5.7}
$$

Equation (5.7) is an equally valid representation of the original system (1.3). We can continue this procedure, but to simplify matters we simply carry out the operations directly on the matrix entries. Thus, the operation that produces (5.7) from (5.5) amounts to subtracting twice the first row from the second in the matrix on the left and in the column vector on the right:

$$
\begin{pmatrix} 1 & 1 & -1 \\ 2 & 1 & -1 \\ 4 & -2 & -3 \end{pmatrix} \begin{pmatrix} x \\ y \\ z \end{pmatrix} = \begin{pmatrix} -3 \\ 0 \\ 4 \end{pmatrix} \xrightarrow{R2 \to R2 - (2 \times R1)} \begin{pmatrix} 1 & 1 & -1 \\ 0 & -1 & 1 \\ 4 & -2 & -3 \end{pmatrix} \begin{pmatrix} x \\ y \\ z \end{pmatrix} = \begin{pmatrix} -3 \\ 6 \\ 4 \end{pmatrix}.
$$

(The notation above the arrow "R2 → R2 − (2 × R1)" means "replace the second row with the original second row minus twice the first row.")

A possible further series of equivalent representations, obtained by the indicated operations on rows, is

$$\begin{pmatrix} 1 & 1 & -1 \\ 0 & -1 & 1 \\ 4 & -2 & -3 \end{pmatrix}\begin{pmatrix} x \\ y \\ z \end{pmatrix} = \begin{pmatrix} -3 \\ 6 \\ 4 \end{pmatrix} \xrightarrow{R3 \rightarrow R3 - (4 \times R1)} \begin{pmatrix} 1 & 1 & -1 \\ 0 & -1 & 1 \\ 0 & -6 & -1 \end{pmatrix}\begin{pmatrix} x \\ y \\ z \end{pmatrix} = \begin{pmatrix} -3 \\ 6 \\ 16 \end{pmatrix}$$

$$\xrightarrow{R3 \rightarrow R3 - (6 \times R2)} \begin{pmatrix} 1 & 1 & -1 \\ 0 & -1 & 1 \\ 0 & 0 & -5 \end{pmatrix}\begin{pmatrix} x \\ y \\ z \end{pmatrix} = \begin{pmatrix} -3 \\ 6 \\ 20 \end{pmatrix}$$

Now, the equivalent system is easier to solve. Writing it out we have

$$\begin{aligned} x + y - z &= -3 \\ -y + z &= 6 \\ -5z &= -20 \end{aligned} \tag{5.8}$$

from which $z = 4$, $y = -2$, $x = 3$.

Note that the final form of the matrix representation involves only zero entries below the diagonal. Such a matrix is said to be *upper-triangular* (meaning the non-zero entries are all in the top right triangle, and on the leading diagonal). This enables us to solve (5.8) easily for z, then y, then x working backwards up the system. In fact, we say that the system of equations (5.8) can be solved by *back-substitution*. The aim of the series of equivalent representations was to achieve this triangular form. Note: reduction to a *lower triangular* form by a similar method would be equally effective.

Rather than use back-substitution we can use row operations to reduce the matrix still further. Continuing, we have

$$\begin{pmatrix} 1 & 1 & -1 \\ 0 & -1 & 1 \\ 0 & 0 & -5 \end{pmatrix}\begin{pmatrix} x \\ y \\ z \end{pmatrix} = \begin{pmatrix} -3 \\ 6 \\ -20 \end{pmatrix}$$

$$\xrightarrow[R3 \;\to\; -R3/5]{R2 \;\to\; (5\times R2)+R3} \begin{pmatrix} 1 & 1 & -1 \\ 0 & -5 & 0 \\ 0 & 0 & 1 \end{pmatrix}\begin{pmatrix} x \\ y \\ z \end{pmatrix} = \begin{pmatrix} -3 \\ 10 \\ 4 \end{pmatrix}$$

$$\xrightarrow[R2 \;\to\; -R2/5]{R1 \;\to\; R1+(R2/5)+R3} \begin{pmatrix} 1 & 0 & 0 \\ 0 & 1 & 0 \\ 0 & 0 & 1 \end{pmatrix}\begin{pmatrix} x \\ y \\ z \end{pmatrix} = \begin{pmatrix} 3 \\ -2 \\ 4 \end{pmatrix}$$

from which the solution can be read off directly.

It is clearly unnecessary to write out the (x, y, z) column and the equals sign at every step, and the right hand side can be carried along as part of the array, as we will see in the following example.

Example 5.2 Solve the system

$$\begin{aligned} x + 2y + 3z &= 10 \\ 2x + 5y + 8z &= 24 \\ 3x + 8y + 14z &= 39 \end{aligned} \tag{5.9}$$

by reduction to triangular form and back-substitution.

We work through a series of equivalent representations to obtain a triangular form:

$$\left(\begin{array}{ccc|c} 1 & 2 & 3 & 10 \\ 2 & 5 & 8 & 24 \\ 3 & 8 & 14 & 39 \end{array}\right) \xrightarrow[R3 \;\to\; R3-(3\times R1)]{R2 \;\to\; R2-(2\times R1)} \left(\begin{array}{ccc|c} 1 & 2 & 3 & 10 \\ 0 & 1 & 2 & 4 \\ 0 & 2 & 5 & 9 \end{array}\right).$$

Start from the left hand column using either of the rows above.

$$\begin{pmatrix} 1 & 2 & 3 & | & 10 \\ 0 & 1 & 2 & | & 4 \\ 0 & 2 & 5 & | & 9 \end{pmatrix} \xrightarrow{R3 \to R3 - (2 \times R2)} \begin{pmatrix} 1 & 2 & 3 & | & 10 \\ 0 & 1 & 2 & | & 4 \\ 0 & 0 & 1 & | & 1 \end{pmatrix}.$$

From the upper triangular form write out the equivalent system and solve by back-substitution.

$$x + 2y + 3z = 10$$
$$y + 2z = 4$$
$$z = 1$$

from which we see $z = 1$, and so $y = 4 - 2 = 2$ and $x = 10 - (2 \times 2) - 3 \times 1 = 3$. Or,

$$z = 1, y = 2, x = 3.$$

This method is a useful way of organizing the working of the solution of linear systems by hand. But, like the simultaneous equations method, it is not a practical algorithm for really large systems or for implementation as a computer program. (For larger systems special methods have been developed.)

Exercise 5.2 Solve the system

$$x + y + 3z = 5$$
$$x + 6y + 10z = 17$$
$$-x + 4y - 5z = 8.$$

5.3. PRODUCTS

In Section 5.1, we learned how to multiply a matrix and a column vector. We multiply two matrices by treating the second one as a set of column vectors. For example, multiplying the same 3×3 matrix by different column vectors we get

$$\begin{pmatrix} 3 & 0 & -1 \\ 0 & 1 & 5 \\ 1 & -2 & 3 \end{pmatrix} \begin{pmatrix} 8 \\ 5 \\ 0 \end{pmatrix} = \begin{pmatrix} 24 \\ 5 \\ -2 \end{pmatrix},$$

$$\begin{pmatrix} 3 & 0 & -1 \\ 0 & 1 & 5 \\ 1 & -2 & 3 \end{pmatrix} \begin{pmatrix} 1 \\ 2 \\ 1 \end{pmatrix} = \begin{pmatrix} 2 \\ 7 \\ 0 \end{pmatrix},$$

$$\begin{pmatrix} 3 & 0 & -1 \\ 0 & 1 & 5 \\ 1 & -2 & 3 \end{pmatrix} \begin{pmatrix} 1 \\ -3 \\ -1 \end{pmatrix} = \begin{pmatrix} 4 \\ -8 \\ 4 \end{pmatrix}.$$

So, putting these together

$$\begin{pmatrix} 3 & 0 & -1 \\ 0 & 1 & 5 \\ 1 & -2 & 3 \end{pmatrix} \begin{pmatrix} 8 & 1 & 1 \\ 5 & 2 & -3 \\ 0 & 1 & -1 \end{pmatrix} = \begin{pmatrix} 24 & 2 & 4 \\ 5 & 7 & -8 \\ -2 & 0 & 4 \end{pmatrix}.$$

Exercise 5.3 Let

$$A = \begin{pmatrix} 3 & 0 & -1 \\ 0 & 1 & 5 \\ 1 & -2 & 3 \end{pmatrix} \text{ and } B = \begin{pmatrix} 8 & 1 & 1 \\ 5 & 2 & -3 \\ 0 & 1 & -1 \end{pmatrix}.$$

Find

(i) **BA** (the product of **B** and **A** in the indicated order)

(ii) **AB**

(iii) Does **AB** = **BA** in this case?

If $AB \neq BA$ for matrices A and B, we say that A and B do not *commute* (with respect to multiplication). In general, matrix multiplication is not commutative: the result depends on the order of multiplication. Therefore, if we are multiplying a matrix equation

$B = C$ by a matrix A we must specify whether this is on the left or the right, giving $AB = AC$ or $BA = CA$. (Both are true but they are different equations if A and B do not commute.)

Exercise 5.4 If A and B are as in Exercise 5.3 and

$$C = \begin{pmatrix} 3 & 0 & 1 \\ 1 & -1 & 1 \\ 0 & 2 & 1 \end{pmatrix},$$

find

(i) $C(AB)$,

(ii) $(CA)B$.

(iii) Verify that $C(AB) = (CA)B$ in this case.

The equality $C(AB) = (CA)B$ is called the *associative law* and is true for all matrices A, B, C. Therefore we can write CAB for this product without ambiguity. Multiplication by a scalar can be done at any stage of taking a matrix product. For example,

$$\lambda(ABC) = (\lambda A)BC = A(\lambda(BC)) \text{ etc.}$$

Exercise 5.5 The following 2×2 matrices are important in quantum mechanics ($i = \sqrt{-1}$ as usual):

$$\sigma_1 = \begin{pmatrix} 0 & 1 \\ 1 & 0 \end{pmatrix}, \quad \sigma_2 = \begin{pmatrix} 0 & -i \\ i & 0 \end{pmatrix}, \quad \sigma_3 = \begin{pmatrix} 1 & 0 \\ 0 & -1 \end{pmatrix}$$

Find a relation between $\sigma_1\sigma_2 - \sigma_2\sigma_1$ and σ_3.

We can also write the product in terms of matrix elements. If A is an $M \times K$ matrix and B is a $K \times N$ matrix, we can compute the elements of the product $C = AB$, which is an $M \times N$ matrix, as follows

$$c_{ij} = \sum_{k=1}^{K} a_{ik}b_{kj}$$

where $i = 1, 2, \ldots, M$ and $j = 1, 2, \ldots, N$. For example for the product of a 2×3 matrix with a 3×2 matrix we have

$$\begin{pmatrix} a_{11} & a_{12} & a_{13} \\ a_{21} & a_{22} & a_{23} \end{pmatrix} \begin{pmatrix} b_{11} & b_{12} \\ b_{21} & b_{22} \\ b_{31} & b_{32} \end{pmatrix} = \begin{pmatrix} c_{11} & c_{12} \\ c_{21} & c_{22} \end{pmatrix}$$

where $c_{11} = a_{11}b_{11} + a_{12}b_{21} + a_{13}b_{31}$ (formed from the 1st row of **A** and the 1st column of **B** as marked).

5.4. THE IDENTITY MATRIX

Definition: the identity matrix. The matrix

$$\mathsf{I} = \begin{pmatrix} 1 & 0 & 0 \\ 0 & 1 & 0 \\ 0 & 0 & 1 \end{pmatrix}$$

is called the (3×3) *identity matrix*. Sometimes it is called the *unit* matrix. The $N \times N$ identity matrix (always square) has ones on the leading diagonal (top left to bottom right) and zeros elsewhere.

$$\mathsf{I}_N = \begin{pmatrix} 1 & 0 & \cdots & 0 \\ 0 & 1 & & \vdots \\ \vdots & & \ddots & 0 \\ 0 & \cdots & 0 & 1 \end{pmatrix}$$

Exercise 5.6

(i) For a general 3×3 matrix

$$A = \begin{pmatrix} a_{11} & a_{12} & a_{13} \\ a_{21} & a_{22} & a_{23} \\ a_{31} & a_{32} & a_{33} \end{pmatrix},$$

show that $IA = AI = A$.

(ii) What is SA where S is the matrix

$$S = \begin{pmatrix} 7 & 0 & 0 \\ 0 & 7 & 0 \\ 0 & 0 & 7 \end{pmatrix} ?$$

5.5. SYMMETRIC AND ANTISYMMETRIC MATRICES

Definition: transpose of a matrix. If A is the matrix

$$A = \begin{pmatrix} a_{11} & a_{12} & a_{13} \\ a_{21} & a_{22} & a_{23} \\ a_{31} & a_{32} & a_{33} \end{pmatrix},$$

the *transpose* of A is the matrix A^T:

$$A^T = \begin{pmatrix} a_{11} & a_{21} & a_{31} \\ a_{12} & a_{22} & a_{32} \\ a_{13} & a_{23} & a_{33} \end{pmatrix},$$

obtained by interchanging rows and columns. We call A *symmetric* if $A^T = A$ and *antisymmetric* if $A^T = -A$.

Exercise 5.7 Let A and B be the matrices defined in Exercise 5.3. Find A^T and B^T and verify that

$$A^TB^T = (BA)^T.$$

Note the order in which the matrices are multiplied. This result is true for general matrices A and B.

Exercise 5.8

(i) The matrix

$$\begin{pmatrix} \cdot & a & b \\ \cdot & \cdot & c \\ \cdot & \cdot & \cdot \end{pmatrix}$$

is antisymmetric. Fill in the missing elements.

(ii) For any matrices A and B, what are $(A^T)^T$ and $(A+B)^T$?

(iii) Verify that if M is an arbitrary matrix, the matrices $\frac{1}{2}(M+M^T)$ and $\frac{1}{2}(M-M^T)$ are respectively symmetric and antisymmetric. Deduce that every matrix is the sum of a symmetric and an antisymmetric matrix.

Part (*i*) of Exercise 5.8 gives another way of looking at an antisymmetric matrix: the terms below the diagonal are the negatives of the corresponding terms above the diagonal and the diagonal terms are zero.

5.6. INVERSES

Definition: Given a matrix A, if a matrix M exists with $AM = MA = I$ then M is the *inverse* of A. We write $M = A^{-1}$. In words: the product of a matrix and its inverse is the identity matrix ($A^{-1}A = I$). (Of course, if M is the inverse of A then A is the inverse of M.)

When we work with scalars we consider $ab = c$ and $a = c/b$ as equivalent ($b \neq 0$). But for matrices "division" does not have such a clear meaning. For matrices we instead work with $\mathbf{AB} = \mathbf{C}$ and $\mathbf{A} = \mathbf{CB}^{-1}$.

Exercise 5.9 Are the following statements (a) true or (b) false?

(i) $\begin{pmatrix} \dfrac{1}{2} & 0 & 0 \\ 0 & -\dfrac{1}{3} & 0 \\ 0 & 0 & 1 \end{pmatrix}$ is the inverse of $\begin{pmatrix} 2 & 0 & 0 \\ 0 & -3 & 0 \\ 0 & 0 & 1 \end{pmatrix}$,

(ii) $\dfrac{1}{25}\begin{pmatrix} 5 & 0 & -5 \\ -6 & 10 & 1 \\ 7 & 5 & 3 \end{pmatrix}$ is the inverse of $\begin{pmatrix} 1 & -1 & 2 \\ 1 & 2 & 1 \\ -4 & -1 & 2 \end{pmatrix}$,

(iii) $25\begin{pmatrix} 1 & -1 & 2 \\ 1 & 2 & 1 \\ -4 & -1 & 2 \end{pmatrix}$ is the inverse of $\begin{pmatrix} 5 & 0 & -5 \\ -6 & 10 & 1 \\ 7 & 5 & 3 \end{pmatrix}$.

In fact if $\mathbf{AM} = \mathbf{I}$ it can be shown that automatically $\mathbf{MA} = \mathbf{I}$, so there is no need to check both conditions for an inverse. We now look at a way to calculate the inverse.

Let $\mathbf{A}$ be the general 3×3 matrix

$$\mathbf{A} = \begin{pmatrix} a_{11} & a_{12} & a_{13} \\ a_{21} & a_{22} & a_{23} \\ a_{31} & a_{32} & a_{33} \end{pmatrix}$$

and let its inverse be

$$\mathbf{A}^{-1} = \begin{pmatrix} x_1 & x_2 & x_3 \\ y_1 & y_2 & y_3 \\ z_1 & z_2 & z_3 \end{pmatrix}.$$

To find A^{-1} explicitly in terms of the components of A we have to solve

$$AA^{-1} = \begin{pmatrix} a_{11} & a_{12} & a_{13} \\ a_{21} & a_{22} & a_{23} \\ a_{31} & a_{32} & a_{33} \end{pmatrix} \begin{pmatrix} x_1 & x_2 & x_3 \\ y_1 & y_2 & y_3 \\ z_1 & z_2 & z_3 \end{pmatrix} = I = \begin{pmatrix} 1 & 0 & 0 \\ 0 & 1 & 0 \\ 0 & 0 & 1 \end{pmatrix}$$

i.e. we must solve the equations

$$\begin{pmatrix} a_{11} & a_{12} & a_{13} \\ a_{21} & a_{22} & a_{23} \\ a_{31} & a_{32} & a_{33} \end{pmatrix} \begin{pmatrix} x_1 \\ y_1 \\ z_1 \end{pmatrix} = \begin{pmatrix} 1 \\ 0 \\ 0 \end{pmatrix},$$

$$\begin{pmatrix} a_{11} & a_{12} & a_{13} \\ a_{21} & a_{22} & a_{23} \\ a_{31} & a_{32} & a_{33} \end{pmatrix} \begin{pmatrix} x_2 \\ y_2 \\ z_2 \end{pmatrix} = \begin{pmatrix} 0 \\ 1 \\ 0 \end{pmatrix}$$

$$\begin{pmatrix} a_{11} & a_{12} & a_{13} \\ a_{21} & a_{22} & a_{23} \\ a_{31} & a_{32} & a_{33} \end{pmatrix} \begin{pmatrix} x_3 \\ y_3 \\ z_3 \end{pmatrix} = \begin{pmatrix} 0 \\ 0 \\ 1 \end{pmatrix}$$

for (x_1, y_1, z_1), (x_2, y_2, z_2) and (x_3, y_3, z_3), respectively. We can do this using the method of Section 5.2 three times, or all in one go as in the following example.

Example 5.3 Find the inverse of $A = \begin{pmatrix} 1 & 2 & 3 \\ 2 & 5 & 8 \\ 3 & 8 & 14 \end{pmatrix}$.

We have to solve

$$\begin{pmatrix} 1 & 2 & 3 \\ 2 & 5 & 8 \\ 3 & 8 & 14 \end{pmatrix} \begin{pmatrix} x_1 \\ y_1 \\ z_1 \end{pmatrix} = \begin{pmatrix} 1 \\ 0 \\ 0 \end{pmatrix} \text{ etc.}$$

To do this we construct a sequence of equivalent representations that reduce the matrix to the identity matrix.

$$\begin{pmatrix} 1 & 2 & 3 & | & 1 & 0 & 0 \\ 2 & 5 & 8 & | & 0 & 1 & 0 \\ 3 & 8 & 14 & | & 0 & 0 & 1 \end{pmatrix} \xrightarrow{\substack{R2-(2\times R1) \\ R3-(3\times R1)}} \begin{pmatrix} 1 & 2 & 3 & | & 1 & 0 & 0 \\ 0 & 1 & 2 & | & -2 & 1 & 0 \\ 0 & 2 & 5 & | & -3 & 0 & 1 \end{pmatrix}$$

$$\xrightarrow{R3-(2\times R2)} \begin{pmatrix} 1 & 2 & 3 & | & 1 & 0 & 0 \\ 0 & 1 & 2 & | & -2 & 1 & 0 \\ 0 & 0 & 1 & | & 1 & -2 & 1 \end{pmatrix}$$

$$\xrightarrow{\substack{R1-(2\times R2) \\ R2-(2\times R3)}} \begin{pmatrix} 1 & 0 & -1 & | & 5 & -2 & 0 \\ 0 & 1 & 0 & | & -4 & 5 & -2 \\ 0 & 0 & 1 & | & 1 & -2 & 1 \end{pmatrix},$$

and so finally we have

$$\xrightarrow{R1+R3} \begin{pmatrix} 1 & 0 & 0 & | & 6 & -4 & 1 \\ 0 & 1 & 0 & | & -4 & 5 & -2 \\ 0 & 0 & 1 & | & 1 & -2 & 1 \end{pmatrix}. \tag{5.10}$$

The equivalent system is now

$$I \begin{pmatrix} x_1 \\ y_1 \\ z_1 \end{pmatrix} = \begin{pmatrix} 6 \\ -4 \\ 1 \end{pmatrix}, \quad I \begin{pmatrix} x_2 \\ y_2 \\ z_2 \end{pmatrix} = \begin{pmatrix} -4 \\ 5 \\ -2 \end{pmatrix}, \quad I \begin{pmatrix} x_3 \\ y_3 \\ z_3 \end{pmatrix} = \begin{pmatrix} 1 \\ -2 \\ 1 \end{pmatrix},$$

from which we read off (x_1, y_1, z_1) etc. It is clear that the matrix to the right of the vertical bar in the final representation (5.10) is the required inverse, i.e.

$$A^{-1} = \begin{pmatrix} 6 & -4 & 1 \\ -4 & 5 & -2 \\ 1 & -2 & 1 \end{pmatrix}.$$

The rule for finding **A−1** is therefore: convert the matrix (**A**|**I**) by row operations of the above type to (**I**|**B**); if this can be done then **A−1** exists and is equal to **B**.

Exercise 5.10 Find the inverse **B** of $\;A = \begin{pmatrix} 1 & 1 & 3 \\ 1 & 6 & 10 \\ -2 & 8 & 10 \end{pmatrix}\;$ by the method indicated. Check that **AB = I**.

In this section, we have found a method by which we can construct the inverse of any given matrix for which an inverse exists. In the next sections we shall find a general formula for the inverse of a square matrix of any size. (Note that matrix inverses are only defined for square matrices.) We begin with 2 × 2 matrices.

5.7. THE INVERSE OF A 2 × 2 MATRIX

We solve the general 2 × 2 system

$$ax + by = h \tag{5.11}$$
$$cx + dy = k \tag{5.12}$$

which can be written in matrix form as

$$\begin{pmatrix} a & b \\ c & d \end{pmatrix} \begin{pmatrix} x \\ y \end{pmatrix} = \begin{pmatrix} h \\ k \end{pmatrix}. \tag{5.13}$$

Example 5.4 Show that if $ad - bc \neq 0$ then the solution of equations (5.11) and (5.12) is

$$x = \frac{dh - bk}{ad - bc}, \quad y = \frac{ak - ch}{ad - bc}. \tag{5.14}$$

To eliminate y between the pair of simultaneous equations (5.11) and (5.12), we multiply equation (5.11) by d, multiply equation (5.12) by b, and then subtract.

$$dax + dby = dh,$$
$$bcx + bdy = bk.$$

So

$$dax - bcx = dh - bk,$$

which leads to an equation for x

$$x = \frac{dh - bk}{ad - bc},$$

as required. Similarly, to eliminate x multiply equation (5.11) by c, multiply equation (5.12) by a, and then subtract.

Alternatively, we can start from the matrix form and manipulate the array of coefficients as in Sections 5.2 or 5.6.

We can write equations (5.14) in a matrix form as

$$x = \frac{1}{ad - bc}(d, -b)\begin{pmatrix} h \\ k \end{pmatrix}, \quad y = \frac{1}{ad - bc}(-c, a)\begin{pmatrix} h \\ k \end{pmatrix}$$

or, combining these, as

$$\begin{pmatrix} x \\ y \end{pmatrix} = \frac{1}{ad - bc}\begin{pmatrix} d & -b \\ -c & a \end{pmatrix}\begin{pmatrix} h \\ k \end{pmatrix} \qquad (5.15)$$

Let

$$\mathsf{A} = \begin{pmatrix} a & b \\ c & d \end{pmatrix}, \mathrm{x} = \begin{pmatrix} x \\ y \end{pmatrix} \quad \text{and} \quad \mathbf{h} = \begin{pmatrix} h \\ k \end{pmatrix}.$$

Then equation (5.13) can be written as $\mathsf{A}\mathrm{x} = \mathbf{h}$ and therefore equation (5.15) is $\mathrm{x} = \mathsf{A}^{-1}\mathbf{h}$. Thus, if we have a matrix

$$A = \begin{pmatrix} a & b \\ c & d \end{pmatrix}, \text{ then we can write } A^{-1} = \frac{1}{ad - bc} \begin{pmatrix} d & -b \\ -c & a \end{pmatrix}.$$

This is the general formula for the inverse of a 2 × 2 matrix.

Note that from this general form we can see that not all 2 × 2 matrices have an inverse. Namely, if $ad - bc = 0$, then A^{-1} will not exist. (This is analogous to the fact that not all numbers have inverses; namely 0 does not have an inverse because there is no number x such that $0x = 1$.)

To extend this to larger matrices we need to introduce the idea of a determinant, which we shall do in the next section.

5.8. DETERMINANTS

Definition: The *determinant*, written det A or $|A|$, of a 2 × 2 matrix A is defined by

$$\det \begin{pmatrix} a & b \\ c & d \end{pmatrix} = ad - bc.$$

Or, using more systematic notation,

$$A = \begin{pmatrix} a_{11} & a_{12} \\ a_{21} & a_{22} \end{pmatrix} \rightarrow \det A = a_{11}a_{22} - a_{12}a_{21}. \tag{5.16}$$

Exercise 5.11 Find the values of

(i) $\det \begin{pmatrix} 1 & 3 \\ -1 & 2 \end{pmatrix}$

(ii) $\det \begin{pmatrix} 3 & 1 \\ 2 & 1 \end{pmatrix}$

(iii) $\det \begin{pmatrix} -3 & -2 \\ -4 & -3 \end{pmatrix}$

Definition: The determinant, det A or |A|, of a 3 × 3 matrix A is defined by

$$\det\begin{pmatrix} a_{11} & a_{12} & a_{13} \\ a_{21} & a_{22} & a_{23} \\ a_{31} & a_{32} & a_{33} \end{pmatrix} = a_{11}\det\begin{pmatrix} a_{22} & a_{23} \\ a_{32} & a_{33} \end{pmatrix} - a_{12}\det\begin{pmatrix} a_{21} & a_{23} \\ a_{31} & a_{33} \end{pmatrix}$$

$$+ a_{13}\det\begin{pmatrix} a_{21} & a_{22} \\ a_{31} & a_{32} \end{pmatrix} \tag{5.17}$$

Note the alternation of signs: in general the sign of the coefficient of a_{ij} is $(-1)^{i+j}$. Note also how the 2 × 2 determinants in (5.17) are formed: for example, we take a_{11}, cross out the row and column that it is in, and this leaves the first 2 × 2 determinant on the right of equation (5.17).

$$\begin{pmatrix} a_{11} & a_{12} & a_{13} \\ a_{21} & a_{22} & a_{23} \\ a_{31} & a_{32} & a_{33} \end{pmatrix} \rightarrow a_{11}\det\begin{pmatrix} a_{22} & a_{23} \\ a_{32} & a_{33} \end{pmatrix}$$

Then we do likewise with a_{12} and a_{13}. The determinants of the (2 × 2) submatrices, obtained by taking out row i and column j, are called the *minors*, M_{ij}, of A.

Example 5.5 If $A = \begin{pmatrix} 1 & -1 & 0 \\ 2 & 4 & -3 \\ -2 & 1 & 1 \end{pmatrix}$, find det A.

Expanding by the first row as in the definition (5.17) gives

$$\det A = 1\times\det\begin{pmatrix} 4 & -3 \\ 1 & 1 \end{pmatrix} - (-1)\times\det\begin{pmatrix} 2 & -3 \\ -2 & 1 \end{pmatrix} + 0\times\det\begin{pmatrix} 2 & 4 \\ -2 & 1 \end{pmatrix}$$

$$= (1\times7) + (1\times-4) + 0 = 3.$$

Exercise 5.12

(i) Let the matrix $\mathbf{B} = \begin{pmatrix} 0 & 1 & 2 \\ -7 & 3 & -4 \\ 1 & 2 & 3 \end{pmatrix}$. Verify that det $\mathbf{B}$ = -17.

(ii) For A as defined in Example 5.5, and $\mathbf{B}$ defined here, find AB and verify that

$$\det \mathbf{AB} = \det \mathbf{A} \cdot \det \mathbf{B}.$$

Exercise 5.13 By induction, generalize the definition of a determinant to $N \times N$ matrices.

Another way to remember how to construct the determinant of a 3×3 matrix is to write the first two columns on the right, and then draw diagonal lines as follows:

$$\begin{array}{ccc|cc} a_{11} & a_{12} & a_{13} & a_{11} & a_{12} \\ a_{21} & a_{22} & a_{23} & a_{21} & a_{22} \\ a_{31} & a_{32} & a_{33} & a_{31} & a_{32} \end{array}$$
$$- \quad - \quad - \quad + \quad + \quad +$$

This is called *Sarrus's rule*. We sum the products along the southeast (top left to bottom right) lines, and subtract the products along the southwest (top right to bottom left) lines. It is a useful way to remember the terms of the determinant of a 3×3 matrix, but it does not extend to larger matrices.

det $\mathbf{A}$

$$= a_{11}a_{22}a_{33} - a_{11}a_{23}a_{32} + a_{12}a_{23}a_{31} - a_{12}a_{21}a_{33} + a_{13}a_{21}a_{32} - a_{13}a_{22}a_{31}$$

5.9. PROPERTIES OF DETERMINANTS

The following properties are valid for determinants of $N \times N$ matrices. (Note that determinants are only defined for square matrices.) We consider proofs only in the 2×2 case, where they involve only elementary algebra. In the following, A is a general $N \times N$ matrix.

Property 1: If **B** is obtained from **A** by multiplying any *one* row or column by λ, then $\det \mathbf{B} = \lambda \det \mathbf{A}$. It follows that $\det(\lambda \mathbf{A}) = \lambda^N \det \mathbf{A}$.

Example 5.6 Prove that $\det(\lambda \mathbf{A}) = \lambda^N \det \mathbf{A}$ for $N = 2$.

$$\lambda \mathbf{A} = \begin{pmatrix} \lambda a_{11} & \lambda a_{12} \\ \lambda a_{21} & \lambda a_{22} \end{pmatrix}.$$

So,

$$\det(\lambda \mathbf{A}) = \lambda a_{11} \lambda a_{22} - \lambda a_{12} \lambda a_{21}$$
$$= \lambda^2 (a_{11} a_{22} - a_{12} a_{21})$$
$$= \lambda^2 \det \mathbf{A}.$$

Property 2: If $\tilde{\mathbf{A}}$ is obtained from **A** by interchanging a pair of rows or a pair of columns, then $\det \tilde{\mathbf{A}} = -\det \mathbf{A}$.

Exercise 5.14 Show that

$$\det \begin{pmatrix} a_{11} & a_{12} \\ a_{21} & a_{22} \end{pmatrix} = -\det \begin{pmatrix} a_{12} & a_{11} \\ a_{22} & a_{21} \end{pmatrix} = -\det \begin{pmatrix} a_{21} & a_{22} \\ a_{11} & a_{12} \end{pmatrix}.$$

In our definition of the determinant we singled out the first row of the matrix. Property 2 shows that there is nothing special about the first row: by Property 2 we can use any row to frame a definition and hence to calculate a determinant. This can be used to simplify the calculation, e.g. by selecting a row containing one or more zeros.

Example 5.7 Find $\det \begin{pmatrix} 2 & 4 & -3 \\ 1 & -1 & 0 \\ 2 & 1 & 1 \end{pmatrix}$.

First, interchange the first and second rows; then expand by the top row. The zero means there are only two minors

(2×2 determinants) to evaluate. We have already evaluated this determinant in Example 5.5.

$$\det\begin{pmatrix} 2 & 4 & -3 \\ 1 & -1 & 0 \\ 2 & 1 & 1 \end{pmatrix} = -\det\begin{pmatrix} 1 & -1 & 0 \\ 2 & 4 & -3 \\ 2 & 1 & 1 \end{pmatrix} = -3.$$

Alternative solution Expand by the second row. This is exactly the same calculation, but arrived at in a different way.

$$\det\begin{pmatrix} 2 & 4 & -3 \\ 1 & -1 & 0 \\ 2 & 1 & 1 \end{pmatrix} = (-1)^{2+1} \cdot (1) \cdot \det\begin{pmatrix} 4 & -3 \\ 1 & 1 \end{pmatrix}$$

$$+ (-1)^{2+2} \cdot (-1) \cdot \det\begin{pmatrix} 2 & -3 \\ 2 & 1 \end{pmatrix}$$

$$= -7 + 4 = -3.$$

Note the signs here: recall from the definition that the coefficient of a_{ij} has a factor $(-1)^{i+j}$.

Property 3: If A has two rows or columns proportional then det A = 0. (In particular det A = 0 if two rows or columns are the same.)

Exercise 5.15 Show that $\det\begin{pmatrix} a_{11} & \lambda a_{11} \\ a_{12} & \lambda a_{12} \end{pmatrix} = 0$ and that $\det\begin{pmatrix} a_{11} & a_{12} \\ \lambda a_{11} & \lambda a_{12} \end{pmatrix} = 0$.

Property 4: A square matrix A and its transpose A^T have the same determinant: det A^T = det A.

Exercise 5.16 Show that

$$\det\begin{pmatrix} a_{11} & a_{12} \\ a_{21} & a_{22} \end{pmatrix} = \det\begin{pmatrix} a_{11} & a_{21} \\ a_{12} & a_{22} \end{pmatrix}.$$

Since transposition interchanges rows and columns, this implies that any statement of a property of the determinant under a row operation must also be valid under the equivalent column operation. Thus we do not have to prove row and column properties separately; e.g. the second result in Exercise 5.15 follows from the first as a consequence of Exercise 5.16. We can also expand a determinant by columns instead of rows.

Property 5: Adding a multiple of one row (or column) to another leaves the determinant unaltered.

Exercise 5.17 Show that

$$\det\begin{pmatrix} a_{11} + \lambda a_{21} & a_{12} + \lambda a_{22} \\ a_{21} & a_{22} \end{pmatrix} = \det\begin{pmatrix} a_{11} & a_{12} \\ a_{21} & a_{22} \end{pmatrix}.$$

Property 6: If the rows (or columns) of A are linearly dependent then $\det A = 0$–i.e. if $\mathbf{a}, \mathbf{b}, \mathbf{c}, \dots$ are the rows of A, regarded as row vectors, and $\lambda\mathbf{a} + \mu\mathbf{b} + \nu\mathbf{c} + \dots = \mathbf{0}$ (with at least one of $\lambda, \mu, \nu, \dots \neq 0$, then $\det A = 0$. (Recall the discussion of linear dependence for vectors in Section 4.2.)

Example 5.8 If two vectors are linearly dependent then

$$\lambda\begin{pmatrix} a_{11} \\ a_{21} \end{pmatrix} + \mu\begin{pmatrix} a_{12} \\ a_{22} \end{pmatrix} = \begin{pmatrix} 0 \\ 0 \end{pmatrix},$$

and we have that

$$\lambda a_{11} + \mu a_{21} = 0 \text{ and } \lambda a_{12} + \mu a_{22} = 0 \qquad (5.18)$$

with λ, μ not both 0. Show that $\det\begin{pmatrix} a_{11} & a_{12} \\ a_{21} & a_{22} \end{pmatrix} = 0$.

Assume $\lambda \neq 0$. Add $(\mu/\lambda) \times$ row 2 to row 1. By Property 5 this does not change the determinant:

$$\det A = \det\begin{pmatrix} a_{11} + \dfrac{\mu}{\lambda}a_{21} & a_{12} + \dfrac{\mu}{\lambda}a_{22} \\ a_{21} & a_{22} \end{pmatrix}.$$

But we know from equations (5.18) that $a_{11} + \frac{\mu}{\lambda}a_{21} = 0 = a_{12} + \frac{\mu}{\lambda}a_{22}$. Therefore

$$\det A = \begin{pmatrix} 0 & 0 \\ a_{21} & a_{22} \end{pmatrix} = 0.$$

If $\mu \neq 0$ we instead add $\lambda/\mu \times$ row 1 to row 2.

Property 7: If A and B are square matrices of the same size, then the determinant of the product is the product of the determinants: det(AB) = det A det B.

> **Exercise 5.18** For general matrices
>
> $$A = \begin{pmatrix} a_{11} & a_{12} \\ a_{21} & a_{22} \end{pmatrix}, \quad B = \begin{pmatrix} b_{11} & b_{12} \\ b_{21} & b_{22} \end{pmatrix},$$
>
> show by explicit calculation of both sides that det AB = det A det B.

Property 8: A matrix A has an inverse if and only if det A $\neq$ 0. (A is then said to be *non-singular*.)

Here we summarize some of the important properties of determinants.

1. If every element of one row (or column) is multiplied by a scalar λ, so is the determinant. If every element of the matrix A ($N \times N$) is multiplied by a scalar λ, the determinant is multiplied by λ^N: det(λA) = λ^N det A.

2. If two rows (or columns) are interchanged, the determinant changes sign (but not magnitude).

3. If two rows (or columns) of A are proportional, then det A = 0.

4. If the rows (or columns) are linearly dependent, then det A = 0.

5. The determinant is unchanged if a constant multiple of one row (or column) is added to another row (or column).

6. The transpose A^T has the same determinant as A: $\det A^T = \det A$.

7. If A and B are square matrices (of the same size) then $\det(AB) = \det A \det$

8. A matrix A has an inverse A^{-1} if and only if $\det A \neq 0$ (meaning it is non-singular).

9. If a row or column is all zeros, then $\det A = 0$.

5.10. SOLUTION OF 2×2 LINEAR SYSTEMS

In section 5.7, we found an expression for the solution of the general 2×2 system:

$$\begin{aligned} ax + by &= h \\ cx + dy &= k \end{aligned} \quad \Leftrightarrow \quad \begin{pmatrix} a & b \\ c & d \end{pmatrix}\begin{pmatrix} x \\ y \end{pmatrix} = \begin{pmatrix} h \\ k \end{pmatrix}.$$

Using determinants we can write the solution (Equation 5.14) as

$$x = \frac{\det\begin{pmatrix} h & b \\ k & d \end{pmatrix}}{\det\begin{pmatrix} a & b \\ c & d \end{pmatrix}}, \quad y = \frac{\det\begin{pmatrix} a & h \\ c & k \end{pmatrix}}{\det\begin{pmatrix} a & b \\ c & d \end{pmatrix}}. \tag{5.19}$$

This alternative form is called *Cramer's rule*.

- If $\det\begin{pmatrix} a & b \\ c & d \end{pmatrix} \neq 0$, the system is said to be *non-singular*.

  In this case Cramer's rule gives the unique solution of the system of equations.

- If $\det\begin{pmatrix} a & b \\ c & d \end{pmatrix} = 0$, the system is said to be *singular*. There

are then two possibilities:

1. The system has no solution, because the contradiction 0 = 1 can be derived.

2. Otherwise there are infinitely many solutions.

Example 5.9 Determine whether the following system has a solution, and if so find every solution:

$$3x + 4y = 5$$
$$6x + 9y = 9$$

First check if the system of equations has a solution:

$$\det\begin{pmatrix} 3 & 4 \\ 6 & 9 \end{pmatrix} = 3 \neq 0,$$

so the system has a unique solution. By Cramer's rule this is

$$x = \frac{\det\begin{pmatrix} 5 & 4 \\ 9 & 9 \end{pmatrix}}{\det\begin{pmatrix} 3 & 4 \\ 6 & 9 \end{pmatrix}} = \frac{9}{3} = 3, \quad y = \frac{\det\begin{pmatrix} 3 & 5 \\ 6 & 9 \end{pmatrix}}{\det\begin{pmatrix} 3 & 4 \\ 6 & 9 \end{pmatrix}} = \frac{-3}{3} = -1.$$

The two simultaneous equations could also be solved by the elementary method of elimination.

Example 5.10 Determine whether the following system has a solution, and if so find every solution:

$$3x + 4y = 5$$
$$6x + 8y = 9$$

First check if the system of equations has a solution. A singular system has either no solution or infinitely many.

$$\det\begin{pmatrix} 3 & 4 \\ 6 & 8 \end{pmatrix} = 0,$$

so the system is singular. Here we shall use standard row and column operations on the coefficients to attempt to solve the system. For the 2 × 2 case we could also use the elementary method of elimination

$$\begin{pmatrix} 3 & 4 & \vline & 5 \\ 6 & 8 & \vline & 8 \end{pmatrix} \quad \underset{\rule{2cm}{0.4pt}}{R2 - (2 \times R1)} \quad \begin{pmatrix} 3 & 4 & \vline & 5 \\ 0 & 0 & \vline & -2 \end{pmatrix}$$

from which $0 = -2$, a contradiction. The system therefore has no solution.

Take another look at the two equations we began with. The left sides are the same except for a factor 2, but the right sides are not; hence they contradict one another.

Example 5.11 Determine whether the following system has a solution, and if so find every solution:

$$3x + 4y = 5$$
$$6x + 8y = 10$$

We have that

$$\det \begin{pmatrix} 3 & 4 \\ 6 & 8 \end{pmatrix} = 0,$$

so the system is singular. Using row reduction:

$$\begin{pmatrix} 3 & 4 & \vline & 5 \\ 6 & 8 & \vline & 10 \end{pmatrix} \quad \underset{\rule{2cm}{0.4pt}}{R2 - (2 \times R1)} \quad \begin{pmatrix} 3 & 4 & \vline & 5 \\ 0 & 0 & \vline & 0 \end{pmatrix}$$

so $0 = 0$ and $3x + 4y = 5$; thus the system is equivalent to the one equation, $3x + 4y = 5$. There are infinitely many solutions. For example, let $x = t$ (an arbitrary number) and $y = \frac{1}{4}(5 - 3t)$.

Take another look at the two equations we began with. The second equation is just twice the first equation, and therefore adds no more information than one equation alone (which then cannot be solved uniquely for two unknowns).

Exercise 5.19 For each of the following systems, determine whether it has a solution and if so find every solution:

(i) $x + 2y = 4$

$2x + 2y = 6$

(ii) $x + 2y = 4$

$2x + 4y = 8$

(iii) $x + 2y = 4$

$2x + 4y = 3$

5.11. SOLUTION OF 3 × 3 SYSTEMS

Cramer's rule: The solution of the general 3 × 3 linear system of equations

$$\begin{pmatrix} a_{11} & a_{12} & a_{13} \\ a_{21} & a_{22} & a_{33} \\ a_{31} & a_{32} & a_{33} \end{pmatrix} \begin{pmatrix} x \\ y \\ z \end{pmatrix} = \begin{pmatrix} h_1 \\ h_2 \\ h_3 \end{pmatrix} \tag{5.20}$$

is

$$x = \frac{\det \begin{pmatrix} h_1 & a_{12} & a_{13} \\ h_2 & a_{22} & a_{23} \\ h_3 & a_{32} & a_{33} \end{pmatrix}}{\det A}, \quad y = \frac{\det \begin{pmatrix} a_{11} & h_1 & a_{13} \\ a_{21} & h_2 & a_{23} \\ a_{31} & h_3 & a_{33} \end{pmatrix}}{\det A}, \quad z = \frac{\det \begin{pmatrix} a_{11} & a_{12} & h_1 \\ a_{21} & a_{22} & h_3 \\ a_{31} & a_{32} & h_3 \end{pmatrix}}{\det A}.$$

$$\tag{5.21}$$

Proof: We can write the first of equations (5.20) as

$$h_1 - a_{11}x - a_{12}y - a_{13}z = 0. \tag{5.22}$$

We verify that (5.21) satisfies this by evaluating the left hand side of (5.22). We get (using property 2 from Section 5.9 to exchange columns, changing signs)

$$\det A \left(h_1 - a_{11}x - a_{12}y - a_{13}z\right)$$

$$= h_1 \det A - a_{11} \det \begin{pmatrix} h_1 & a_{12} & a_{13} \\ h_2 & a_{22} & a_{23} \\ h_3 & a_{32} & a_{33} \end{pmatrix} + a_{12} \det \begin{pmatrix} h_1 & a_{11} & a_{13} \\ h_2 & a_{21} & a_{23} \\ h_3 & a_{31} & a_{33} \end{pmatrix}$$

$$- a_{13} \det \begin{pmatrix} h_1 & a_{11} & a_{12} \\ h_2 & a_{21} & a_{22} \\ h_3 & a_{31} & a_{32} \end{pmatrix}$$

$$= \det \begin{pmatrix} h_1 & a_{11} & a_{12} & a_{13} \\ h_1 & a_{11} & a_{12} & a_{13} \\ h_2 & a_{21} & a_{22} & a_{33} \\ h_3 & a_{31} & a_{32} & a_{33} \end{pmatrix}$$

by expanding the determinant by the first row,

$$= 0,$$

since two rows of the determinant are equal. The other two equations in (5.20) can be verified similarly. This completes the proof.

- If $\det A \neq 0$, the system is said to be *non-singular*. In this case Cramer's rule gives the unique solution of the system of equations.

- If $\det A = 0$, the system is said to be *singular*. There are then two possibilities:

  1. The system has no solution, because the contradiction $0 = 1$ can be derived.

  2. Otherwise there are infinitely many solutions.

Exercise 5.20

(i) Determine which of the following systems has a unique solution

$$2x - 3y + 3z = -2$$
$$2x - 2y + 6z = 1$$
$$x + \lambda z = k$$

for (a) $\lambda = 6, k = 1$, (b) $\lambda = 3, k = 3$, (c) $\lambda = 6, k = 7/2$.

(ii) Use Cramer's rule to find the unique solution.

(iii) Find all the solutions in case (c).

5.12. HOMOGENEOUS SYSTEMS

If $(h_1, h_2, h_3) = (0, 0, 0)$ then (5.20) can be written

$$\mathbf{A}x = 0.$$

We refer to this as a *homogeneous* linear system.

- If a homogeneous system is *non-singular* (i.e. if det $\mathbf{A} \neq 0$) Cramer's rule gives $\mathbf{x} = \mathbf{0}$ (i.e. $x = 0$, $y = 0$, $z = 0$). We say that the system has only the *trivial solution*.

- If a homogeneous system is singular (i.e. if det $\mathbf{A} = 0$) then there will be *infinitely many solutions*. (One might guess this since Cramer's rule gives 0/0 for each of the x, y, and z. A proper proof can be constructed from the converse of property 6 of determinants: that the rows of a singular matrix are linearly dependent; but we shall not stop to do this). We shall see an important application of this in Section 5.14.

We have stated the theory in terms of 3×3 systems, but the argument applies to any $n{\times}n$ homogeneous system (i.e. any system of the form $\mathbf{A}x = 0$ with $\mathbf{A}$ an $n \times n$ matrix. In particular if $\mathbf{A}$ is a number, a say (a 1×1 matrix) then $ax = 0$ has only the trivial solution $x = 0$ if $a \neq 0$ and infinitely many solutions if $a = 0$. The statements above are just the (important!) generalizations of this result to simultaneous equations.

5.13. A FORMULA FOR THE INVERSE MATRIX

Let

$$\mathbf{A} = \begin{pmatrix} a_{11} & a_{12} & a_{13} \\ a_{21} & a_{22} & a_{23} \\ a_{31} & a_{32} & a_{33} \end{pmatrix}.$$

In Section 5.8, we defined the minor of an element of $\mathbf{A}$. For example,

- the minor of a_{11} is det $\begin{pmatrix} a_{22} & a_{23} \\ a_{32} & a_{33} \end{pmatrix}$,

- the minor of a_{12} is $\det \begin{pmatrix} a_{21} & a_{23} \\ a_{31} & a_{33} \end{pmatrix}$

and so on. In general, to obtain the minor of a_{ij} we cross out the ith row and jth column from A and take the determinant of what remains. We now define the cofactor of a_{ij}, call it c_{ij}:

$$c_{ij} = (-1)^{i+j} \times M_{ij} \text{ where } M_{ij} \text{ is the } (i, j) \text{ minor of } A.$$

So

$$c_{11} = \det \begin{pmatrix} a_{22} & a_{23} \\ a_{32} & a_{33} \end{pmatrix}, \quad c_{12} = -\det \begin{pmatrix} a_{21} & a_{23} \\ a_{31} & a_{33} \end{pmatrix}.$$

and so on.

Using cofactors the determinant for a 3×3 matrix can be written

$$\det A = a_{11}c_{11} + a_{12}c_{12} + a_{13}c_{13}$$

where

$$c_{12} = (-1)^{1+2} M_{12} \text{ with minor } M_{12} = \det \begin{pmatrix} a_{21} & a_{23} \\ a_{31} & a_{33} \end{pmatrix}$$

and so on. The determinant of any $N \times N$ matrix can be written similarly as the sum of the products of the elements of any row (or column) with their corresponding cofactors.

Finally, we define the adjoint matrix to A, adj A, as the matrix constructed from the cofactors of A:

$$\text{adj } A = \begin{pmatrix} c_{11} & c_{21} & c_{31} \\ c_{12} & c_{22} & c_{32} \\ c_{13} & c_{23} & c_{33} \end{pmatrix}.$$

Note the arrangement of rows and columns – it is the transpose of what you might guess.

Exercise 5.21 Verify that if A is any 2×2 matrix whatsoever, then

$$A(\text{adj } A) = (\det A)I.$$

In fact this result holds for any square matrix. From it we get a formula for A^{-1}:

$$A^{-1} = \frac{1}{\det A} \text{adj } A. \qquad (5.23)$$

It is clear from this that A^{-1} exists provided A is non-singular ($\det A \neq 0$) (property 8).

Consider now the linear system

$$Ax = \mathbf{h}.$$

If $\det A \neq 0$ we can form A^{-1}. Multiplying through on the left by A^{-1} we get

$$A^{-1}Ax = A^{-1}\mathbf{h}.$$

Since $A^{-1}Ax = |x = x$, it follows that

$$x = A^{-1}\mathbf{h}$$

as the solution of the system. Thus, we see again that the system has a unique solution if $\det A \neq 0$. This is what we set out to understand in Section 5.1 (Equation 5.6).

5.14. EIGENVALUES AND EIGENVECTORS

Consider the linear system of equations for x

$$Ax = \lambda x \qquad (5.24)$$

where A is a given matrix and λ some real number. For example, if A is a 2×2 matrix, this system would be explicitly

$$a_{11}x + a_{12}y = \lambda x$$
$$a_{21}x + a_{22}y = \lambda y \qquad (5.25)$$

or equivalently

$$a_{11}x + a_{12}y - \lambda x + 0y = 0$$
$$a_{21}x + a_{22}y + 0x - \lambda y = 0. \qquad (5.26)$$

Putting equations (5.26) back in matrix form gives

$$\begin{pmatrix} a_{11} & a_{12} \\ a_{21} & a_{22} \end{pmatrix}\begin{pmatrix} x \\ y \end{pmatrix} + \begin{pmatrix} -\lambda & 0 \\ 0 & -\lambda \end{pmatrix}\begin{pmatrix} x \\ y \end{pmatrix} = \begin{pmatrix} 0 \\ 0 \end{pmatrix}$$

which we can write as

$$\mathbf{Ax} - \lambda\mathbf{Ix} = 0$$

or

$$(\mathbf{A} - \lambda\mathbf{I})\mathbf{x} = 0, \qquad (5.27)$$

where $\mathbf{I}$ is the identity matrix. Equation (5.24) can be written as equation (5.27) for any $N \times N$ matrix $\mathbf{A}$ and the $N \times N$ identity matrix $\mathbf{I}$.

From Section 5.12, we know that if $\det(\mathbf{A} - \lambda\mathbf{I}) = 0$ then equation (5.27) has a unique solution, namely $\mathbf{x} = \mathbf{0}$. This is not of any interest. To obtain non-trivial solutions of equation (5.24) we must impose the condition

$$\det(\mathbf{A} - \lambda\mathbf{I}) = 0. \qquad (5.28)$$

Since $\mathbf{A}$ is supposed given, equation (5.28) can be satisfied only if λ takes appropriate values.

Definition: A value of λ that satisfies equation (5.28) is called an *eigenvalue* of the matrix $\mathbf{A}$. If λ is an eigenvalue, equation (5.24) has infinitely many non-zero solutions for $\mathbf{x}$. These solutions are called *eigenvectors* of the matrix $\mathbf{A}$ (corresponding to the eigenvalue λ). Not all of the (infinite number of) eigenvectors will be independent. In fact, if A is an $n \times n$ matrix it will have at most n independent eigenvectors.

Example 5.12 Find the eigenvalues of the matrix

$$A = \begin{pmatrix} 0 & 1 & -1 \\ 1 & 2 & 1 \\ -1 & 1 & 2 \end{pmatrix}.$$

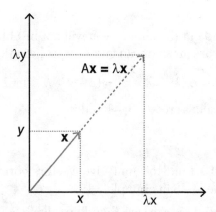

FIGURE 5.1: Illustration of an eigenvector. The vector **x** is (left) multiplied by the matrix A and the result is another vector in the same direction as **x** but scaled in magnitude by a factor λ (its eigenvalue). If **x** is an eigenvector of A (with eigenvalue λ), then so is any multiple of this, i.e. α**x** is also an eigenvector (and it has the same eigenvalue). We often use unit eigenvectors, which are scaled to have unit length (**x** · **x** = 1).

The eigenvalues are solutions of $\det(\mathbf{A} - \lambda\mathbf{I}) = 0$:

$$\det\left(\begin{pmatrix} 0 & 1 & -1 \\ 1 & 2 & 1 \\ -1 & 1 & 2 \end{pmatrix} - \begin{pmatrix} \lambda & 0 & 0 \\ 0 & \lambda & 0 \\ 0 & 0 & \lambda \end{pmatrix}\right)$$

$$= \det\begin{pmatrix} -\lambda & 1 & -1 \\ 1 & (2-\lambda) & 1 \\ -1 & 1 & (2-\lambda) \end{pmatrix} = 0.$$

Write out the determinant by expanding out the top row, i.e.

$$0 = (-\lambda)[(2 - \lambda)^2 - 1] - (1)[(2 - \lambda) + 1] + (-1)[1 + (2 -\lambda)];$$

$(2 - \lambda)^2 - 1$ is a "square minus a square" so can be factorized:

$$0 = -\lambda[(2 - \lambda) + 1][(2 - \lambda) - 1] - [(2 - \lambda) + 1] - [1 + (2 - \lambda)];$$

$(2 - \lambda + 1) = (3 - \lambda)$ is a common factor, so we take it outside:

$$0 = - \lambda(3 - \lambda)(1 - \lambda) - (3 - \lambda) - (3 - \lambda) = -(3 - \lambda)[\lambda(1 - \lambda) + 2].$$

Do not expand out the cubic – you will not be able to factorize it (unless you are given or can guess one solution). Thus,

$$0 = (3 - \lambda)(\lambda^2 - \lambda - 2) = (3 - \lambda)(\lambda + 1)(\lambda - 2).$$

So the eigenvalues are $\lambda = 3, -1$ and 2.

Example 5.13 Find the unit eigenvectors corresponding to the eigenvalue $\lambda = 3$ in example 5.12.

The eigenvectors (x, y, z) are found by solving equation (5.27): $(A - \lambda I)x = 0$, or

$$\begin{pmatrix} -\lambda & 1 & -1 \\ 1 & (2-\lambda) & 1 \\ -1 & 1 & (2-\lambda) \end{pmatrix} \begin{pmatrix} x \\ y \\ z \end{pmatrix} = \begin{pmatrix} 0 \\ 0 \\ 0 \end{pmatrix}.$$

The eigenvectors corresponding to $\lambda = 3$ therefore satisfy

$$\begin{pmatrix} -3 & 1 & -1 \\ 1 & -1 & 1 \\ -1 & 1 & -1 \end{pmatrix} \begin{pmatrix} x \\ y \\ z \end{pmatrix} = \begin{pmatrix} 0 \\ 0 \\ 0 \end{pmatrix},$$

which we can write as

$$-3x + y - z = 0,$$
$$x - y + z = 0,$$
$$-x + y - z = 0.$$

The last two equations are the same, so rearranging we get

$$y = x + z.$$

Substitute for y in the first equation, and we obtain:

$$-3x + (x + z) - z = 0,$$

from which $x = 0$, and hence $y = z$. A vector (x, y, z) with $x = 0$ and $y = z$ must be of the form $(0, k, k)$, so the eigenvectors are $(0, k, k)$, $k \neq 0$.

To find the unit eigenvectors we must chose k such that the vector length (magnitude) is unity. The length of the vector $(0, k, k)$ is $\sqrt{2k^2}$; therefore choosing $k = \pm 1/\sqrt{2}$ will give a vector of unit length. The unit eigenvectors are $\pm\left(1/\sqrt{2}\right)(0, 1, 1)$.

Note what is going on here: We have just chosen λ such that the equations do not have a unique solution. Therefore there is no point in trying to find one. A non-unique solution must contain an arbitrary parameter! Usually one chooses this to be one of x, y or z (whichever is most convenient) but any combination will do. Whatever choice is made, the same pair of equal and opposite unit eigenvectors will be obtained.

Example 5.14 Find the remaining eigenvectors following from example 5.12.

For $\lambda = 2$ we have similarly

$$(A - I)x = (A - 2I)x = 0,$$

i.e.

$$\begin{pmatrix} -2 & 1 & -1 \\ 1 & 0 & 1 \\ -1 & 1 & 0 \end{pmatrix} \begin{pmatrix} x \\ y \\ z \end{pmatrix} = \begin{pmatrix} 0 \\ 0 \\ 0 \end{pmatrix},$$

which gives

$$-2x + y - z = 0,$$
$$x + z = 0,$$
$$-x + y = 0,$$

from which $z = -x, y = x$. Thus, the eigenvectors are of the form $(k, k, -k)$, $k \neq 0$. The unit eigenvectors are $\pm\left(1/\sqrt{3}\right)(1,1,-1)$.

For $\lambda = -1$, similarly, the corresponding eigenvectors are found to be of the form $(2k, -k, k)$. The unit eigenvectors are $\pm\left(1/\sqrt{6}\right)(2,-1,1)$.

Note: an $N \times N$ real *symmetric* matrix has N eigenvalues which are all real numbers (but not necessarily distinct if the eigenvalue equation has repeated roots). This can provide a useful check on your work.

Exercise 5.22 Show that $\lambda = 1$ is an eigenvalue of the matrix

$$\begin{pmatrix} 1 & -4 & 4 \\ -4 & -1 & 0 \\ 4 & 0 & 3 \end{pmatrix}.$$

Find the other eigenvalues and the corresponding eigenvectors.

5.15. MATRICES AS TRANSFORMATIONS

We can view the application of a matrix A to a vector x as either effecting a transformation from the vector x to the vector $x' = Ax$ (an *active* transformation) or as giving the coordinates $x' = Ax$ of a fixed vector x relative to a new coordinate system (a *passive* transformation). Depending on the context, different choices of A correspond to different active transformations or to different changes of coordinate system.

Rotation in the Plane

Suppose that (x, y) axes in the plane are rotated through an angle θ to new axes (x', y'). A point P acquires new coordinates (x', y') which, by Figure 5.2, are related to its original coordinates (x, y) by

$$x' = OC = OD + DC = OD + AQ = x\cos(\theta) + y\sin(\theta)$$
$$y' = OE = PQ - QC = PQ - AD = y\cos(\theta) - x\sin(\theta)$$

Then the change of coordinates from (x, y) to (x', y') is

$$x' = x\cos(\theta) + y\sin(\theta)$$
$$y' = -x\sin(\theta) + y\cos(\theta)$$

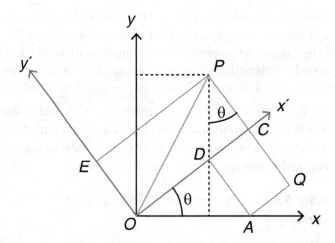

FIGURE 5.2: Rotation of coordinates in the plane. A point P given in coordinates (x, y). The same point is at coordinates (x', y') in axes that are rotated (counterclockwise) by an angle θ from the original axes.

and the change of coordinates can be written in matrix form as

$$\mathbf{x'} = \mathsf{R}\,\mathbf{x}, \tag{5.29}$$

where R is the matrix

$$\mathsf{R} = \begin{pmatrix} \cos(\theta) & \sin(\theta) \\ -\sin(\theta) & \cos(\theta) \end{pmatrix}.$$

Alternatively, we may think of this as a transformation of the point P with coordinates (x, y) to a new point P' with coordinates (x', y') in a fixed coordinate system by rotation about the origin through the angle $-\theta$:

$$\begin{pmatrix} x \\ y \end{pmatrix} \rightarrow \begin{pmatrix} x' \\ y' \end{pmatrix}.$$

Exercise 5.23

(i) Show that $\mathsf{R}\mathsf{R}^\mathsf{T} = \mathsf{I}$ and $\mathsf{R}^\mathsf{T}\mathsf{R} = \mathsf{I}$, i.e. $\mathsf{R}^{-1} = \mathsf{R}^\mathsf{T}$.

(ii) What matrix would represent the transformation of a point by rotation about the origin through an angle $+\theta$ in a fixed coordinate system?

A square matrix satisfying $R^{-1} = R^T$ is said to be *orthonormal* or *unitary*. Rotations are represented by orthonormal matrices. The converse is slightly more complicated: any transformation effected by an orthonormal matrix with positive determinant is a rotation. (Negative determinant gives a rotation followed, or preceded, by a reflection.)

Note that if the rows of a 3×3 matrix are $\mathbf{r}_1$, $\mathbf{r}_2$ and $\mathbf{r}_3$ then the requirement of orthogonormality can be seen equivalently to be that $\mathbf{r}_i \cdot \mathbf{r}_j = 0$ for $i \neq j$ (when $\mathbf{r}_1$, $\mathbf{r}_2$ and $\mathbf{r}_3$ are considered as vectors) and that $\mathbf{r}_1$, $\mathbf{r}_2$, and $\mathbf{r}_3$ are unit vectors ($\mathbf{r}_1 \cdot \mathbf{r}_1 = 1$ etc.).

Exercise 5.24 For the matrix R below, compute

(i) R^T and

(ii) RR^T.

(iii) Explain why the matrix is orthogonal.

(iv) Find the corresponding angle of rotation if this represents a rotation of coordinate axes:

$$R = \begin{pmatrix} 1/2 & \sqrt{3}/2 \\ -\sqrt{3}/2 & 1/2 \end{pmatrix}.$$

Rotation in Three Dimensions

A 3×3 orthonormal matrix (i.e. a matrix R such that $RR^T = I$) with determinant $+1$ effects a rotation in 3D. The axis of rotation is the only real eigenvector of the matrix. The angle of rotation θ is obtained from $1 + 2 \cos(\theta) = R_{11} + R_{22} + R_{33}$ (although we shall not prove this here).

Exercise 5.25 Verify that the matrix

$$\begin{pmatrix} 2/3 & -1/3 & 2/3 \\ 2/3 & 2/3 & -1/3 \\ -1/3 & 2/3 & 2/3 \end{pmatrix}$$

is orthogonal. Show that $\lambda = 1$ is an eigenvalue of the matrix and hence find the axis and angle of rotation.

Projection. The matrix transformation $\mathbf{x}' = \mathbf{Px}$ where

$$P = \begin{pmatrix} 1 & 0 & 0 \\ 0 & 1 & 0 \\ 0 & 0 & 0 \end{pmatrix}$$

effects a projection onto the (x, y) plane, since it gives

$$x' = x, \, y' = y, \, z' = 0.$$

So any point (x, y, z) becomes

$$\begin{pmatrix} x \\ y \\ z \end{pmatrix} \rightarrow \begin{pmatrix} x' = x \\ y' = y \\ z' = 0 \end{pmatrix}.$$

Note that $\det P = 0$. A transformation represented by a singular matrix P is a projection if $P^2 = P$ (i.e. repetition of the transformation has no further effect).

Figure 5.3 shows an example using the projection matrix

$$P = \begin{pmatrix} 1 & 0 & 0 \\ 0 & 1 & 0 \\ 0 & 0 & 0 \end{pmatrix}.$$

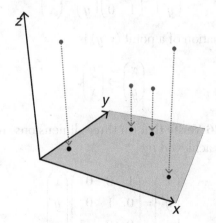

FIGURE 5.3: Left: an example of orthogonal projection. Points are projected onto the (x, y) plane using the matrix P.

Figure 5.4 shows an example using the projection matrix

$$Q = \begin{pmatrix} 1 & -1 \\ 0 & 0 \end{pmatrix}.$$

Reflection. Reflection about a plane is another transformation that can be achieved by application of a matrix. In the case of reflection, the determinant of the corresponding matrix is always -1. For example, reflection about the line $y = x$ is achieved by the matrix

$$R = \begin{pmatrix} 0 & 1 \\ 1 & 0 \end{pmatrix}$$

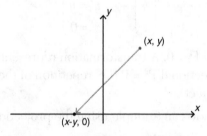

FIGURE 5.4: In general a projection will not be orthogonal. For example the matrix Q represents the projection in two dimensions as shown in the figure.

since then

$$\begin{pmatrix} x' \\ y' \end{pmatrix} = \begin{pmatrix} 0 & 1 \\ 1 & 0 \end{pmatrix} \begin{pmatrix} x \\ y \end{pmatrix} = \begin{pmatrix} y \\ x \end{pmatrix}$$

so the transformation of a point (x, y) is

$$\begin{pmatrix} x \\ y \end{pmatrix} \rightarrow \begin{pmatrix} y \\ x \end{pmatrix}.$$

Exercise 5.26 Verify that (in three dimensions) reflection in the (x, y) plane is achieved by

$$\begin{pmatrix} x' \\ y' \\ z' \end{pmatrix} = \begin{pmatrix} 1 & 0 & 0 \\ 0 & 1 & 0 \\ 0 & 0 & -1 \end{pmatrix} \begin{pmatrix} x \\ y \\ z \end{pmatrix}.$$

5.16.   EXTENSION: TRANSFORMATION GROUPS

Consider again the rotations in the plane from section (5.15). Rotation clockwise through the angle θ (an active transformation) is effected by the matrix

$$\begin{pmatrix} \cos(\theta) & -\sin(\theta) \\ \sin(\theta) & \cos(\theta) \end{pmatrix}$$

acting on the vector (x, y) (since an active transformation clockwise through an angle θ is equivalent to a passive transformation counter-clockwise through θ.) A further rotation through θ' will be given by the product

$$\begin{pmatrix} \cos(\theta') & -\sin(\theta') \\ \sin(\theta') & \cos(\theta') \end{pmatrix} \begin{pmatrix} \cos(\theta) & -\sin(\theta) \\ \sin(\theta) & \cos(\theta) \end{pmatrix} = \begin{pmatrix} \cos(\theta+\theta') & -\sin(\theta+\theta') \\ \sin(\theta+\theta') & \cos(\theta+\theta') \end{pmatrix}$$

acting on (x, y), which, as we expect, is a rotation through an angle $\theta + \theta'$. The product of two rotation matrices is therefore always another rotation matrix with a different parameter ($\theta+\theta'$ in this example). The set of rotation matrices also contains an identity element which leaves a vector unchanged, namely the rotation matrix with parameter $\theta = 0$, and a matrix corresponding to the inverse of any rotation, through θ say, that returns us to the original vector, namely the matrix with parameter $-\theta$.

These conditions, together with the fact that matrix multiplication is associative (i.e. **A(BC) = (AB)C**), define the conditions for the set of matrices to form a *group*. We can now work backwards. Any set of matrices that have the property of constituting a group will effect a rotation in some space. Thus we can use a group of transformations to define "vectors" in the corresponding space. The simplest example would be to go from three dimensions to n-dimensional vectors transforming under the $n \times n$ rotation matrices. Another example is provided by four-vectors in relativity, which transform under a set of matrices representing the Lorentz group. (These matrices are different from four-dimensional rotations, so relativistic 4-vectors are not the same as "ordinary" vectors in four dimensions.) We can also define objects representing transformations in some kind of internal spaces, such as the two (complex) dimensional spin space of the electron, or the three-dimensional color space of quarks.

The components of the transformation matrices we have considered so far are differentiable functions of the one or more parameters (representing what are called Lie groups). We can also extend groups of transformations, such the reflections we considered above, which have finite numbers of elements. These represent discrete symmetries such as one finds, for example, in crystal structures.

Revision Notes

After completing this chapter you should be able to

- Solve a system of linear equations by reducing the matrix representation to *triangular form*

- Find the *inverse* of a matrix by row or column operations. Perform *algebraic operations* on matrices

- Find the *determinant* of a matrix, using the properties of determinants to simplify the calculation

- Use det **AB** = det **A** det **B** to find the determinant of a *product*

- Use the determinant of coefficients to determine whether a system of equations has one, many or no solutions

- Define and compute *eigenvalues* and *eigenvectors*

- Construct matrix representations of simple *transformations*

- Determine whether a given matrix is *orthogonal*

5.17. EXERCISES

1. If $\begin{pmatrix} 1 & 0 & 2 \\ 0 & 1 & 3 \\ -1 & 1 & 1 \end{pmatrix}$ find det **A**, and hence (or otherwise) show

that the equations
$$x + 2z = 0$$
$$y + 3z = 0$$
$$-x + y + z = 0$$

have non-trivial solutions.

2. Solve the equations

$$2x + 3y + 5z = 7$$
$$x + 7y + z = 6$$
$$3x + y + 2z = 9.$$

3. Find the values of λ and μ for which the equations

$$x + 2y = \mu$$
$$2x + \lambda y = 2$$

have (i) a unique solution, (ii) no solution, (iii) an infinite number of solutions. Give expressions for the solutions where possible.

4. For what value(s) of λ does the system

$$x - 2y + z = 0$$
$$x + \lambda z = 0$$
$$3x + y + 2z = 0$$

have non-trivial solutions?

5. Show that the product of the two matrices $\begin{pmatrix} 2 & 0 & -1 \\ 0 & 2 & 1 \\ 2 & -1 & 0 \end{pmatrix}$

and $\begin{pmatrix} 1 & 1 & 2 \\ 2 & 2 & -2 \\ -4 & 2 & 4 \end{pmatrix}$ is 6I, where I is the unit matrix. Hence

(or otherwise) solve the set of linear equations

$$x + y + 2z = 1$$
$$2x + 2y - 2z = 3$$
$$-4x + 2y + 4z = 5.$$

6. Given $A = \begin{pmatrix} 1 & 0 & -1 \\ 2 & 1 & 0 \\ -1 & 2 & -1 \end{pmatrix}$ and $B = \begin{pmatrix} 3 & -2 & 1 \\ 3 & 0 & 1 \\ 2 & 0 & 1 \end{pmatrix}$ find

det(AB). What is det(BA)?

7. Evaluate the determinant $\begin{vmatrix} 0 & -1 & 2 & 0 \\ 4 & 0 & 0 & 1 \\ 0 & 2 & 2 & -1 \\ -1 & -1 & 1 & 1 \end{vmatrix}$.

8. If $A = \begin{pmatrix} 1 & 0 & -1 \\ 2 & 2 & 5 \\ 3 & -1 & 2 \end{pmatrix}$ and $B = \begin{pmatrix} 0 & 1 & 3 \\ -1 & 1 & 1 \\ -2 & 1 & 0 \end{pmatrix}$ what are AB, A^T,

    $(A^T B^T)^T$ where A^T denotes the transpose of A?

9. If $A = \begin{pmatrix} 1 & 0 & 4 \\ -1 & 1 & 0 \\ 3 & 1 & -1 \end{pmatrix}$ and $B = \begin{pmatrix} 1 & -1 & -2 \\ 0 & 0 & 1 \\ 4 & 0 & 1 \end{pmatrix}$ find AB and

    verify by explicit calculation of both sides that det AB = det A det B.

10. What are the eigenvalues of the matrix $\begin{pmatrix} 1 & 2 \\ 2 & 3 \end{pmatrix}$?

11. Verify that $\begin{pmatrix} 1 \\ 1 \end{pmatrix}$ is an eigenvector of $\begin{pmatrix} 2 & 3 \\ 3 & 2 \end{pmatrix}$ and find a

    second linearly independent eigenvector.

12. The matrix A and its eigenvalues, λ, and eigenvectors, $\mathbf{x}$, satisfy $A\mathbf{x} = \lambda\mathbf{x}$. Deduce the eigenvalues of A^{-1}.

13. A linear transformation is represented by

    $$\begin{pmatrix} x' \\ y' \end{pmatrix} = \begin{pmatrix} -1 & 1 \\ 2 & 3 \end{pmatrix} \begin{pmatrix} x \\ y \end{pmatrix}.$$

    Find the inverse transformation and hence the equation of the line $2y + x = 0$ in the (x', y') coordinate system.

14. Find the eigenvalues of the matrix $Q = \begin{pmatrix} 1 & 0 & -1 \\ 0 & 2 & 0 \\ -1 & 0 & -1 \end{pmatrix}$ and find

    the eigenvector corresponding to the eigenvalue $\lambda = 2$.

15. Show that $\lambda = 1$ is an eigenvalue of the matrix

$$Q = \begin{pmatrix} 1 & 0 & -1 \\ -2 & 1 & 0 \\ 2 & 0 & 4 \end{pmatrix}.$$ Find the corresponding eigenvector and

the other eigenvalues.

5.18. PROBLEMS

1. Show that the eigenvalues of the matrix

$$A = \begin{pmatrix} 1 & 3 & 0 \\ 3 & -2 & -1 \\ 0 & -1 & 1 \end{pmatrix}$$

are 1, 3, and -4, and find corresponding eigenvectors. Verify that these eigenvectors are mutually orthogonal.

If the eigenvectors are (x_1, x_2, x_3), (y_1, y_2, y_3), (z_1, z_2, z_3) and a matrix P is defined as

$$P = \begin{pmatrix} x_1 & x_2 & x_3 \\ y_1 & y_2 & y_3 \\ z_1 & z_2 & z_3 \end{pmatrix}$$

show that PAP^T is a diagonal matrix.

2. Reduce to triangular form the linear equations

$$4x + 3y + 2z = 1$$
$$w + 4y + 3z = 2$$
$$2w + x + 4z = 3$$
$$3w + 2x + y + \lambda z = u$$

where λ, u are any real numbers. For what values of λ, u will the equations have a solution? When is the solution unique? Taking $u = 4$, what is the solution for the cases $\lambda = 0$ and $\lambda = 5$?

3. Solve the linear equations

$$x + 3y + 2z = 4$$
$$2x + y + 3z = 5$$
$$3x + 2y + (1 + \mu)z = 0$$

where μ is a constant, and hence show that there is a single value of μ for which the equations have no solution. Find the value of μ which makes $y = 0$, and find the corresponding values for x and z.

4. The coordinates (x, y, z) of a point, P, satisfy the determinantal equation

$$\begin{vmatrix} x & y & z & 1 \\ 1 & 0 & 1 & 1 \\ 2 & 1 & 0 & 1 \\ 1 & 3 & 2 & 1 \end{vmatrix} = 0.$$

By working out the determinant to express the equation in a simpler form, show that P must lie on a certain plane. Verify that the point with coordinates $(1, 3, 2)$ lies on this plane, and explain how this could be shown directly using the determinantal form of the equation.

5. Show that the matrix $A = \begin{pmatrix} 1 & 0 & 2 \\ 0 & -1 & 0 \\ 3 & 0 & 1 \end{pmatrix}$ satisfies the equation

$$(A - \lambda_1 I)(A - \lambda_2 I)(A - \lambda_3 I) = 0$$

where λ_1, λ_2 and λ_3 are the eigenvalues of A and I is the unit matrix.

By expanding the equation for A above into a polynomial of the form $A^3 + aA^2 + bA + cI = 0$, find the inverse of A.

6. Show explicitly that the matrix

$$A = \begin{pmatrix} 2 & 1 & 1 \\ 0 & 2 & 3 \\ 0 & 0 & 5 \end{pmatrix}$$

satisfies the matrix equation

$$A^3 - 9A^2 + 24A - 20I = 0.$$

Hence or otherwise find A^{-1}.

6

DIFFERENTIAL EQUATIONS 1

Differential equations are fundamental to physical science. The behavior of any system continuously evolving in time (the motion of a body subject to a force, a chemical reaction, ...) is governed by one or more differential equations. So is the behavior of continuous systems in space. Solving the equations means finding out how the body moves given the forces acting, how the reaction proceeds given the reagents etc. Only the simplest of equations can be solved exactly. But these are important in developing a physical intuition for the sort of behavior to expect. (Conversely a good physical insight will provide a basis for the mathematical solution.) They are also used later in the derivation of approximations to the solutions of more complex equations. You need to develop a good understanding of the equations in this chapter.

6.1. WHAT ARE DIFFERENTIAL EQUATIONS?

A differential equation is an equation for an unknown function, $y(t)$ say, of an independent variable (in this case t) that involves one or more derivatives of y with respect to t.

Example 6.1 Here are some examples of differential equations, from different fields.

A simple model for population growth (with $\lambda > 0$), or radioactive decay (with $\lambda < 0$)

$$\frac{dy}{dt} = \lambda y. \qquad (6.1)$$

Simple harmonic motion, ω a constant

$$\frac{d^2 y}{dt^2} = -\omega^2 y. \qquad (6.2)$$

Governor of steam turbine

$$\frac{d^3 y}{dt^3} - 3\frac{d^2 y}{dt^2} + 4\frac{dy}{dt} + 2y = 40. \qquad (6.3)$$

Motion of a planet in general relativity

$$\frac{d^2 u}{d\theta^2} + u = \kappa + 3u^2. \qquad (6.4)$$

Economics growth model, A, B constants

$$\frac{dr}{dt} = \left(Ar^{1/2} + 1\right)\left(Br^{1/2} + 1\right). \qquad (6.5)$$

Geophysics, $f(x)$ a given function of x

$$\frac{d}{dx}\left\{x^5 f(x)\sqrt{1+y}\right\} = 5f(x)x^4. \qquad (6.6)$$

The *order* of a differential equation is the number of the highest derivative appearing in it.

Example 6.2 What are the orders of equations (6.1) and (6.3)?

Equation (6.1) is first order because it contains only first derivatives (i.e. dy/dt).

Equation (6.3) is third order because the number of the highest derivative is 3 (i.e. d^3y/dt^3).

Exercise 6.1 What are the orders of the differential equations (6.2) and (6.4) to (6.6)?

It might be helpful to clarify the notation that is used for differential equations. If a function y depends on time t we often write $\dot{y}$ instead of dy/dt, $\ddot{y}$ for d^2y/dt^2 etc. If the independent variable is not time but position, x, we may use "prime" symbols like y', y'', ... y^n for dy/dx, d^2y/dx^2, ..., d^ny/dx^n. For example, equation (6.4) can be written as

$$u'' + u = \kappa + 3u^2.$$

Yet another notation is to use superscript numbers to signify the order of the derivative; e.g. $f^{(3)}(x)$ means d^3f/dx^3 and so on. In treating the dynamics of a point particle, we often use $x(t)$ to denote the position as a function of time, with x here as the dependent variable. For most of this chapter we shall use x or t as the independent variable.

6.2. SOLVING DIFFERENTIAL EQUATIONS

So the simplest differential equations are those of the form $y' = f(x)$, or $y'' = f(x)$, ..., $y^{(n)} = f(x)$. To solve the equation means to determine the unknown function $y(x)$ which will turn the equation into an identity upon substitution. All we need to do to solve these types of differential equations is integrate the appropriate number of times (not forgetting the correct number of constants of integration).

Definition: We call the solution of a differential equation that contains the correct number of arbitrary constants the *general solution, $y(x)$*.

Example 6.3 Solve for $y(x)$ the differential equation $\dfrac{dy}{dx} = x^2$

Step 1: In this simple case we can integrate both sides directly to find $y(x)$

$$y(x) = \int x^2 dx = \frac{x^3}{3} + C$$

where C is a constant.

Step 2: We can then write our solution to the differential equation as

$$y(x) = \frac{x^3}{3} + C$$

Do not forget the constants when solving differential equations.

Step 3: We can check that this is indeed a solution by back-substitution:

$$\frac{dy(x)}{dx} = \frac{d}{dx}\left(\frac{x^3}{3} + C\right) = x^2$$

Do not omit step 3 to save time. On the contrary, it can save a lot of wasted work if you pick up a mistake at this point.

Exercise 6.2 Solve the following, and check your solutions:

(i) $\dfrac{dy}{dx} = e^x$,

(ii) $\dfrac{dy}{dt} = \sin(t)$.

6.3. FIRST ORDER SEPARABLE EQUATIONS

A first order differential equation is called separable if it can be written in the form

$$\frac{dy}{dx} = f(x)g(y), \tag{6.7}$$

where $f(x)$ is a function of x only and $g(y)$ is a function of y only, and this includes cases where f or g are simply constants. Equations of this form can be rearranged so that all terms depending on x and all terms depending on y appear on opposite sides of the equation (this is the *separation* of the variables), and then integrated

$$\int \frac{dy}{g(y)} = \int f(x)dx. \tag{6.8}$$

It should then be possible to find the solution $y(x)$ that satisfies the differential equation.

Note that we cannot write equation (6.7) as $y = g(y) \int f(x)dx$ as many begining students are tempted to do, because y and hence $g(y)$ is a function of x.

Example 6.4 Solve for $y(x)$ the differential equation $yy' = x$.

Step 1: We *separate the variables* in the equation in such a way that the left hand side becomes a function of y only, and the right side a function only of x.

$$\int y \frac{dy}{dx} dx = \int x \, dx,$$

or

$$\int y \, dy = \int x \, dx.$$

In this form both sides can be directly integrated.

Step 2: There is no need to include two constants of integration here: any constant on the left can be taken over to the right and included in C.

$$y^2/2 = x^2/2 + C.$$

Step 3: Write the *general solution.* Hence

$$y(x) = (x^2 + 2C)^{1/2}.$$

As noted in the example above, constants should be combined where possible so that the total number of arbitrary constants in the solution is immediately visible. The solution here contains only one arbitrary constant, however arrived at. In fact, the factor 2 in $2C$ is of no significance ($2C$, like C, can be any arbitrary number), so we would normally write the solution to example 6.4 just as

$$y(x) = (x^2 + C)^{1/2}.$$

The only difference between Example 6.4 and those in Section 6.2 is that on integration we obtain a function of y ($\frac{y^2}{2}$ here) on the left hand side rather than y itself. So equations that can be expressed in the form we are considering are known as *first order separable equations.*

Key point: Rearrange *first order separable* equations so that you have

Function of y, y' only = Function of x only

(x does not appear explicitly)   (y does not appear explicitly)

i.e. the LHS is a function of y only, and the RHS is a function of x only. Some manipulation may be required to put the equation in this form. For example $dy/dx = x/y$ is equivalent to Example 6.4. To see this, just multiply through by y.

Exercise 6.3 Solve the following first order separable equations;

(i) $yy' = a$ (a given constant) for $y(x)$,

(ii) $y' = (1 - y^2)^{1/2} \xi^{1/2}$ for $y(\xi)$,

(iii) $t(\dot{y}) + 2 = y$ for $y(t)$.

Note that each of the solutions in exercise (6.3) contains one arbitrary constant (and no more). This is a general rule: the solution to a first order equation contains at most one arbitrary constant.

Thus a first order equation is solved by some formula $y = f(x, C)$ involving an arbitrary constant C. For all possible values of C we get all of the possible *particular* solutions. We will see later that solutions to second order equations have at most two arbitrary constants. Third order equations require three arbitrary constants, and so on.

Key point: The general solution to an nth order equation *must* contain n arbitrary constants.

We will come back to particular solutions later on, in Section 6.9, and in Chapter 8.

6.4. LINEARITY VERSUS NON-LINEARITY

Suppose that an nth order differential equation has x as the independent variable and y as the unknown function. Recall that the nth derivative of y can be written as $y^{(n)}$. If the equation can be written in the form

$$A_0 y^{(n)} + A_1 y^{(n-1)} + \ldots + A_{n-1} y + A_n = 0, \qquad (6.9)$$

where $A_0, A_1, \ldots, A_n$ may be constants or functions of x only, then it is called a *linear* differential equation. The left side of equation (6.9) is a linear combination of $y, y^{(1)}, y^{(2)}, \ldots, y^{(n)}$.

However, if the $A_0, A_1, \ldots, A_n$ terms themselves contain any of $y, y^{(1)}, y^{(2)}$ or higher derivatives, products of y (e.g. y^2), powers of y

(e.g. $y^{1/2}$) or any functions of y (e.g. $\sin(y)$ or e^y), then the equation is *non-linear.*

Example 6.5 State if the following equations are linear or nonlinear, and why

(i) $\dot{y} = \lambda t$, (ii) $y'' = -(\sin(x))y$, and (iii) $yy' = x$.

    (i) $\dot{y} = \lambda t$ is linear because the unknown function $\dot{y}$ appears linearly.

    (ii) $y'' = -(\sin(x))y$ is linear because the unknown functions ($y(x)$ and their derivatives) appear linearly, even though x does not.

    (iii) $yy' = x$ is non-linear because it involves a product of unknown functions, y and y'.

Exercise 6.4 Classify the equations shown in Example 6.1 as linear or non-linear.

Linear equations are important examples of differential equations because:

(i) They can often be solved.

(ii) Approximations to interesting physical problems can often be obtained as linear equations.

(iii) When $A_n = 0$ in equation (6.9) solutions can be added to give new solutions (see Section 6.5).

When faced with a linear equation in the form of equation (6.9) we generally take A_n over to the right hand side and also divide through by the coefficient A_0 of the highest order derivative $y^{(n)}$. Thus first and second order equations in this form can alternatively be written as

$$y' + P(x)y = Q(x), \tag{6.10}$$

and

$$y'' + F(x)y' + G(x)y = H(x), \qquad (6.11)$$

where $P(x)$, $Q(x)$, $F(x)$, $G(x)$ and $H(x)$ are referred to as the *coefficients* of y and its derivatives, and are functions of x only (or in some cases are replaced by constants; see Section 6.7). For most of the rest of this chapter the equations we shall look at will be linear and usually of first or second order.

6.5. HOMOGENEOUS VERSUS INHOMOGENEOUS

The linear differential equation $y' + P(x)y = Q(x)$ is called *homogeneous* if $Q(x) = 0$ and *inhomogeneous* otherwise. Similarly, the linear differential equation $y'' + F(x)y' + G(x)y = H(x)$ is called *homogeneous* if $H(x) = 0$ and *inhomogeneous* otherwise. Some authors call these *unforced* (homogeneous) and *forced* (inhomogeneous) because the term $Q(x)$ is often thought of as a forcing term for a physical system.

For example,

$$y'' + \omega^2 y = 0 \text{ is homogeneous}, \qquad (6.12)$$
$$y'' + \omega^2 y = 2x \text{ is inhomogeneous}. \qquad (6.13)$$

If two solutions $y = y_1(x)$ and $y = y_2(x)$ of a homogeneous linear equation are known then any linear combination $y = Ay_1(x) + By_2(x)$ (with A and B constants) is itself another solution. This is not true for inhomogeneous equations.

To see how this works, let us work through an example.

Example 6.6 Verify that $y = \sin(\omega x)$ and $y = \cos(\omega x)$ are solutions of equation (6.12) and that, if A and B are arbitrary constants, then $y = A\sin(\omega x) + B\cos(\omega x)$ is also a solution.

We verify that something is a solution by substitution in the given equation. If $y = \sin(\omega x)$ then

$$y' = \omega \cos(\omega x),$$

$$y'' = -\omega^2 \sin(\omega x).$$

Substituting into equation (6.12) we see that $-\omega^2 \sin(\omega x) + \omega^2 \sin(\omega x) = 0$, and we have indeed verified that $y = \sin(\omega x)$ is a valid solution.

Following a similar method we can easily show that $y = \cos(\omega x)$ is also a solution.

Finally, we can show that the sum of the two solutions $y = A \sin(\omega x) + B \cos(\omega x)$ is also a solution by direct substitution into equation (6.12)

$$\frac{d^2}{dx^2}\{A\sin(\omega x) + B\cos(\omega x)\} + \omega^2\left(A\sin(\omega x) + B\cos(\omega x)\right)$$

$$= A\left(\frac{d^2}{dx^2}\{\sin(\omega x)\} + \omega^2\sin(\omega x)\right) + B\left(\frac{d^2}{dx^2}\{\cos(\omega x)\} + \omega^2\cos(\omega x)\right)$$

$$= A\left(-\omega^2\sin(\omega x) + \omega^2\sin(\omega x)\right) + B\left(-\omega^2\cos(\omega x) + \omega^2\cos(\omega x)\right)$$

$$= 0$$

Therefore $y = A\sin(\omega x) + B\cos(\omega x)$ is also a solution.

Exercise 6.5

(i) Verify that $y = 1$ and $y = t$ are two solutions of $d^2y/dt^2 = 0$ and that $y = A + Bt$ is a solution (A and B constants).

(ii) Let $y = y_1(x)$ and $y = y_2(x)$ be two solutions of $y'' + F(x)y' + G(x)y = 0$. Show that $Ay_1 + By_2$ is also a solution for arbitrary constants A, B.

(iii) Given that e^x and e^{-x} are solutions of $d^2y/dx^2 - y = 0$, explain why it follows that $\sinh(x)$ and $\cosh(x)$ are also solutions.

(iv) Verify that $y = 2x/\omega^2$ is a solution of equation (6.13) but that $y = 2x/\omega^2 + 2x/\omega^2 = 4x/\omega^2$ is not.

(v) If $y = y_1(x)$ and $y = y_2(x)$ are two solutions of the inhomogeneous linear equation $y'' + F(x)y' + G(x)y = H(x)$, what differential equation is satisfied by $Ay_1 + By_2$, where A and B are constants?

6.6. FINDING SOLUTIONS

We now return to Example 6.3: $dy/dx = x^2$. The solution is the integral of x^2. Suppose that we remember the integral of x^2 is something like x^3 but have forgotten whether it is $3x^3$ or $x^3/3$ (this example is simple, but if it were more complex we might not remember the correct factor). We can guess a more general form, $y = Ax^3$, and then solve for A. Substitution in the differential equation gives

$$\frac{d}{dx}\{Ax^3\} = 3Ax^2. \tag{6.14}$$

This is a solution if $3Ax^2 = x^2$, i.e. if $A = 1/3$. (Of course, the general solution still involves a constant of integration C.) Similarly, if you guess that the integral is some power of x but cannot remember which one, you could try $y = Ax^n$ where both A and n are to be determined by substitution.

Example 6.7 Solve $dy/dx = x^2$ by trying a solution of the form $y = Ax^n + C$, with A, n and C constants.

Step 1: We let y be of the proposed form and substitute it into the given differential equation.

$$y' = nAx^{n-1} = x^2.$$

Step 2: Work out the conditions under which the proposed form leads to a solution. If this is not possible, then there is no solution of the proposed form, and an alternative guess is required. The above condition is true if both

$$A = 1/n \text{ and } n - 1 = 2,$$

that is, when

$$n = 3, \text{ and } a = 1/3$$

Step 3: Finally we can write down our general solution as

$$y = x^3/3 + C.$$

We can use this same approach not only where we have forgotten an exact integral but where we perhaps never knew one, as long as we can guess the correct *form* of the solution. This may seem rather mysterious the first time you see it, but it is not really mystery – it is experience. You will gain this experience by working through this chapter. If you are mathematically minded you will soon remember what sort of functions satisfy what sort of equations. If you are physically minded you will use physical intuition to guess what sort of behavior to expect from a given system (as represented by a differential equation).

But what happens if we make the wrong guess? Let us see. Suppose that, instead of trying $y = Ax^3$ in equation (6.14), we guess $y = Ax^4$ with A a constant. Then $dy/dx = 4Ax^3 = x^2$ if $A = 1/4x$. This is impossible. (We differentiated y assuming A was a constant and it has turned out not to be.) So the guess is wrong and we stop here. The failure tells us that we must start again with a better guess.

The final problem is that the "guesswork" does not feel like "real math" and we would naturally prefer to be given an algorithm to follow in order to solve these types of equation (i.e. a set of instructions that always works). There is no algorithm. That is okay; after all integration is a process of educated guesswork too, and we have learned to cope with that. Solving a differential equation involves a similar process of educated guesswork, based on a correct identification of the type of equation to be solved.

Exercise 6.6

(i) By trying a solution of the form $y = A\cos(px)$ solve $y' = \sin(2x)$ for $y(x)$.

(ii) By trying a solution of the form $y = (Ax+B)e^{2x}$ solve $y' = 3e^{2x}$ for $y(x)$.

However, suppose we try to solve $y' = e^{-x^2}\ln(x)$. We want to be able to solve the indefinite integral $\int e^{-x^2}\ln(x)\,dx$ – but this cannot be expressed in terms of known functions. Therefore not all differential equations have solutions expressible in terms of known functions. In such cases information about the solution must be obtained by either numerical or approximate methods, but we will not tackle that here.

6.7. HOMOGENEOUS LINEAR EQUATIONS

The following equation is similar to equation (6.10):

$$y' + py = 0,$$

but here we have a *constant coefficient* of y, p = constant (instead of a known function of x, $P(x)$). In addition, it is *first order* (it contains y differentiated just once), and is *linear* (the left hand side is a linear combination of y and y'). Also, unlike equation (6.10) it is *homogeneous* (there is no term independent of y, i.e. $Q(x) = 0$). It is also *separable* (it can be written as $dy/y = pdx$). We can therefore integrate it directly to obtain the general solution, as in Section 6.3.

Alternatively, we can solve it by using physical or mathematical intuition as the basis of a guessed solution. Let us rewrite the equation as $y' = -py$. In this form you may recognize that the equation looks like the radioactive decay law. So we could reasonably expect an exponential type solution, say $y = Ae^{\lambda x}$, where A is a constant.

Alternatively, the equation looks something like $y' = y$ which we know is satisfied if $y = e^x$. So again we are led to try $y = Ae^{\lambda x}$.

Differentiating, we find $y' = \lambda A e^{\lambda x}$ and therefore, to satisfy the equation, we need $\lambda A e^{\lambda x} + p A e^{\lambda x} = 0$. This is true if $\lambda = -p$ but with no restriction on A. Thus the general solution in this example, containing one arbitrary constant, is $y = A e^{-px}$.

Let us take a look at a complete worked example.

Example 6.8 Solve $y'' - 4y = 0$.

Step 1: This equation is second order, linear, homogeneous and has constant coefficients. But it is *not* separable. We therefore have to call on some other experience to solve it. Since it looks similar to the example we discussed above we can try an initial guess

$$y = A e^{\lambda x},$$

and it follows that

$$y' = A \lambda e^{\lambda x} \text{ and } y'' = A \lambda^2 e^{\lambda x}.$$

Step 2: To test if our trial function really is a solution we substitute y and y'' above into the original differential equation, which gives

$$y'' - 4y = (A \lambda^2 e^{\lambda x}) - 4(A e^{\lambda x}) = 0.$$

Step 3: Divide through by $A e^{\lambda x}$ to get $\lambda^2 - 4 = 0$, from which we can see that $\lambda^2 = 4$ and $\lambda = \pm 2$. We have therefore found two solutions

$$y_1 = A e^{2x} \text{ and } y_2 = A e^{-2x}.$$

Step 4: The general solution must contain two *independent* arbitrary constants though, A and B say, and therefore the *general solution* is written down as the sum of the two solutions as follows:

$$y = A e^{2x} + B e^{-2x},$$

where A and B are arbitrary constants. This follows because a linear combination of solutions of a linear equation gives another solution. This can be checked by substituting back into the differential equation.

We will now look at another example where we will solve a similar second order differential equation.

Example 6.9 Solve $y'' - y' - 6y = 0$.

Step 1: The equation is second order, linear, homogeneous with constant coefficients, so our experience so far tells us what form to try. So we try as our initial guess

$$y = Ae^{\lambda x},$$

and therefore

$$y' = A\lambda e^{\lambda x} \text{ and } y'' = A\lambda^2 e^{\lambda x}.$$

Step 2: To test the trial function substitute into the differential equation to be satisfied. Then cancel $Ae^{\lambda x}$ to get the auxiliary equation.

$$y'' - y' - 6y = A\lambda^2 e^{\lambda x} - A\lambda e^{\lambda x} - 6Ae^{\lambda x} = 0,$$

$$\lambda^2 - \lambda - 6 = 0.$$

This removes the x dependence so our guess was a good one. The resulting equation in λ is called the *auxiliary* equation.

Step 3: Solve the auxiliary equation for λ

$$(\lambda - 3)(\lambda + 2) = 0,$$

$$\lambda = -2 \text{ or } \lambda = 3.$$

Step 4: Substitute the values of λ to form the solutions for y. The two solutions are

$$y_1 = Ae^{-2x} \text{ and } y_2 = Ae^{3x}.$$

Remember, the general solution involves two arbitrary constants since we are solving a second order equation. Hence the *general solution* is written as the sum of y_1 and y_2, but with two independent arbitrary constants

$$y = Ae^{-2x} + Be^{3x},$$

where A and B are arbitrary constants.

It is very important to remember this method of solving differential equations with constant coefficients. Let us work through one more example, following the same steps to first form the auxiliary equation. We can then solve for λ and write the general solution.

Example 6.10 Solve $y'' - 3y' - 4y = 0$.

Step 1: First check if the equation is linear (no products of $y, y', \ldots$)? Is it homogeneous (there is no term independent of y)? Does it have constant coefficients $(1, -3, -4)$? This method works only if you answer *yes* to all three questions. With experience you can do the differentiation and substitution in your head. For example, we try as our initial guess

$$y = Ae^{\lambda x},$$

$$y' = A\lambda e^{\lambda x} \text{ and } y'' = A\lambda^2 e^{\lambda x}.$$

Step 2: Form the auxiliary equation by substituting our initial guess into the differential equation and canceling $Ae^{\lambda x}$.

$$\lambda^2 - 3\lambda - 4 = 0.$$

Step 3: Solve the auxiliary equation. And hence

$$(\lambda - 4)(\lambda + 1) = 0,$$

$$\lambda = 4 \text{ or } -1$$

Step 4: The *general solution* is the sum

$$y = Ae^{4x} + Be^{-x}$$

where A and B are independent arbitrary constants.

Note that in these cases where one of the roots of the auxiliary equation is positive the general solution is unstable since the solution will tend to infinity as $x \to \infty$.

Exercise 6.7

(i) Show that the equation $y'' + 7y' + 12y = 0$ has the general solution $y = Ae^{-4x} + Be^{-3x}$.

(ii) Show that this can also be written as $y = e^{-7x/2}(P \cosh(x/2) + Q \sinh(x/2))$ with $P = A + B$, $Q = B - A$.

Exercise 6.8 Assuming $b^2 - 4ac > 0$, find the general solution of $ay'' + by' + cy = 0$ (a, b, c constants).

Provided the differential equation is linear, homogeneous and with constant coefficients, that is all there is to finding the general solution. Do not try to use this method for equations which are nonlinear, inhomogeneous or which do not have constant coefficients.

6.8. AUXILIARY EQUATIONS WITH REPEATED ROOTS

In the previous section, we looked at solving the auxiliary equation for examples where λ has two different values, λ_1 and λ_2 say.

Here we are going to look at the special case where we find λ has two equal values when we solve the auxiliary equation, or *repeated roots*. Let us look at an example, following the same steps as previously.

Example 6.11 Solve $y'' + 4y' + 4y = 0$.

Step 1: Is the equation linear, homogeneous and does it have constant coefficients? We can answer *yes* to all of these questions and thus we begin by trying

$$y = Ae^{\lambda x}.$$

Step 2: Substitute y, y' and y'' in the differential equation, and form the auxiliary equation

$$\lambda^2 + 4\lambda + 4 = 0.$$

Step 3: Solve the auxiliary equation

$$(\lambda + 2)(\lambda + 2) = 0 \text{ and thus } \lambda = -2, -2.$$

Step 4: Write the solution

$$y = Ae^{-2x} + Be^{-2x}$$

which just combines to give $y = Ce^{-2x}$.

But this cannot be the *general solution* because it involves only one arbitrary constant.

We know that every second order differential equation must have a solution which contains two arbitrary constants, so there must be another term that contains the second constant. If we proceed systematically with this example we can try to find a second solution of the differential equation when one solution is known. We will see how to solve these *repeated roots* examples in general.

The trick is to try y = (unknown function) × (known solution). We try

$$y = u(x)e^{-2x},$$

and then calculate y' and y''

$$y' = u'(x)e^{-2x} - 2u(x)e^{-2x},$$

$$y'' = u''(x)e^{-2x} - 2u'(x)e^{-2x} - 2u'(x)e^{-2x} + 4u(x)e^{-2x}.$$

Now we have to find $u(x)$. By substituting y, y' and y'' into the differential equation and canceling e^{-2x} we can find $u(x)$:

$$y''+4y'+4y = [u''(x)-4u'(x)+4u(x)]+4[u'(x)-2u(x)]+4[u(x)] = 0.$$

Simplifying we are left with $u''(x) = 0$, so that $u'(x) = A$, and $u(x) = Ax + B$ where A and B constants. Hence the general solution is

$$y = (Ax + B)e^{-2x}.$$

In fact, you do not need to go through the whole process each time, if you can remember the form of the solution; you can just use it directly.

Key point: If the auxiliary equation has λ as a real and *repeated root*, we can try a solution of the form $y = (Ax + B)e^{\lambda x}$.

Exercise 6.9 Find the general solution of $y'' + 6y' + 9y = 0$.

6.9. PARTICULAR INTEGRALS

Having looked at how to solve homogeneous differential equations (with constant coefficients), we are now going to see what happens when the right hand side of equation (6.9) is not zero. Consider a linear inhomogeneous equation of the form

$$y'' + fy' + gy = H(x), \tag{6.15}$$

where $H(x)$ is a known function of x (similar to Equation 6.11, but here we have constant coefficients f and g instead of known functions of x, $F(x)$ and $G(x)$). If we can find a solution of equation (6.15) which has no arbitrary constants, this is called a *particular integral* (PI). Systematic methods do exist for finding PIs. However, in simple cases, the best method is again to use educated guesswork. Roughly speaking, we are going to try a PI of similar form to $H(x)$. The cases that follow are those that occur most commonly. Remember, we assume throughout that f and g are constants. Thus this section explains how to find a PI for inhomogeneous linear equations with constant coefficients. The examples below will all be for second order equations, but the method works similarly if the order is higher.

Case 1 Find a PI for $y'' + fy' + gy = H(x)$ where $H(x)$ is a polynomial in x.

Method Try a PI of the form y_{PI} = polynomial of the same degree.

Example 6.12 Find a PI for

$$y'' + 7y' + 12y = 24x + 26.$$

Step 1: First we try a polynomial of the same form as $H(x)$

$$y_{PI} = Cx + D,$$

$$y'_{PI} = C \text{ and } y''_{PI} = 0.$$

Step 2: Substitute in the differential equations

$$0 + 7C + 12(Cx + D) = 24x + 26.$$

Step 3: Since this must hold for all x we can compare coefficients of x^0 (= 1) and coefficients of x^1 (= x). So we have that

$$12C = 24, \text{ and}$$

$$7C + 12D = 26.$$

Solving these we find

$$C = 2, \text{ and } D = 1.$$

Step 4: Write down the particular integral solution

$$y_{PI} = 2x + 1.$$

The method fails if $g = 0$ in equation (6.15). In this case substituting $y''_{PI} = 0$ and $y'_{PI} = A$ in the left hand side in the example above would not work. The way to proceed under these circumstances is to multiply through by x in the original trial y_{PI} and try again. The correct guess in this case is $y_{PI} = Cx^2 + Dx$.

Exercise 6.10 Find a PI of the differential equations

   (i) $y'' - 3y' - 4y = -16x - 40$,

   (ii) $y'' - 2y' + y = x^2 - 3$.

Case 2 Find a PI for $y'' + f y' + gy = H(x)$ where $H(x) = Ce^{mx}$ (C and m constants).

Method Try a PI of the form $y_{PI} = De^{mx}$.

Example 6.13 Find a particular integral for

$$y'' + 7y' + 12y = 5e^{2x}.$$

Step 1: Guess a PI of the same form as $H(x)$ in the differential equation.

$$y_{PI} = De^{2x}.$$

Step 2: Differentiate y_{PI} twice

$$y'_{PI} = 2De^{2x} \text{ and } y''_{PI} = 4De^{2x}.$$

Step 3: Substitute into the differential equation and solve for D.

$$(4D + 14D + 12D)e^{2x} = 5e^{2x}.$$

Because we have guessed a solution of the right form the factor e^{2x} cancels, leaving an equation for D. Therefore

$$30D = 5 \quad \text{or} \quad D = \frac{1}{6}$$

Step 4: Write out the PI

$$y_{PI} = \frac{1}{6}e^{2x}.$$

As with Case 1, sometimes this method will fail. The left hand side equals zero if e^{mx} is a solution of the homogeneous equation (i.e. the solution to $y'' + 7y' + 12y = 0$, found by solving the auxiliary equation), in which case we cannot solve for D.

Key point: If using $y_{PI} = De^{mx}$ makes the left hand side = 0, then the next guess is $y_{PI} = Dxe^{mx}$. If this fails in the same way, the correct approach is to multiply through by x and try again. So, next we would try $y_{PI} = Dx^2e^{mx}$, and so on until the method works.

Exercise 6.11 Find a PI for each of the differential equations

(i) $y'' - 2y' + 3y = e^{2x}$,

(ii) $y'' + 2y' - 3y = e^x$.

Case 3 Find a PI for $y'' + fy' + gy = H(x)$ where $H(x) = (a + \beta x)e^{mx}$, and a, β and m are constants.

Method Try a PI of the form $y_{PI} = (C + Dx)e^{mx}$.

Again the method may fail (if substituting $y_{PI} = Ce^{mx}$ into the LHS gives 0). The correct guess in that case is then to try $y_{PI} = (Cx + Dx^2)e^{mx}$.

Case 4 Find a PI for $y'' + fy' + gy = H(x)$ where $H(x) = a \cos(mx) + \beta \sin(mx)$, and a, β and m are constants.

Method Try a PI $y_{PI} = C \cos(mx) + D \sin(mx)$. (In Chapter 8 we shall look at a more efficient method.)

Example 6.14 Find a PI for $y'' + 7y' + 12y = -3 \sin(3x)$.

Step 1: Choose an appropriate form for y_{PI} based on the form of $H(x)$.

$$y_{PI} = C \cos(3x) + D \sin(3x).$$

Step 2: Differentiate y_{PI} twice and substitute into the differential equation

$$y'_{PI} = -3C \sin(3x) + 3D \cos(3x),$$

$$y''_{PI} = -9C \cos(3x) - 9D \sin(3x).$$

Substituting into the differential equation we get

$$(-9C \cos(3x) - 9D \sin(3x)),$$

$$+7(3D \cos(3x) - 3C \sin(3x)),$$

$$+12(C \cos(3x) + D \sin(3x)) = -3\sin(3x).$$

Step 3: Since the equation holds for all x we can equate coefficients of $\cos(x)$ and $\sin(x)$

$$-9C + 21D + 12C = 0,$$

$$-9D - 21C + 12D = -3.$$

Therefore $C = -7D$ from the first equation, so $150D = -3$ from the second. From which

$$D = \frac{1}{50}, \quad C = \frac{7}{50}.$$

Step 4: Write out the PI

$$y = \frac{7}{50}\cos(3x) - \frac{1}{50}\sin(3x).$$

Note that even though $H(x)$ in the example above involves only $\sin(3x)$ (and with integer coefficient -3) we find that the PI involves both $\cos(3x)$ *and* $\sin(3x)$ (and unwieldy fractions).

Exercise 6.12 Find a PI of the differential equations

(i) $y'' + 7y' + 12y = -150 \cos(3x)$

(ii) $y'' - 2y' + 3y = 34 \sin(2x)$.

Case 5 Find a PI for $y'' + fy' + gy = H(x)$ where $H(x) =$ (sum of functions of different types).

Method We solve the functions separately (as per the examples above) and add the solutions together.

Example 6.15 Find a PI for

$$y'' + 7y' + 12y = 5e^{2x} + 24x + 26.$$

We have already found PIs for when the right hand side is $24x + 26$ and $5e^{2x}$. The solutions have been found in Examples 6.12 and 6.13. The required PI is therefore the sum of the two y_{PI} solutions for those examples

$$y_{PI} = 2x + 1 + \frac{1}{6}e^{2x}$$

Exercise 6.13 Find a particular integral for each of the following:

(i) $y'' + y = 2e^x + x + 14$

(ii) $y'' + 3y' + 2y = \sin(4x)$

(iii) $y'' - 2y' - 3y = 2e^{-x}$.

It is often useful to think of the above differential equations as determining the response of a physical system, described by $y(x)$ (or $y(t)$), to a "driving force" given by the known functions of x or t on the right hand side. The rules show that an oscillating force produces an oscillating response, an exponential force causes an exponential response, and so on.

6.10. INHOMOGENEOUS LINEAR EQUATIONS

In Section 6.9, we found how to obtain a PI of an equation of the form

$$y'' + fy' + gy = H(x), \tag{6.16}$$

with f and g constants and $H(x)$ a known function of x. In Sections 6.7 and 6.8, we saw how to obtain the general solution of equations of the form

$$y' + gy = 0, \tag{6.17}$$

and

$$y'' + fy' + gy = 0. \tag{6.18}$$

Key point: The general solution of an inhomogeneous differential equation has two components:

(i) Solving the homogeneous part (i.e. $H(x) = 0$) gives the *complementary function* (CF), $y_{CF}(x)$.

(ii) Solving the inhomogeneous term (i.e. $H(x) \neq 0$) gives the *particular integral* (PI), $y_{PI}(x)$.

The full general solution of the equation

$$y'' + fy' + gy = H(x)$$

is given by the sum of the complementary function and the particular integral

$$y_{GS} = y_{CF} + y_{PI}.$$

Let us now take a look at some examples of how to do this, i.e. solving the homogeneous part of the differential equation and adding it to the particular integral to find the *general solution*.

Example 6.16 Find the general solution of

$$y'' + 7y' + 12y = 24x + 26$$

Step 1: Find the complementary function (from Exercise 6.7)

$$y_{CF} = Ae^{-4x} + Be^{-3x}.$$

Step 2: Find the particular integral (from Example 6.12)

$$y_{PI} = 2x + 1.$$

Step 3: Write down the general solution. The general solution is the sum of the CF and the PI, and hence we can write y_{GS} as

$$y_{GS} = y_{CF} + y_{PI} = Ae^{-4x} + Be^{-3x} + 2x + 1.$$

Here is a more detailed example.

Example 6.17 Find the general solution (CF+PI) of $y'' + 4y' + 4y = 2e^{-2x}$.

Step 1: Find the CF by solving the auxiliary equation. Begin with a guess

$$y = Ae^{\lambda x},$$
$$\therefore y' = A\lambda e^{\lambda x} \text{ and } y'' = A\lambda^2 e^{\lambda x},$$
$$\lambda^2 + 4\lambda + 4 = 0.$$

Now we have the auxiliary equation we solve it

$$(\lambda + 2)^2 = 0,$$
$$\lambda = -2.$$

The auxiliary equation has repeated roots. We know that for the case of repeated roots we can write down the solution directly as $y_{CF} = (Ax + B)e^{\lambda x}$. This gives us the complementary function

$$y_{CF} = (Ax + B)e^{-2x}.$$

Step 2: Find the PI by guessing a solution of the same form as the RHS. If we try $y = Ce^{-2x}$ for a PI we will get zero when we substitute y_{PI}, y'_{PI} and y''_{PI} back into the differential equation. This arises because Ce^{-2x} is part of the CF so it cannot be a PI. As we have a repeated roots example, this will also be true if we multiply by x, as Cxe^{-2x} is also part of the CF. So we multiply by x again and try

$$y_{PI} = Cx^2 e^{-2x},$$
$$y'_{PI} = 2Cxe^{-2x} - 2Cx^2 e^{-2x},$$
$$y''_{PI} = 2Ce^{-2x} - 4Cxe^{-2x} - 4Cxe^{-2x} + 4Cx^2 e^{-2x}.$$

We now need to find the constant, C. We substitute y_{PI} y'_{PI} and y''_{PI} into the equation to be solved, collect the coefficients of x and solve for C

$$y'' + 4y' + 4y = 2Ce^{-2x} - 8Cxe^{-2x}$$
$$+ 4(2Cxe^{-2x} - 2Cx^2e^{-2x}) + 4Cx^2e^{-2x},$$
$$= 2e^{-2x}.$$

Collecting terms, first coefficients of x^2, then x^1, then x^0

$$4C - 8C + 4C = 0,$$
$$-8C + 8C = 0,$$
$$2C = 2.$$

Therefore $C = 1$, and we have the particular integral

$$y_{PI} = x^2e^{-2x}.$$

Step 3: Write down the general solution, which is the sum of the CF and PI

$$y_{GS} = (x^2 + Ax + B)\,e^{-2x}.$$

Note the strategy used above: we proceed systematically until we find something that works (i.e. $C \neq 0$) for our particular integral. But be sensible: if things start to get really complicated it is probably due to having made a mistake early on.

Here is an example from electrical circuit theory:

Example 6.18 The equation

$$R\frac{dq}{dt} + q/C = V$$

describes the time dependent response of the current and the charge in an electical circuit comprising a resistance R, a capacitor C and a battery, voltage V_0. Since this is a linear equation with constant coefficients we look for a CF of the form $q_{CF}(t) = ae^{pt}$ with a a constant. This gives

$$Rp + 1/C = 0$$

hence $p = -1/RC$. For the PI we try $q_{PI}(t) = A$. This gives $A = CV_0$. Putting this together, the general solution is $q(t) = ae^{-t/RC} + V_0/C$.

Exercise 6.14 Find the general solution of each of the following:

(i) $y'' - y = x$

(ii) $y'' + 4y' + 3y = e^{-x}$

(iii) $y'' + 2y' + y = xe^{-x}$.

Exercise 6.15 This is another example from electrical circuit theory. If we apply an alternating voltage to a circuit with a resistance R and capacitance C the charge on the capacitor is governed by

$$R\frac{dq}{dt} + q/C = V_0 \cos(\omega t). \tag{6.19}$$

Find the general solution for $q(t)$.

6.11. INTEGRATING FACTOR METHOD

In this section, we will introduce the *integrating factor* method for solving any first order linear differential equations.

We start with any first order differential equation of the form

$$y' + P(x)y = Q(x), \tag{6.20}$$

where $P(x)$ and $Q(x)$ are known functions of x as in equation (6.10). We define the integrating factor as $R(x) = e^{\int P(x)\,dx}$, and will use this to solve our differential equation. First we will look at a worked example, and then discuss how the method works.

Example 6.19 The equation $xy' - 2y = x^2$ is first order and linear. Find its general solution.

Step 1: First we divide through by the coefficient of y' so that we have our equation in the correct form

$$y' - \frac{2}{x}y = x, \quad \text{so} \quad P(x) = -\frac{2}{x}.$$

Step 2: Next, we use $P(x)$ to find the *integrating factor, R(x)*

$$R(x) = e^{\int P(x)dx} = e^{\int -\frac{2}{x}dx} = e^{-2\int \frac{1}{x}dx},$$

therefore

$$R(x) = e^{-2\ln(x)} = e^{\ln(x^{-2})} = \frac{1}{x^2}.$$

Step 3: Now, we multiply through by the integrating factor $R(x)$

$$\frac{1}{x^2}y' - \frac{2}{x^3}y = \frac{1}{x}.$$

Step 4: We note that the left hand side is the derivative of a product, namely $\frac{d}{dx}\left(\frac{1}{x^2}y\right)$, and hence we can write

$$\frac{d}{dx}\left(\frac{1}{x^2}y\right) = \frac{1}{x}.$$

Step 5: Finally, we integrate both sides to obtain the general solution. Do not forget the constant of integration, and then rearrange for y_{GS}

$$\frac{1}{x^2}y = \int \frac{1}{x}dx,$$
$$= \ln(x) + C,$$

and therefore $\qquad y_{GS} = x^2 \ln(x) + Cx^2.$

The method works for every first order linear equation. The key to the method is remembering how to find the integrating factor.

Proof: We need to verify that multiplying through by the integrating factor $R(x)$ leads to a left hand side that can be put in the form of the derivative of a product. According to the rule to be proved, we first take the integrating factor

$$R(x) = e^{\int P(x)\, dx}.$$

We then need to take the derivative of $R(x)$. However, $R(x)$ is a composite of two functions $h(x)$ and $g(x)$, say, so we need to use the chain rule to find $R'(x)$. So if we let

$$h(g(x)) = e^{g(x)} \text{ where } g(x) = \int P(x)\, dx,$$

then we can write the derivative of $R(x) = h(g(x))$ as

$$R'(x) = \frac{dh}{dg}\frac{dg}{dx},$$

$$R'(x) = \left(e^{\int P(x)dx} \right) \cdot \left(\frac{d}{dx} \int P(x)dx \right) = R(x)P(x).$$

So we have found that the derivative of the integrating factor is simply the product of the integrating factor itself and the coefficient of y in equation (6.20).

Now let us multiply equation (6.20) by $R(x)$.

$$R(x)y' + R(x)P(x)y = R(x)Q(x).$$

We notice that (using the product rule)

$$\frac{d}{dx}(R(x)y) = R(x)y' + R'(x)y,$$

$$= R(x)y' + R(x)P(x)y.$$

This is just the left hand side of equation (6.20) multiplied by the integrating factor $R(x)$, as required. Note that the left hand side becomes $\frac{d}{dx}(R(x)y)$ every time, so there is no need to work this out each time. You can use the result directly as in the following example.

Example 6.20 Find the general solution of $x^2y' + xy = 1/x$.

Step 1: Divide by the coefficient of y' (which in this case is x^2)

$$y' + \frac{1}{x}y = \frac{1}{x^3}.$$

Step 2: Find the integrating factor

$$R(x) = e^{\int \frac{1}{x}dx} = e^{\ln(x)} = x.$$

Step 3: Multiply through by $R(x)$

$$xy' + y = \frac{1}{x^2}.$$

Step 4: Recall that the left side is the derivative of the product $R(x)y$

$$\frac{d}{dx}(xy) = \frac{1}{x^2}.$$

Step 5: Integrate both sides, and rearrange to find the general solution y_{GS}

$$xy = -\frac{1}{x} + c,$$

$$\therefore y_{GS} = -\frac{1}{x^2} + \frac{c}{x}.$$

Exercise 6.16 Solve by this method

(i) $y' + y = 2$

(ii) $x^{-1}y' - 2y = 1$

6.12. EXTENSION: SPECIAL FUNCTIONS

In this section, we look briefly at a way to derive solutions of general linear equations with coefficients that are not constant, but instead functions of the independent variable. An example would be

$$x^2 \frac{d^2 y}{dx^2} + x \frac{dy}{dx} + \left(x^2 - a^2\right) y = 0$$

with a a constant. This introduces the topic of special functions.

We know that the general solution to the equation

$$y'' + y = 0 \tag{6.21}$$

is $y = a \cos(x) + b \sin(x)$. Expanding sin and cos in a Taylor series, we can express the general solution as a power series:

$$y = a \left(1 - \frac{x^2}{2} + \frac{x^4}{4!} + \dots \right) + b \left(x - \frac{x^3}{3!} + \frac{x^5}{5!} + \dots \right).$$

Now, if we had not already known about trigonometric functions, could we have derived these series solutions? To this end, let us suppose that we seek a solution of (6.21) of the form

$$y = x^v \sum_{n=0}^{\infty} a_n x^n$$

where we have introduced the factor x^v because we are assuming that we do not yet know that the solution can be expanded in integer powers. By differentiating the series we must have that

$$y'' + y = \sum_{n=0}^{\infty} (n+v)(n+v-1) a_n x^{n+v-2} + \sum_{n=0}^{\infty} a_n x^n = 0.$$

To find v and the coefficients a_n we compare powers of x. From x^{v-2} we get

$$v(v-1)a_0 = 0$$

from which $v = 0$ or $v = 1$, since we assume $a_0 \neq 0$. We take $v = 0$; taking $v = 1$ turns out to give nothing more. With $v = 0$ we compare in turn the powers of x^0, x^1 and x^n giving

$$x^0 : 2a_2 + a_0 = 0$$

$$x^1 : 3 \times 2a_3 + a_1 = 0$$

$$x^n : (n+2)(n+1)a_{n+2} + a_n = 0$$

Thus

$$a_{n+2} = -\frac{a_n}{(n+2)(n+1)} = \frac{a_{n-2}}{(n+2)(n+1)n(n-1)} = \cdots$$

$$= \begin{cases} \dfrac{(-1)^{n/2} a_0}{(n+2)!} & n \text{ even} \\[2ex] \dfrac{(-1)^{(n-1)/2} a_1}{(n+2)!} & n \text{ odd} \end{cases}$$

We deduce that

$$y = a_0 \left(1 - \frac{x^2}{2!} + \frac{x^4}{4!} + \cdots\right) - a_1 \left(x - \frac{x^3}{3!} + \frac{x^5}{5!} + \cdots\right),$$

which is of the form $y = a \cos(x) + b \sin(x)$, where we could now *define* $\sin(x)$ and $\cos(x)$ by the series.

Of course, defining sin and cos in this way we now have to derive their other properties, which is a major undertaking. However, for more general linear differential equations, where we do not already have known solutions to hand, the approach via solution in series gives us a way of defining new functions and determining their properties.

The most common functions studied in this way are called *special functions*. This characterization clearly leaves open which functions are included, but we'll give a snapshot of some of the most common ones.

Bessel functions: The defining equation is

$$x^2 \frac{d^2 y}{dx^2} + x \frac{dy}{dx} + (x^2 - n^2) y = 0$$

which is satisfied by

$$y(x) = J_n(x) = \sum_{m=0}^{\infty} \frac{(-1)^m}{\Gamma(m+1)\Gamma(m+n+1)} \left(\frac{x}{2}\right)^{2m+n}.$$

If n is not an integer, then J_{-n} is a second independent solution. If n is an integer, then the second solution involves $\ln(x)$ so cannot be obtained by the series method directly.

A special case occurs when the series terminates (because a coefficient – and hence all succeeding ones – equals zero). In that case we obtain a polynomial solution rather than a new function, but because of the importance of these cases, the specific polynomials are given names.

Hermite polynomials, $H_n(x)$: The defining differential equation is

$$\frac{d^2 H_n}{dx^2} + (2n + 1 - x^2)H_n = 0$$

which, for integer n, has the polynomial solutions

$$H_0(x) = 1, H_1(x) = 2x, H_{n+1}(x) = 2xH_n(x) - 2nH_{n-1}(x).$$

Laguerre polynomials, $L_n(x)$: The defining differential equation is

$$x\frac{d^2 L_n}{dx^2} + (1 - x)\frac{dL_n}{dx} + nL_n = 0$$

which, for integer n, has the polynomial solutions

$$L_0 = 1, \quad L_1(x) = 1 - x, \quad L_{n+1}(x) = \frac{2n + 1 - x}{(n+1)}L_n(x) - \frac{n}{n+1}L_{n-1}(x).$$

Legendre polynomials, $P_l(x)$: The defining differential equation is

$$(1 - x^2)\frac{d^2 P_l}{dx^2} - 2x\frac{d^2 P_l}{dx} + l(l+1)P_l(x) = 0,$$

where l is an integer. (The use of l instead of n is conventional, arising from the association of Legendre polynomials with angular momentum, which is customarily denoted by l in quantum mechanics.) The polynomial solutions are

$$P_0(x)=1, \quad P_1(x)=x, \quad P_l(x)=\frac{(2l-1)}{l}xP_{l-1}(x)-\frac{(l-1)}{l}P_{l-2}(x).$$

Before the advent of computers series solutions were used extensively to elucidate the properties of solutions of differential equations and to compute numerical solutions. Nowadays, this aspect has been superseded by more direct numerical methods. Nevertheless, many of these examples (and others) are of theoretical importance in quantum mechanics and other areas of physics that involve linear differential equations.

Revision Notes

After completing this chapter you should be able to

- Define the terms order, linear, homogeneous, constant coefficients with reference to a differential equation

- Explain the terms complementary function (CF) and particular integral (PI)

- State the number of constants in the general solution

- Find a PI and CF for a linear equation with constant coefficients by trial solutions of the appropriate form, and hence find the general solution as PI + CF

- Solve a linear equation of the first order by the integrating factor method

Exercises and problems on differential equations can be found at the end of Chapter 8.

COMPLEX NUMBERS

Complex numbers, which involve the square root of -1, are an essential tool in physics and mathematics. It is necessary to use complex numbers in many areas of study, from electronics, to quantum mechanics, to economics. They arise most directly from studying the solution to polynomial equations, but they allow us to see deep connections between algebra and geometry, and between the exponential, trigonometric and hyperbolic functions. In this chapter, we will introduce you to some of the fundamental results and applications of complex numbers.

7.1. INTRODUCTION OF COMPLEX NUMBERS

Within the system of integers it is not possible to divide certain (in fact, most) pairs of numbers (because the result m/n is usually not an integer for m and n integers). We therefore introduce the system of *rational* numbers (fractions). In the system of rational numbers the quadratic equation $x^2 = 2$ does not have a solution. We therefore introduce some extra numbers, the *irrational* numbers, which cannot be expressed as fractions. The number $\sqrt{2}$ is defined such that $\left(\sqrt{2}\right)^2 = 2$ (You may be able to write down any number of more or less accurate approximations to $\sqrt{2}$, as fractions or finite decimals, but you can't say what $\sqrt{2}$ really *is* except through this definition.) The two solutions of $x^2 = 2$ are then $\pm\sqrt{2}$. The irrational numbers (including those that, like π but unlike $\sqrt{2}$, are not solutions of algebraic equations with integer coefficients), together with the rationals, make up the *real* number system.

Now to the main point. The real number system is not large enough to allow us to solve every quadratic equation. For example, the quadratic equation

$$z^2 + 1 = 0$$

has no real roots (see Figure 7.1). This is the same as saying the equation

$$z^2 = -1 \tag{7.1}$$

cannot be solved by a real number. We therefore introduce another new type of number, call it $\sqrt{-1}$ or i.

The "number" i is defined such that $i^2 = -1$.

You can't say what i really *is* except through this definition. The two solutions of equation (7.1) are then $z = \pm i$. It turns out that the introduction of i enables us to take the square root of any number, hence to solve any quadratic equation. (In fact, the introduction of i is sufficient to solve *any* polynomial equation, not just quadratics, as we shall see later.)

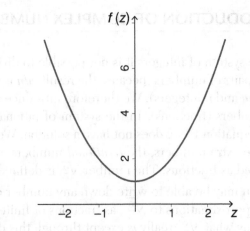

FIGURE 7.1: Graph of the quadratic function $f(z) = z^2 + 1$. As this does not cross the z-axis, there are no real z values for which $f(z) = 0$; the quadratic equation $z^2 + 1 = 0$ has no real roots.

Example 7.1 Find the solutions of $z^2 = -4$.

Write z^2 as the product of two numbers with known square roots.

$$z^2 = 4 \times (-1).$$

So,

$$z = \pm\sqrt{4} \times \sqrt{(-1)} = \pm 2 \times i = \pm 2i.$$

Example 7.2 Find the solutions of $z^2 - 2z + 5 = 0$.

Complete the square to get

$$z^2 - 2z + 5 = (z-1)^2 + 4 = 0,$$

which is true if

$$(Z-1)^2 = -4$$
$$Z - 1 = \pm 2i$$
$$z = 1 \pm 2i.$$

An alternative method is, of course, to use the formula for the roots of a quadratic equation: i.e. if

$$z = \left[-(-2) \pm \sqrt{(4-20)} \right]/2$$
$$= \left[2 \pm \sqrt{(-1)} \times \sqrt{16} \right]/2$$
$$= 1 \pm 2i$$

Exercise 7.1 Find, in the form $x + yi$, all solutions of the quadratic equations

  (i) $z^2 - 6z + 25 = 0$

  (ii) $2z^2 + 2z + 5 = 0$.

An expression of the form $x + yi$ is called a *complex* number. The complex number $x + i0$ is the same as the *real* number x. The complex number $0 + yi$ is called *pure imaginary* and is written as yi. For a complex number $z = x + yi$, the real number x is called the "real part of z," while the real number y is called the "imaginary part of z." Note that we refer to the imaginary part as y and not iy. This way of writing complex numbers – as $x + yi$ – is called the *standard form*.

If $z = x + yi$ we write

$$x = \operatorname{Re}(z),$$

(or sometimes $x = \Re(z)$, with an "R" in the Fraktur typeface), the real part of z. And we write

$$y = \operatorname{Im}(z)$$

(or sometimes $y = \Im(z)$, with an "I" in the Fraktur typeface), the imaginary part of z.

Two complex numbers $z_1 = x_1 + y_1 i$ and $z_2 = x_2 + y_2 i$ are equal if, and only if, their real and imaginary parts are equal (i.e. if $x_1 = x_2$ and $y_1 = y_2$). This is called equating real and imaginary parts.

Example 7.3 What are u and v (assumed real) if $u^2 + 2uvi = 25 + 30i$?

Equate the real and imaginary parts

$$u^2 = 25 \Rightarrow u = \pm 5$$
$$2uv = 30 \Rightarrow v = \pm 3.$$

Put only the corresponding values together

$$u = 5, v = 3 \quad \text{or} \quad u = -5, v = -3.$$

7.2. OPERATIONS ON COMPLEX NUMBERS

The result of any combination of algebraic operations on complex numbers is a complex number because it can always be reduced to an expression of the form $x + yi$.

Arithmetic

Let $z_1 = x_1 + y_1 i$ and $z_2 = x_2 + y_2 i$ be two complex numbers. In the following list we put the result of an algebraic operation on z_1 and z_2 in standard form ($x + yi$).

1. *Addition:* $z_1 + z_2 = (x_1 + x_2) + (y_1 + y_2)i$

2. *Subtraction:* $z_1 - z_2 = (x_1 - x_2) + (y_1 - y_2)i$

3. *Negative:* $-z_1 = -x_1 - y_1 i = -x_1 + (-y_1)i$

4. *Multiplication:* $z_1 \times z_2 = (x_1 + y_1 i)(x_2 + y_2 i) = (x_1 x_2 - y_1 y_2) + (x_1 y_2 + y_1 x_2)i$

The basic rule of complex algebra is: treat complex numbers as if real but replace i^2 by -1.

There is no need to memorize the formula for $(x_1 + y_1 i)(x_2 + y_2 i)$, since it comes immediately from multiplying out (to give $x_1 x_2 + x_1 y_2 i + y_1 x_2 i + y_1 y_2 i^2$) putting $i^2 = -1$, and collecting real and imaginary parts.

Exercise 7.2 Write in standard form ($x + yi$) each of

(i) $(2 + 3i) + (7 - 6i)$,

(ii) $(9 - 3i) - (7 - 6i)$,

(iii) $(2 + 3i)(7 - 6i)$.

Exercise 7.3 Work out $[z - (2 + 3i)] \times [z - (2 - 3i)]$ and hence find the solutions of $z^2 - 4z + 13 = 0$.

Example 7.4 By working out the real and imaginary parts of both sides, show that $z_1 z_2 = z_2 z_1$ (so complex numbers may be multiplied in any order). We say that multiplication of complex numbers is "commutative."

Let $z_1 = x_1 + y_1 i$, $z_2 = x_2 + y_2 i$. Then

$$z_1 z_2 = (x_1 + y_1 i)(x_2 + y_2 i).$$

Multiply out and put $i^2 = -1$ to get

$$= x_1 x_2 - y_1 y_2 + (x_1 y_2 + y_1 x_2)i.$$

Interchange the order, since reals can be multiplied in any order.

$$= x_2 x_1 - y_2 y_1 + (x_2 y_1 + y_2 x_1)i$$

$$= z_2 z_1.$$

The final equality follows by symmetry: there is no need to multiply out $z_2 z_1$ explicitly because it can be obtained from $z_1 z_2$ by interchanging 1 and 2 throughout.

Exercise 7.4 By working out the real and imaginary parts of both sides, show that $z_3(z_1 + z_2) = z_3 z_1 + z_3 z_2$ (multiplication of complex numbers is "distributive").

Complex Conjugate

If $z = x + yi$ we define the complex conjugate of z to be

$$z^* = x - yi. \tag{7.2}$$

Thus, to obtain the complex conjugate of a complex number, simply replace i by $-i$. Other notations you may come across are $\overline{z}$ and $z^\dagger$. The complex conjugate of a complex number is an important concept that occurs frequently.

Example 7.5 Verify that if $z = x + yi$, then $zz^* = x^2 + y^2$. Use equation (7.2) to write the complex conjugate, and then multiply out, putting $i^2 = -1$.

$$zz^* = (x + yi)(x - yi)$$

$$= x^2 - (yi)^2 + xyi - xyi$$

$$= x^2 - (y)^2(-1)$$

$$= x^2 + y^2.$$

We see that a complex number multiplied by its complex conjugate gives a (positive) real number. The complex conjugate of a real number x is simply x (real).

Exercise 7.5 By considering the real and imaginary parts show that

(i) $\left(z_1 + z_2\right)^* = z_1^* + z_2^*$,

(ii) $\left(z_1 z_2\right)^* = z_1^* z_2^*$.

These are useful results: the conjugate of a sum of two complex numbers is the sum of their conjugates; the conjugate of the product of two complex numbers is the product of the two conjugates.

Exercise 7.6 Write each of the following in standard form $(x + yi)$, and find their complex conjugates.

(i) $(1-i)^4$

(ii) $\left(-\dfrac{1}{2} + \dfrac{\sqrt{3}}{2}i\right)^3$

(iii) $\left(-\dfrac{1}{2} + \dfrac{\sqrt{3}}{2}i\right)^3$

Division

The procedure for division of two complex numbers is similar to that for multiplication. We write the complex numbers in standard form, $z_1 = x_1 + y_1 i$ and $z_2 = x_2 + y_2 i$, take the ratio and manipulate the result into a simpler form.

$$\frac{z_1}{z_2} = \frac{x_1 + y_1 i}{x_2 + y_2 i} = \frac{x_1 + y_1 i}{x_2 + y_2 i} \times \frac{x_2 - y_2 i}{x_2 - y_2 i}.$$

The trick is to multiply top and bottom by $z_2^* = x_2 - iy_2$, as in the expression on the right, which will give an expression with a real number as the denominator

$$= \frac{x_1 x_2 + y_1 y_2}{x_2^2 + y_2^2} + \frac{y_1 x_2 - x_1 y_2}{x_2^2 + y_2^2} i$$

provided $z_2 \neq 0$. The method is important. It enables you to write a complex fraction in the standard form $x + yi$. Rather than remembering this formula it is better to derive the result directly in each particular case as shown next.

Example 7.6 Express $\dfrac{5+i}{3+2i}$ in the standard form $x + yi$.

We multiply top and bottom by the complex conjugate of the bottom

$$\frac{5+i}{3+2i} = \frac{5+i}{3+2i} \times \frac{3-2i}{3-2i}$$

$$= \frac{15 + 3i - 10i - 2i^2}{3^2 + 2^2}$$

$$= \frac{17 - 7i}{13} = \frac{17}{13} - \frac{7}{13} i.$$

Exercise 7.7 Write $\dfrac{2-5i}{2+3i}$ in the standard form $x + yi$.

Exercise 7.8 Show that $\left(z_1 / z_2\right)^* = z_1^* / z_2^*$ if $z_2 \neq 0$.

Now we know how to multiply and divide complex numbers we can compute the inverse of a complex number. If $z \neq 0$, then

$$\frac{1}{z} = \frac{1}{z}\frac{z^*}{z^*} = \frac{x}{x^2+y^2} + \left(\frac{-y}{x^2+y^2}\right)i. \tag{7.3}$$

Hence every complex number z ($\neq 0$) has an inverse. From this it follows that if $z_1 z_2 = 0$, then $z_1 = 0$ or $z_2 = 0$. (Proof: suppose $z_1 \neq 0$; then we can multiply through by $1/z_1$ to get $z_2 = 0$.)

Exercise 7.9 Verify that the product of a complex number with its inverse is unity: $(1/z) \times z = 1$.

Now we can summarize the rules of algebra for complex numbers.

1. *Addition:* $z_1 + z_2 = (x_1 + x_2) + (y_1 + y_2)i$

2. *Subtraction:* $z_1 - z_2 = (x_1 - x_2) + (y_1 - y_2)i$

3. *Multiplication by real number:* $az_1 = ax_1 + ay_1 i$

4. *Multiplication:* $z_1 \times z_2 = (x_1 x_2 - y_1 y_2) + (x_1 y_2 + y_1 x_2)i$

5. *Conjugation:* $z^* = x - yi$.

6. *Division:* $\dfrac{z_1}{z_2} = \dfrac{z_1 z_2^*}{z_2 z_2^*} = \dfrac{x_1 x_2 + y_1 y_2}{x_2^2 + y_2^2} + \left(\dfrac{y_1 x_2 - x_1 y_2}{x_2^2 + y_2^2}\right)i$

7. *Inverse:* $\dfrac{1}{z} = \dfrac{1}{z}\dfrac{z^*}{z^*} = \dfrac{x}{x^2+y^2} + \left(\dfrac{-y}{x^2+y^2}\right)i$.

7.3.   QUADRATIC EQUATIONS

We have seen how to solve any quadratic equation, $az^2 + bz + c = 0$, with real coefficients, a, b, and c, by means of the formula

$$z = \frac{-b \pm \sqrt{(b^2 - 4ac)}}{2a}, \tag{7.4}$$

and if $4ac > b^2$ you use i to take the square root of the negative number.

Suppose, however, that one or more of the coefficients a, b or c is complex. We can still use this formula. But is the square root of a complex number a complex number, or do we have to introduce further new numbers into our scheme? Let us start with the simplest example, the square root of i itself.

Example 7.7 Solve $z^2 = i$ (i.e. find the square root of i).

Let $z = x + yi$ with x and y real (assuming that the result will be a complex number)

$$z^2 = (x + yi)^2$$
$$= x^2 - y^2 + 2xyi = i$$

Equate real and imaginary parts

$$x^2 - y^2 = 0, \quad \text{and} \quad 2xy = 1$$

These simultaneous equations have two real solutions,

$$x = +\sqrt{1/2}, \quad \text{and} \quad y = +\sqrt{1/2}, \qquad \text{or}$$
$$x = -\sqrt{1/2}, \quad \text{and} \quad y = -\sqrt{1/2}.$$

Therefore

$$z = \frac{1}{\sqrt{2}} + \frac{i}{\sqrt{2}} \quad \text{or} \quad z = -\frac{1}{\sqrt{2}} - \frac{i}{\sqrt{2}}$$

We have therefore shown that the square root of i is another complex number, not something new. Taking the square root of a general complex number is more difficult. The following shows two ways of doing it; we shall give a systematic method for finding all roots later on (and they will all turn out to be complex numbers).

Example 7.8 Solve $z^2 = -5 + 12i$ (i.e. find the square root of $-5 + 12i$).

Let $z = x + yi$, with x and y real

$$z^2 = (x + yi)^2$$

$$= x^2 - y^2 + 2xyi = -5 + 12i$$

Equate real and imaginary parts and solve for x and y

$$x^2 - y^2 = -5 \quad \text{and} \quad 2xy = 12$$

$$\therefore y = 6/x \quad \text{and} \quad x^2 - 36/x^2 = -5$$

Multiply through by x^2 and factorize the resulting quadratic for x^2

$$x^4 + 5x^2 - 36 = 0,$$
$$(x^2 + 9)(x^2 - 4) = 0.$$
$$\therefore x^2 = -9 \text{ or } x^2 = 4$$

$x^2 = -9$ is impossible for x real. Hence $x = +2$, and $y = 6/x = +3$, or $x = -2$ and $y = 6/x = -3$. Therefore

$$z = 2 + 3i \quad \text{or} \quad z = -2 - 3i.$$

Alternative Solution: Proceed as before to find $x^2 - y^2$ and xy by equating real and imaginary parts.

$$x^2 - y^2 = -5 \quad \text{and} \quad 2xy = 12$$

We can solve these by a trick which avoids the need to solve a quadratic equation. The trick is to find $x^2 + y^2$ in terms of the already known $x^2 - y^2$ and xy.

$$\left(x^2+y^2\right)^2 =\left(x^2-y^2\right)^2 +\left(2xy\right)^2$$
$$=(-5)^2+(12)^2=169$$
$$\left(x^2+y^2\right)=\sqrt{169}=13$$

Now we can solve for x^2 or y^2. Adding the above equation to $(x^2 - y^2) = -5$ we get

$$2x^2 = 13 - 5 = 8, \therefore x = \pm2$$

And so, using $2xy = 12$ we get

$$y = 6/x = \pm3$$

Putting this all together

$$z = 2 + 3i \quad \text{or} \quad z = -2 - 3i.$$

Exercise 7.10 Use one of these methods to find the square root of $5 + 12i$.

Since we now know how to find the square roots of a complex number we can now solve any quadratic by using the standard formula. We know $az^2 + bz + c = 0$ has roots

$$z=\frac{-b\pm\sqrt{\left(b^2-4ac\right)}}{2a}. \tag{7.5}$$

Example 7.9 Solve $z^2 - (2 - i)z + (2 - 4i) = 0$.

The solution of the quadratic equation is

$$z=\frac{(2-i)\pm\sqrt{\left(2-i\right)^2-4(1)(2-4i)}}{2}$$
$$=\frac{(2-i)\pm\left(-5+12i\right)^{1/2}}{2}.$$

To obtain $(-5 + 12i)^{1/2}$ we have to solve $w^2 = -5 + 12i$ (see example 7.8):

$$(-5 + 12i)^{1/2} = \pm(2 + 3i).$$

Therefore

$$z = \frac{(2-i) \pm (2+3i)}{2}$$

$$z = 2 + i \quad \text{or} \quad z = -2i.$$

Exercise 7.11 Deduce from the solution of Example 7.9 that

$$z^2 - (2 - i)z + (2 - 4i) = (z - (2 + i))(z + 2i)$$

and verify the result by multiplying out the right side.

Exercise 7.12 Solve the quadratic equations

(i) $z^2 - (2 + 2i)z + (2i - 16) = 0$

(ii) $z^2 - (4 + i)z + (5 + 5i) = 0.$

7.4. THE ARGAND DIAGRAM

The complex number $z = x + yi$ can be represented by the point (x, y) in the plane. This graphical representation of complex numbers is called the *Argand diagram*. Essentially, we use the real and imaginary parts of a complex number as coordinates. See Figure 7.2.

Example 7.10 Indicate the position of $3 - 4i$ in the Argand diagram.

This will be the point $(3, -4)$ shown in Figure 7.2.

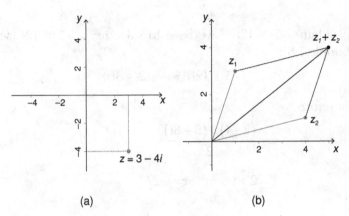

FIGURE 7.2: (a) Argand diagram for the complex number $z = 3 - 4i$. (b) Argand diagram showing the addition of two complex numbers $z_1 + z_2$. Note that the real part is along the horizontal axis ($x = \text{Re}(z)$), the imaginary part is along the vertical axis ($y = \text{Im}(z)$).

Figure 7.2(b) shows an Argand diagram for the sum of two complex numbers. The real and imaginary parts each add to form a third complex number, just as the components of (Cartesian) vectors add in two-dimensional space. Figure 7.3(a) shows an Argand diagram for a complex number and its conjugate. When we form the conjugate we swap the sign of the imaginary part, which is a reflection about the x-axis in the Argand diagram. Figure 7.3(b) shows an Argand diagram for a complex number z and the same number multiplied by i repeatedly: z, zi, $-z$ and $-zi$. Each multiplication by i rotates the point in the Argand diagram by 90 degrees ($\pi/2$ radians) about the origin, in the counterclockwise direction. Clearly there is a close connection between complex numbers and two-dimensional geometry, especially rotations and reflections.

Exercise 7.13 Indicate the positions in the Argand diagram of the following complex numbers. (You will need to rewrite the complex numbers in the standard form $x + yi$; refer to Exercise 7.6.)

(i) $2 + 5i$

(ii) $(1 - i)^4$

(iii) $\left(-\dfrac{1}{2} + \dfrac{\sqrt{3}}{2} i \right)^2$

(iv) $\left(-\dfrac{1}{2} + \dfrac{\sqrt{3}}{2} i \right)^3$

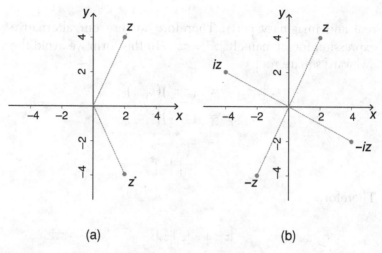

FIGURE 7.3: (a) Argand diagram for the complex number $z = 2 + 4i$ and its conjugate $z^* = 2 - 4i$. (b) Argand diagram showing the complex number $z = 2 + 4i$ and the complex numbers obtained by repeated multiplication by i: $z, zi, -z$ and $-zi$.

7.5. MODULUS AND ARGUMENT OF A COMPLEX NUMBER

The *modulus* of the complex number $z = (x + yi)$ is

$$|z| = (x^2 + y^2)^{1/2}. \tag{7.6}$$

The modulus of z is the distance of (x, y) from the origin in the Argand diagram. Alternatively, we have

$$|z| = (z^*z)^{1/2}, \tag{7.7}$$

from Example 7.5, and hence

$$|z|^2 = z^*z. \tag{7.8}$$

This can be very useful in manipulations.

Example 7.11 Show that $|z_1 z_2| = |z_1||z_2|$.

This can be shown by putting $z_1 = x_1 + y_1 i$, $z_2 = x_2 + y_2 i$. However, the result involves only the complex numbers and not their real and imaginary parts. Therefore we use our alternative expression for $|z|$, namely $|z|^2 = zz^*$. (In this form we avoid the awkward square root.)

$$
\begin{aligned}
|z_1 z_2|^2 &= (z_1 z_2)(z_1 z_2)^* \\
&= (z_1 z_2)(z_1^* z_2^*) \\
&= z_1 z_1^* z_2 z_2^* \\
&= |z_1|^2 |z_2|^2.
\end{aligned}
$$

Therefore

$$|z_1 z_2| = |z_1||z_2|.$$

Exercise 7.14 Show that, for any complex numbers z_1 and z_2,

$$\left|\frac{z_1}{z_2}\right| = \frac{|z_1|}{|z_2|}.$$

Exercise 7.15 Find $|z|$ if

(i) $z = -\sqrt{3} + i$,

(ii) $z = \left(-\sqrt{3} + i\right)^2$,

(iii) $z = \dfrac{1 + 3i}{1 + i}$.

If the complex number $z \neq 0$ is represented by the point $P = (x, y)$ in the Argand diagram, then the *argument* of z, $\arg(z)$, is the angle from the x-axis to the line OP joining point P with the origin O. See Figure 7.4:

$$\arg(z) = \tan^{-1}\left(\frac{y}{x}\right). \tag{7.9}$$

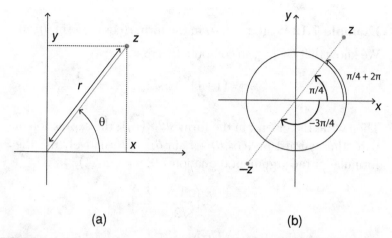

(a)   (b)

FIGURE 7.4: (a) A complex number z in the Argand diagram, showing its modulus and argument (r, θ) and its real and imaginary parts (x, y), which are polar coordinates and Cartesian coordinates in the Argand diagram, respectively. (b) The Argand diagram for $z = 1 + i$ and $-z = -1 - i$. The argument is defined up to multiples of 2π. The argument of z is $\pi/4$ or $\pi/4 + 2\pi = 9\pi/4$ etc.; the argument of $-z$ is $-3\pi/4$ or $-3\pi/4 + 2\pi = 5\pi/4$ etc.

This formula needs to be used carefully to obtain a correct angle from the inverse tan. For example, $\arg(1 + i) = \tan^{-1}(1) = \pi/4$, but $\arg(-1 - i) = \tan^{-1}(1) = -3\pi/4$. (See Figure 7.4(b).) Note also that $\arg(z)$ is many-valued: we can add arbitrary multiples of 2π and get the same argument. If $z = 0$ then $\arg(z)$ is not defined.

Exercise 7.16 Find the modulus and argument of $1 - i$ and $-\sqrt{3} - i$.

From Figure 7.4(a) we see that $(|z|, \arg(z))$ are the polar coordinates of z in the Argand diagram. So write $r = |z|$, $\theta = \arg(z)$. Then $x = r\cos(\theta)$, $y = r\sin(\theta)$ and we get the *polar representation* of a complex number

$$z = r(\cos\theta + i\sin\theta). \tag{7.10}$$

This is important. We can use the polar (r, θ) or Cartesian (x, y) representation, or both, as convenient.

Example 7.12 Write $z = 1 + i$ in the form $r(\cos\theta + i\sin\theta)$. We know that $r = |z|$, so compute this first.

$$|z| = (1+1)^{1/2} = \sqrt{2}.$$

Then write the given z in the form $\sqrt{2} \times (...)$; the expression in parentheses must be $(\cos(\theta) + i\sin(\theta))$. Alternatively use the formula for the argument (equation 7.9) $\theta = \tan^{-1}(1)$.

$$
\begin{aligned}
z &= \sqrt{2}\left(\frac{1}{\sqrt{2}} + \frac{i}{\sqrt{2}}\right), \\
&= \sqrt{2}\left(\cos\left(\frac{\pi}{4}\right) + i\sin\left(\frac{\pi}{4}\right)\right).
\end{aligned}
$$

Exercise 7.17 Indicate the positions of the complex numbers i, $1 - i\sqrt{3}$, $-1 - i\sqrt{3}$ on the Argand diagram and write each of them in the form $r(\cos(\theta) + i\sin(\theta))$.

7.6. THE COMPLEX EXPONENTIAL

We shall show that for real θ:

$$e^{i\theta} = \cos(\theta) + i\sin(\theta). \tag{7.11}$$

This result relates the exponential and trigonometric functions through complex numbers, and is enormously useful in physics and mathematics. It is known as *Euler's equation*. (The equation is also true for complex θ.)

Proof: We have (from Chapter 2), that

$$e^x = 1 + x + \frac{x^2}{2!} + \frac{x^3}{3!} + \frac{x^4}{4!} + \dots$$

So, putting $x = i\theta$,

$$e^{i\theta} = 1 + i\theta - \frac{\theta^2}{2!} - \frac{i\theta^3}{3!} + \frac{\theta^4}{4!} + \dots$$

We collect the real terms and the imaginary terms together

$$= \left(1 - \frac{\theta^2}{2!} + ...\right) + i\left(\theta - \frac{\theta^3}{3} + ...\right),$$

and notice these are the series expansions of cos(θ) and sin(θ).

$$= \cos(\theta) + i\,\sin(\theta).$$

Example 7.13 Show that $\cos(\theta) = \frac{1}{2}\left(e^{i\theta} + e^{-i\theta}\right)$.
$e^{-i\theta}$ is obtained from $e^{i\theta}$ either by complex conjugation (replacing i by $-i$) or by letting $\theta \to -\theta$.

$$e^{i\theta} = \cos(\theta) + i\,\sin(\theta),$$

$$e^{-ie} = \cos(\theta) - i\,\sin(\theta).$$

Hence

$$e^{i\theta} + e^{-ie} = 2\cos(\theta)$$

$$\therefore \quad \cos(\theta) = \frac{1}{2}\left(e^{i\theta} + e^{-i\theta}\right)$$

Exercise 7.18 Show that $\sin(\theta) = (e^{i\theta} - e^{-i\theta})/2i$.

We therefore have expressions for sin and cos in terms of complex exponentials

$$\cos(\theta) = \frac{1}{2}\left(e^{i\theta} + e^{-i\theta}\right)$$

$$\sin(\theta) = \frac{1}{2i}\left(e^{i\theta} - e^{-i\theta}\right). \tag{7.12}$$

Exercise 7.19 What, in their simplest forms, are

(i) $e^{i\pi/2}$,

(ii) $e^{i\pi}$,

(iii) $e^{2\pi n i}$ (n = 0, ±1, ±2,...),

(iv) $|e^{i\theta}|$.

Exercise 7.20 Show that $e^{i\theta_1} = e^{i\theta_2}$ if (and only if) $\theta_2 = \theta_{11}+2n\pi$ $(n = 0, \pm1, \pm2 ...)$.

Exercise 7.21 Show that

$$\cosh(\theta) = \cos(i\theta), \sinh(\theta) = -i\sin(i\theta),$$

and that

$$\cos(\theta) = \cosh(i), \sin(\theta) = -i\sinh(i\theta).$$

This explains the similarities (and differences) between the identities involving the trigonometric and hyperbolic functions.

Comparing equations (7.11) and (7.10) we see that any complex number can be written in the form

$$z = re^{i\theta} \qquad\qquad (7.13)$$

where $r = |z|$ and $\theta = \arg(z)$. This form is very useful for taking products and roots of complex numbers as the following examples show.

Example 7.14 If $z = re^{i\theta}$ $(z \neq 0)$, find z^{-1}:

$$z^{-1} = (re^{i\theta})^{-1} = r^{-1}e^{-i\theta}.$$

Example 7.15 If $z_1 = r_1 e^{i\theta_1}$ and $z_2 = r_2e^{i\theta_2}$ find z_1z_2, $|z_1 z_2|$ and $\arg(z_1z_2)$.

$$z_1z_2 = r_1r_2e^{i\theta_1} e^{i\theta_2} = r_1r_2e^{i\theta_1+ i\theta_2} = r_1r_2e^{i(\theta_1+\theta_2)}$$

So $|z_1z_2| = r_1r_2$ and $\arg(z_1z_2) = \theta_1 + \theta_2$.

We see from this example that to form the product of two complex numbers we multiply the moduli and add the arguments.

Exercise 7.22 If $z_1 = r_1 e^{i\theta_1}$ and $z_2 = r_2 e^{i\theta_2}$ $(z_2 \neq 0)$ find $\dfrac{z_1}{z_2}$, $\left|\dfrac{z_1}{z_2}\right|$ and arg $\left(\dfrac{z_1}{z_2}\right)$

Example 7.16 If $z = re^{i\theta}$ find $z^{1/2}$ and indicate the positions of the square roots of z on the Argand diagram.

$$z^{1/2} = (re^{i\theta})^{1/2} = \pm r^{1/2} e^{i\theta/2}.$$

See Figure 7.5. (Remember the ±: there are two square roots!)

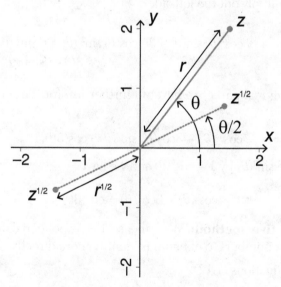

FIGURE 7.5: The Argand diagram showing a complex number z and its two square roots, $z^{1/2}$.

Exercise 7.23 Let $z = 1 + \sqrt{3}i$. By expressing z in the form $re^{i\theta}$, find $z^{1/2}$ and show your result on the Argand diagram.

7.7. DE MOIVRE'S THEOREM

Since $(\cos(\theta) + i \sin(\theta))^n = (e^{i\theta})^n = e^{in\theta} = \cos(n\theta) + i \sin(n\theta)$, we have another fundamental result, known as de Moivre's theorem,

$$(\cos(\theta) + i \sin(\theta))^n = \cos(n\theta) + i \sin(n\theta). \qquad (7.14)$$

valid for $n = 0, \pm1, \pm2, \ldots$. This can be used to give a proof of various complicated-looking identities.

Example 7.17 Show that $4 \cos^3(\theta) = \cos(3\theta) + 3 \cos(\theta)$.

Since the left hand side involves $\cos^3(\theta)$ we try putting $n = 3$ in de Moivre's theorem.

$$(\cos(\theta) + i \sin(\theta))^3 = \cos(3\theta) + i \sin(3\theta).$$

Then multiply out the left side

$$\cos^3(\theta) + 3i \cos^2(\theta) \sin(\theta) + 3i^2 \cos(\theta) \sin^2(\theta) + i^3 \sin^3(\theta)$$
$$= \cos(3\theta) + i \sin(3\theta). \quad (7.5)$$

Equating real parts (we have no further use for the imaginary parts)

$$\cos^3(\theta) - 3\cos(\theta) \sin^2(\theta) = \cos(3\theta).$$

Replace $\sin^2(\theta)$ by $1 - \cos^2(\theta)$ and we have

$$4\cos^3(\theta) - 3\cos(\theta) = \cos(3\theta).$$

Alternative method: We can use the exponential form of complex numbers to obtain this identity more directly.

From equations (7.12)

$$4\cos^3(\theta) = 4\left(\frac{1}{2}\left[e^{i\theta} + e^{-i\theta}\right]\right)^3$$

$$= \frac{1}{2}\left(e^{i\theta} + e^{-i\theta}\right)^3.$$

Now expand this using the binomial theorem

$$4\cos^3(\theta) = \frac{1}{2}\left[\left(e^{i\theta}\right)^3 + 3\left(e^{i\theta}\right)^2\left(e^{-i\theta}\right) + 3\left(e^{i\theta}\right)\left(e^{-i\theta}\right)^2 + \left(e^{-i\theta}\right)^3\right]$$

$$= \frac{1}{2}\left[e^{3i\theta} + 3e^{i\theta} + 3e^{-i\theta} + e^{-3i\theta}\right]$$

$$= \frac{1}{2}\left[e^{3i\theta} + e^{-3i\theta}\right] + \frac{3}{2}\left[e^{i\theta} + e^{-i\theta}\right]$$

$$= \cos(3\theta) + 3\cos(\theta).$$

Exercise 7.24 Show that $\sin^5(\theta) = (\sin(5\theta) - 5\sin(3\theta) + 10\sin(\theta))/16$.

Exercise 7.25 Let C_n stand for the sum $1 + \cos(\theta) + \cos(2\theta) + \dots + \cos[(n-1)\theta]$. The aim of this question is to find a closed-form expression for the sum of this series to n terms.

(i) Use de Moivre's theorem to show that C_n is the real part of

$$1 + e^{i\theta} + \left(e^{i\theta}\right)^2 + \dots + \left(e^{i\theta}\right)^{n-1}.$$

(ii) This is a geometric progression. Identify the common ratio and hence obtain its sum.

(iii) Deduce that C_n is the real part of $\left\{\dfrac{e^{in\theta}-1}{e^{i\theta}-1}\right\}$. Show that

$$\frac{e^{in\theta}-1}{e^{i\theta}-1} = \frac{e^{in\theta/2}\left(e^{in\theta/2}-e^{-in\theta/2}\right)}{e^{i\theta/2}\left(e^{i\theta/2}-e^{-i\theta/2}\right)}$$

and hence that

$$1 + \cos(\theta) + \cos(2\theta) + \dots + \cos((n-1)\theta) = \cos\left(\frac{(n-1)}{2}\theta\right)\frac{\sin(n\theta/2)}{\sin(\theta/2)}$$

This section has shown that one of the many uses of complex numbers is in obtaining real trigonometrical identities. These can also be established without the use of complex numbers, but often not so simply.

7.8.   THE ROOTS OF UNITY

We are familiar with the idea that the solution to $z^2 = 1$ is $z = \pm 1$ so 1 has two square roots. Now that we know about complex numbers we can solve the more general equation $z^n = 1$. The complete set of n complex solutions are the nth roots of unity.

Example 7.18 Show that $z^3 = 1$ is solved by $z = 1, \omega, \omega^2$ where $\omega = e^{2\pi i/3}$. (These are the cube roots of unity.)

Begin by noting that $z^3 = 1 = e^{2\pi i} = e^{4\pi i}$ are three equivalent ways of writing 1. They give *different* cube roots. Any other way (e.g. $e^{6\pi i}$) will repeat one of these. So

$$z = 1, e^{2\pi i/3}, e^{4\pi i/3}.$$

The cube roots are written as 1, ω and ω^2 where $\omega = e^{2\pi i/3}$, and $\omega^2 = e^{4\pi i/3}$.

Exercise 7.26 Show that $z^4 = 1$ is solved by $z = 1, \omega, \omega^2, \omega^3$, where $\omega = e^{\pi i/2}$, and that $z^n = 1$ is solved by $z = 1, \omega, \omega^2,..., \omega^{n-1}$ where $\omega = e^{2\pi i/n}$.

Plot the cube roots of 1 and the fourth roots of 1 in the Argand diagram and describe where the nth roots lie.

7.9.   ROOTS OF REAL POLYNOMIALS

If $P(z)$ is a polynomial with real coefficients and if z_0 is a root, i.e. if $P(z_0) = 0$, then z_0^* is also a root, i.e. $P(z_0^*) = 0$. Thus:

The roots of a real polynomial are either real or occur in complex conjugate pairs

Proof:

Let

$$P(z) = a_n z^n + a_{n-1} z^{n-1} + ... + a_0,$$

where $a_0, a_1, \dots a_n$ are real coefficients. Taking the complex conjugate of this equation gives

$$\left(P(z)\right)^* = a_n^* \left(z^n\right)^* + a_{n-1}^* \left(z^{n-1}\right)^* + \dots + a_0^*,$$

since the conjugate of a sum of complex numbers is the sum of the conjugates of the summands (see Exercise 7.5). Now, noting that $(q^m)^* = (q^*)^m$ for integer m we can write this as

$$= a_n^* \left(z^*\right)^n + a_{n-1}^* \left(z^*\right)^{n-1} + \dots + a_0^*.$$

And since all the a_m's are real $\left(a_m^* = a_m\right)$ we have

$$= a_n(z^*)^n + a_{n-1}(z^*)^{n-1} + \dots + a_0$$
$$= P\left(z^*\right)$$

But $P(z_0) = 0$, so $(P(z_0))^* = 0$, and hence $P\left(z_0^*\right) = 0$ This means that z_0^* is a root if z_0 is a root.

Exercise 7.27 Why does this proof fail if $a_n, a_{n-1}, \dots, a_0$ are not real?

7.10.   ROOTS OF COMPLEX POLYNOMIALS

Remainder theorem: Let $Q(z)$ be a polynomial with complex coefficients. (Of course, this includes the case when the coefficients are in fact real.) Then

z_0 is a root of $Q(z) = 0$ if and only if $z - z_0$ is a factor of $Q(z)$.

Proof: If $(z - z_0)$ is a factor of $Q(z)$ then $Q(z) = (z - z_0)P(z)$ for some polynomial $P(z)$ and hence $Q(z_0) = 0$. Conversely, if $Q(z_0) = 0$ then $Q(z) = Q(z) - Q(z_0)$ and each pair of terms of the same degree is divisible by $(z - z_0)$.

Fundamental theorem of algebra: There is at least one complex number, z_0 say, such that $Q(z_0) = 0$; i.e. every complex polynomial has a root.

This is a remarkable result (and not easy to prove rigorously). Turn back to example 7.1 to remind yourself that it is not true for real numbers. But once we have introduced the solution, i, of one real quadratic equation, $z^2+1 = 0$, we can, in principle, solve *all* polynomial equations. The "in principle" is important – no explicit formulae exist for general equations higher than quartic, even though we know the roots always exist.

Exercise 7.28 Deduce that every polynomial has a complete factorization into linear factors,

$$Q(z) = a_n(z - z_1)(z - z_2) \dots (z - z_n)$$

and that every real polynomial has a complete factorization into some product of real quadratic factors and real linear factors.

Exercise 7.29 Given that $z = 1 + i$ is a root of the equation

$$z^4 - 8z^3 + 24z^2 - 32z + 20 = 0$$

find the other three roots.

7.11. EXTENSION: COMPLEX VARIABLE

We have already met some very simple examples of functions of a complex variable; z^2 and $\sqrt{z}$ are examples. If $z = x + iy$ as usual, we can write $z^2 = (x^2 - y^2) + 2xyi$, that is in the form $f(z) = u(x, y) + iv(x, y)$ where u and v are real functions of the real variables x and y. To extend the calculus from real to complex functions, we need to be able to differentiate $f(z)$ with respect to z. To see that this restricts the possible choices of u and v consider the function F obtained by putting $u = x$ and $v = 0$. If we now let z vary keeping x constant, we get $dF / dz = \lim_{\delta z \to 0}\left(F(z+\delta z) - F(z)\right)/\delta z = 0$ if instead we let z vary keeping y constant, we obtain, similarly, $dF/dz = 1$. To ensure that the limit is independent of how we approach it, it can be shown that u and v must satisfy

$$\frac{\partial u}{\partial x} = \frac{\partial v}{\partial y} \quad \text{and} \quad \frac{\partial u}{\partial y} = -\frac{\partial v}{\partial x}.$$

These are called the Cauchy-Riemann relations. You can check that they are satisfied by $f(z) = z^2$ but not by $F(z) = x$. (Informally, we can see that F is not a function of z only, but of both z and z^* since $x = (z + z^*)/2$.)

Differentiable functions of a complex variable are called analytic functions. They have some remarkable properties. For example, if $f(z)$ is an analytic fucntion in some region Ω, it satisfies Cauchy's theorem

$$\oint f(z)dz = 0$$

where the integral is taken around a circle in Ω with center z_0. (To do the integral around a circle of radius R put $z = z_0 + Re^{i\theta}$, $dz = iRe^{i\theta}d\theta$ and integrate with respect to θ from 0 to 2π. The theorem is true for any closed curve, not just circles, but the technique for evaluating the integral is then more complicated.)

The theorem can be used to prove Cauchy's integral formula for an analytic function:

$$f(z_0) = \frac{1}{2\pi i}\oint \frac{f(z)}{z - z_0}dz.$$

A function analytic in an annulus,[1] A, centered on a point z_0, can be expanded as a Laurent series (the generalization of Taylor series for analytic functions):

$$f(z) = a_0 + a_1(z - z_0) + a_1(z - z_0)^2 + \ldots + \frac{b_1}{z - z_0} + \frac{b_2}{(z - z_0)^2} + \ldots$$

where z lies within the annulus. (If $f(z)$ is analytic within a disc centered on z_0, that includes z_0, then the b_n will all be zero.) Cauchy's formula can be used to show that

$$a_n = \frac{1}{2\pi i}\oint \frac{f(z)}{(z - z_0)^{n+1}}, \quad b_n = \frac{1}{2\pi i}\oint \frac{f(z)}{(z - z_0)^{-n+1}},$$

---

[1] An annulus is the region between two concentric circles.

where the integrals are taken around closed curves within the annulus A, and that

$$\oint f(z)dz = 2\pi i b_1. \tag{7.16}$$

The quantity b_1 is called the residue of $f(z)$ at z_0. Equation (7.16) can be used to evaluate integrals which are often difficult to obtain by other means. (The technique is called contour integration.)

Revision Notes

After completing this chapter you should be able to

- Add, subtract, multiply and divide complex numbers
- Find the complex conjugate of an algebraic expression
- Find the real and imaginary parts of a complex algebraic expression
- Solve quadratic equations with real or complex coefficients
- Find the argument and modulus of a complex number and indicate it on the Argand diagram
- Express a complex number in the form
$$z = r(\cos(\theta) + i\sin(\theta)) = re^{i\theta}$$
- Recall $\cos(\theta) = \dfrac{1}{2}\left(e^{i\theta} + e^{-i\theta}\right), \quad \sin(\theta) = \left(e^{i\theta} - e^{-i\theta}\right)/2i$
- Quote de Moivre's theorem and use it to derive trigonometric identities
- Find the nth roots of unity

7.12. EXERCISES

1. Write in the form $x + iy$, where x and y are real, the complex number u/v, where $u = 1 + 7i$, $v = 1 + i$, and verify that (i) $|x + iy| = |u| / |v|$, (ii) $x - iy = u^* / v^*$.

2. Write in the form $x + iy$, where x and y are real, each of the complex numbers $w = (2 + 3i) / (1 + i)$, $z = 3e^{i\pi/4}$, and their conjugates w^*, z^*.

3. Write the complex number $w = -1 + i\sqrt{3}$ in the form $re^{i\theta}$ where $r > 0$ and $-\pi < \theta < \pi$, and indicate its position on the Argand diagram.

4. Show the positions of the circle $|z| = 5$ and of the particular complex number $z_0 = -3 + 4i$ and of z_0^* in the Argand diagram, and also indicate the approximate positions of the two square roots of z_0.

5. Show the positions of the circle $|z| = 5$ and of the particular complex number $z_0 = 5 - 12i$ in the Argand diagram, write down the values of $|z_0|$ and $\tan(\arg z_0)$, and indicate the approximate positions of the two square roots of z_0.

6. Indicate the positions of the complex numbers $z_1 = 1 + i$ and $z_2 = \sqrt{2}i$ in the Argand diagram, verify that they are at an equal distance from the origin, and explain why $\arg(z_1 + z_2)$ must be $3\pi/8$.

7. Write the complex number $z = -1 + 2i$ in the form $re^{i\theta}$ where $r = |z| > 0$ and $-\pi < \theta < \pi$, expressing θ in terms of the number $\alpha = \tan^{-1} 2$. Show that $z^* = 5 / z$. Also indicate the positions of z, z^* and z^2 on the Argand diagram.

8. Express each of the complex numbers $z_1 = 1 - i$ and $z_2 = i$ in the form $re^{i\theta}$ where $r > 0, -\pi < \theta < \pi$, and indicate their positions in the Argand diagram. What is z_3, if $0, z_1, z_2, z_3$ form a parallelogram?

9. Express the complex number $(1 + i) / (3 + 4i)$ in the form $x + iy$ where x and y are real. Show that if it is written in the form $re^{i\theta}$ where $r > 0$ and $-\pi < \theta \le \pi$ then $r = \sqrt{2} / 5$, and determine θ correct to two decimal places.

10. Given that $z = x + iy$ (where x and y are real) and $z^2 = 3 + 4i$, find the values of $x^2 - y^2$ and xy. Hence show that $X = x^2$ satisfies the quadratic equation $X^2 - 3X - 4 = 0$ and find the two possible values for x.

11. Find the three cube roots of 2 and indicate their positions in the Argand diagram.

12. From the binomial expansion of $(e^{i\theta} + e^{-i\theta})^4$ deduce that

$$8\cos^4(\theta) = \cos(4\theta) + 4\cos(2\theta) + 3.$$

7.13. PROBLEMS

1. Solve the quadratic equation $z^2 - 5iz - 7 + i = 0$, expressing each of the two roots in the form $x + iy$ with real x and y, and hence factorize the left hand side of the equation.

2. (i) Describe how the successive coefficients in the binomial expansion of $(1 + x)^n$ can be calculated; in particular write down the expansion of $(1 + x)^5$ and hence that of $(u + v)^5$.

(ii) Derive the formula $e^{i\theta} = \cos(\theta) + i\sin(\theta)$ from the power series expansions of $e^{i\theta}$, $\cos(\theta)$ and $\sin(\theta)$. Hence show that $(e^{i\theta} + e^{-i\theta}) / 2 = \cos(\theta)$.

(iii) From the last formula, or otherwise, deduce that

$$\cos^5(\theta) = \frac{1}{16}\left(\cos(5\theta) + 5\cos(3\theta) + 10\cos(\theta)\right).$$

3. Show that if $z = re^{i\theta} = r\cos(\theta) + ir\sin(\theta)$ then

$$\text{Im}(e^z) = e^{r\cos(\theta)}\sin(r\sin(\theta)) \qquad (7.17)$$

and also give the real part of e^z. By taking the imaginary part of the expansion of e^z in ascending powers of z, deduce from (7.17) that

$$e^{r\cos(\theta)}\sin(r\sin(\theta)) = r\sin(\theta) + \frac{r^2\sin(2\theta)}{2!} + \frac{r^3\sin(3\theta)}{3!} + \dots$$
$$(7.18)$$

Regarding θ as constant, use (7.18) to evaluate the limit

$$\lim_{r \to 0^+} \frac{1}{r} e^{r\cos(\theta)} \sin\left(r\sin(\theta)\right),$$

and also verify that the same answer is obtained when l'Hôpital's rule is applied instead.

4. Show that if $(a + bi)^2 = 5 - 12i$, where a and b are real, then $a^2 - b^2 = 5$ and $a^2 + b^2 = 13$, and find the two square roots of $5 - 12i$. Hence solve the quadratic equation

$$z^2 - 3z + 1 + 3i = 0,$$

and also factorize the left side. How could it have been easily noticed that $z - i$ is a factor?

5. Let θ be an arbitrary real number, and $\phi = \theta/2$.

   (i) Verify that $1 + e^{i\theta}$ can be written as $2e^{i\phi} \cos(\phi)$.

   (ii) By treating the expression

$$C_n = 1 + n\cos(\theta) + \binom{n}{2}\cos(2\theta) + \binom{n}{3}\cos(3\theta)$$

$$+ \ldots + \cos(n\theta)$$

where n is a positive integer, as the real part of a binomial expansion, show that

$$C_n = (2\cos(\phi))^n \cos(n\phi).$$

   (iii) If $\dfrac{2\pi}{3} < \theta < \dfrac{4\pi}{3}$ what is the behavior of C_n as $n \to \infty$?

6. Verify that if x and y are real numbers with $y > 0$, and

$$p = \frac{1}{\sqrt{2}}\left[x + \left(x^2 + y^2\right)^{1/2}\right]^{1/2} \quad \text{and} \quad q = \frac{1}{\sqrt{2}}\left[-x + \left(x^2 + y^2\right)^{1/2}\right]^{1/2}$$

then $p^2 - q^2 = x$ and $2pq = y$ and therefore $p + iq$ is a square root of $x + iy$. Hence show that the two square roots of

$z = -2 + 2i\sqrt{3}$ are $\pm\left(1 + i\sqrt{3}\right)$. Verify this by writing z in the form $re^{i\theta}$ and finding $r^{1/2}e^{i\theta/2}$.

Solve the quadratic equation $z^2 + \sqrt{2}z + \left(1 - \dfrac{\sqrt{3}}{2}i\right) = 0$.

7. Verify that $z = i$ is a root of the equation

$$z^3 + (2 + i)z^2 - (1 + 4i)z - 2 + 3i = 0,$$

and hence find the other roots. Indicate the positions of the roots in the Argand diagram.

8. Evaluate the sum $\displaystyle\sum_{n=0}^{N-1} \exp\left(2\pi ipn / N\right)$ where p and N are integers (≥ 1), for the cases when $p \neq 0$ and $p = 0$. By putting $p \neq k - k'$ and $p = k + k'$ where k, k' are integers and $k \neq 0$, $k' \neq 0$, show that

$$\sum_{n=0}^{N-1} \cos\left(\frac{2\pi kn}{N}\right)\cos\left(\frac{2\pi k'n}{N}\right) = \begin{cases} \dfrac{N}{2}, & k = k', \\ 0 & \text{otherwise} \end{cases} \qquad (7.19)$$

$$\sum_{n=0}^{N-1} \sin\left(\frac{2\pi kn}{N}\right)\sin\left(\frac{2\pi k'n}{N}\right) = \begin{cases} \dfrac{N}{2}, & k = k', \\ 0 & \text{otherwise} \end{cases} \qquad (7.20)$$

$$\sum_{n=0}^{N-1} \sin\left(\frac{2\pi kn}{N}\right)\cos\left(\frac{2\pi k'n}{N}\right) = 0. \qquad (7.21)$$

What are the values of these sums when $k = k' = 0$?

9. Given that $\exp(z) = \displaystyle\sum_{n=0}^{\infty} \frac{z^n}{n!}$, where z is a complex number, write down the series expansion of $\exp(xe^{i\theta})$ where x and θ are real. Hence sum the series $\displaystyle\sum_{n=0}^{\infty} \frac{x^n}{n!}\cos(n\theta)$ and $\displaystyle\sum_{n=0}^{\infty} \frac{x^n}{n!}\sin(n\theta)$. By equating the coefficients of a suitable power of x, use your results to express $\cos(4\theta)$ and $\sin(4\theta)$ as sums of powers of $\sin(\theta)$ and $\cos(\theta)$.

CHAPTER 8

DIFFERENTIAL EQUATIONS 2

In Chapter 6, we learned different methods for solving some first order and second order differential equations. Having worked through the topic of complex numbers in Chapter 7 we can now see what happens if we are faced with complex roots in an auxiliary equation. We will also demonstrate the use of complex numbers to simplify the solution of certain types of differential equations. First though, we will show how to constrain the physical problem which a given differential equation represents, by setting initial and/ or boundary conditions. By doing this we fix the value of the arbitrary constants that appear in the general solution of a differential equation.

8.1. BOUNDARY AND INITIAL CONDITIONS

We have learned in Chapter 6 that the general solution to an nth order differential equation contains n arbitrary constants. In physical applications of the differential equation these constants are then fixed (and found) by imposing realistic conditions on the solution and / or its derivatives at specified values of the independent variable (say position x or time t). This can be done if we understand how the system that is represented by the differential equation behaves at a given time or position, e.g. at $t = 0$ or $x = 0$.

Example 8.1 Find the solution for $y(t)$ of $\ddot{y} + \omega^2 y = 0$ such that at $t = 0$, $y(0) = 0$ and $\dot{y}(0) = 0$, where $\dot{y}$ means dy/dt and $\ddot{y}$ means d^2y / dt^2 and so on. (In a physical application t might be time and the conditions would represent the given position and speed at an initial instant.)

First we need to find the *general solution*. In this particular case we have found the solution in Section 6.5 of Chapter 6, with t here replacing x.

$$y_{GS} = A\sin(\omega t) + B\cos(\omega t).$$

(This is the equation for simple harmonic motion.) Now we can apply the given *initial* conditions. These are specified at $t = 0$, so we must find the value of our general solution y_{GS} evaluated at $t = 0$.

$$y_{GS}(0) = A\sin(0) + B\cos(0) = A \times 0 + B \times 1 = B.$$

We are given $y(0) = 1$ and therefore, $B = 1$. Now we must differentiate the general solution to get $\dot{y}_{GS}$,

$$\dot{y}_{GS}(t) = A\omega\cos(\omega t) - B\omega\sin(\omega t).$$

We are given $\dot{y}(0) = 0$; therefore at $t = 0$

$$\dot{y}_{GS}(0) = A\omega \times 1 - B \times 0 = A\omega = 0,$$

hence,

$$A = 0.$$

Finally, we substitute $A = 0$ and $B = 1$ in the general solution. This is then the required solution which satisfies the differential equation and also the extra conditions imposed at $t = 0$ (it is often useful to check this). Thus

$$y = A\cos(\omega t).$$

Note that the first step is to obtain the general solution, by any of the methods that you have learned so far (or will learn in future). Once you have done this, you can then apply the initial or boundary conditions.

Key point: When the conditions imposed are values of the solution and its derivatives for the *same* value of the independent variable then these are called *initial conditions*. This is the case in examples 8.1–8.3 and in Exercise 8.1. The independent variable is often time, and the conditions are often imposed at $t = 0$, as in Example 8.1 above. When the conditions are values of the solution for *different* values of the independent variable, as in Exercise 8.2, then these are called *boundary conditions*.

Let us now look at another example of how to apply initial conditions where we have already previously obtained the general solution in the exercises in Chapter 6.

Example 8.2 Solve $y'' + 7y' + 12y = 19 + 12x$ subject to the *initial* conditions $y(0) = 1$ and $y'(0) = 2$, given the general solution $y_{GS}(x) = Ae^{-4x} + Be^{-3x} + x + 1$.

For this example we are able to start from the given general solution. This is the same as the example already solved in Section 6.7.

The general solution, y_{GS}, is the sum of the CF and the PI,

$$y_{GS}(x) = y_{CF+}y_{PI},$$

$$y_{GS}(x) = Ae^{-4x} + Be^{-3x} + x + 1.$$

We are going to need y'_{GS}, to get $y'_{GS}(0)$, and

$$y'_{GS} = -4Ae^{-4x} - 3Be^{-3x} + 1.$$

Now we can apply the initial conditions. We substitute $x = 0$ in the general solution and set $y_{GS}(0) = 1$,

$$y_{GS}(0) = (A \times 1) + (B \times 1) + 0 + 1 = A + B + 1.$$

Then,

$$y_{GS}(0) = A + B + 1 = 1 \text{ or } A = -B.$$

Now apply the condition that $y'_{GS}(0) = 2$ in a similar way, and solve for A and B:

$$y'_{GS}(0) = (-4A \times 1) - (3B \times 1) + 1,$$
$$= -4A - 3B + 1 = 2.$$

This gives

$$-4A - 3B = 1.$$

From above, $A = -B$, so $B = 1$, and $A = -1$. Finally, we substitute the values of A and B into the general solution. Thus our solution is

$$y = -e^{-4x} + e^{-3x} + x + 1.$$

We are now going to look at a complete example, where we will begin by finding the general solution of a second order, linear, inhomogeneous equation.

Example 8.3 Solve $y'' - y = e^x$ subject to the conditions $y_{GS}(0) = 0$ and $y'_{GS}(0) = 1$.

First we need to find the general solution. We do this by first finding the CF when

$$y'' - y = 0.$$

As this is linear and homogeneous with constant coefficients we try

$$y_{CF} = Ae^{\lambda x},$$

then

$$y'_{CF} = \lambda A e^{\lambda x},$$
$$y''_{CF} = \lambda^2 A e^{\lambda x}.$$

Next we substitute back into the differential equation, and form the auxiliary equation

$$\lambda^2 A e^{\lambda x} - A e^{\lambda x} = 0.$$

Then, canceling $A e^{\lambda x}$,

$$\lambda^2 - 1 = 0,$$
$$\Rightarrow \lambda = \pm 1.$$

Finally we can write the complementary function

$$y_{CF} = A e^x + B e^{-x}.$$

Now, to find a PI our first thought is to try $y_{PI} = C e^x$, but unfortunately this will not work because it is part of the CF (try it and see what happens). Recall from Section 6.9 that when this happens we multiply by x and try again.

$$y_{PI} = C x e^x,$$
$$y'_{PI} = C e^x + C x e^x,$$
$$y''_{PI} = 2 C e^x + C x e^x.$$

Substituting this into the differential equation we can solve for C, and hence find y_{PI}. As $y'' - y = e^x$, we have

$$(2 C e^x + C x e^x) - (C x e^x) = e^x,$$
$$2 C e^x = e^x.$$

Therefore we find that $C = 1/2$, and we can write the PI as follows:

$$y_{PI} = \frac{1}{2}xe^x.$$

We can write the general solution as the sum of y_{CF} and y_{PI}:

$$y_{GS} = Ae^x + Be^{-x} + \frac{1}{2}xe^x.$$

To apply the initial conditions we also need y'_{GS} so let us derive that next:

$$y'_{GS} = Ae^x - Be^{-x} + \frac{1}{2}e^x + \frac{1}{2}xe^x.$$

We are given that $y(0) = 0$ and $y'(0) = 1$. Substitute $x = 0$ in y_{GS}, and then in y'_{GS}:

$$y_{GS}(0) = (A \times 1) + (B \times 1) = A + B = 0,$$
$$y'_{GS}(0) = (A \times 1) - (B \times 1) + \frac{1}{2} \times 1 = A - B + \frac{1}{2} = 1.$$

Finally we can solve for A and B and substitute them back into the general solution, y_{GS}. From above we have $A + B = 0$ and $A - B = 1/2$. Adding these together we get

$$2A = \frac{1}{2}, \text{ leading to } A = \frac{1}{4} \text{ and } B = -A = -\frac{1}{4}.$$

Finally, substituting into the GS gives

$$y = \frac{1}{4}e^x - \frac{1}{4}e^{-x} + \frac{1}{2}xe^x.$$

Key point: Note that the initial or boundary conditions are applied to the general solution as a whole, and not just to the CF or PI separately. The first step in solving a differential equation is always to find the *general solution*.

Since the general solution to an nth order equation contains n arbitrary constants, n initial or boundary conditions are required to find the solution.

Exercise 8.1 Solve the equation $y'' - y = x$ for $y(x)$, subject to the initial conditions $y = 0$ and $y'' = 0$ at $x = 0$.

Exercise 8.2 Solve the equation $\ddot{z} = -g$ (where g is a constant) subject to the following boundary conditions:

(i) $z = 5$ at $t = 1$ and $z = 4$ at $t = 2$,

(ii) $\dot{z} = 0$ at $t = 2$ and $z = 0$ at $t = 3$.

Exercise 8.3 Solve the following for $y(x)$ subject to the specified conditions. (See Exercise 6.14).

(i) $y'' - y = x$ subject to $y(0) = 1$, $y'(0) = 2$,

(ii) $y'' + 4y' + 3y = e^{-x}$ subject to $y(0) = 0$, $y'(0) = 2$,

(iii) $y'' + 2y' + y = xe^{-x}$ subject to $y(0) = 0$, $y(1) = 1/e$.

In the examples that we have looked at so far, we have applied initial or boundary conditions to second order differential equations. The method we have discussed here however applies to any type of differential equation, and is the method to follow in general. In the next example, we shall see how to apply the conditions to a first order differential equation, where we will begin by using the integrating factor method to find the general solution.

Example 8.4 Solve $y' + \dfrac{2}{x}y = 1$ subject to $y = 0$ at $x = 1$.

The equation is first order but not separable, and does not have constant coefficients, so we will use the integrating factor method here to find the general solution. We begin by writing down the integrating factor, $R(x)$.

$$R(x) = \exp\left(\int \frac{2}{x}dx\right),$$

$$= e^{2\ln(x)} = e^{\ln(x^2)} = x^2.$$

We then multiply the differential equation by the integrating factor, $R(x)$,

$$x^2 y' + x^2 \frac{2}{x}y = x^2,$$

and recognize that the left hand side is equivalent to the derivative of the product $(R(x) \times y)$ (see Section 6.11):

$$(x^2 y)' = x^2.$$

We now integrate both sides to obtain our general solution, $y_{GS}(x)$. We have

$$x^2 y = \frac{x^3}{3} + C,$$

so that the general solution, which contains one arbitrary constant, C, is

$$y = \frac{x}{3} + \frac{C}{x^2}.$$

We can now impose the boundary condition: at $x = 1$, $y_{GS}(1) = 0$, so

$$y_{GS}(1) = \frac{1}{3} + \frac{C}{1^2} = 0$$

$$\text{and } C = -\frac{1}{3}.$$

The solution to this differential equation for the given boundary condition is therefore

$$y = \frac{x}{3} - \frac{1}{3x^2} = \frac{x^3 - 1}{3x^2}.$$

▌ Exercise 8.4 Solve $y' - 2xy = x$ subject to $y = 0$ at $x = 0$.

We are often interested in how physical systems behave for very large values of the independent variable. In such circumstances we might set the boundary conditions at $\pm\infty$, where our general solution approaches a given value(its asymptotic value). Let us look at an example of how this is done.

Example 8.5 Find the solution of $y' - \frac{y}{x} = \frac{1}{x^2}$ such that $y(x) \to 0$ as $x \to \infty$.

The equation is first order but not separable, and does not have constant coefficients, so we will again use the integrating factor method to solve the equation. We begin by writing down the integrating factor, $R(x)$.

$$R(x) = \exp\left(\int \frac{-1}{x}dx\right),$$

$$= e^{-\ln(x)} = e^{\ln(x^{-1})} = \frac{1}{x}.$$

We then multiply through by the integrating factor

$$\frac{1}{x}y' - \frac{y}{x^2} = \frac{1}{x^3},$$

which can be written as

$$\left(\frac{y}{x}\right)' = \frac{1}{x^3}.$$

Integrating both sides and rearranging gives

$$y_{GS} = -\frac{1}{2x} + Cx.$$

In order that $y \to 0$ as $x \to \infty$ we must have $C = 0$, so the required solution is

$$y = -\frac{1}{2x}.$$

Let us try one more example of this before we move on.

Example 8.6 Find the general solution of $\ddot{y} - y = 0$ such that $y \to 0$ as $t \to +\infty$ and hence find the solution satisfying $y(0) = 1$ that remains bounded (i.e. does not take arbitrarily large values) for all $t \geq 0$.

The equation is linear, homogeneous and has constant coefficients so we try $y = Ae^{\lambda t}$. Then

$$\dot{y} = \lambda A e^{\lambda t},$$
$$\ddot{y} = \lambda^2 A e^{\lambda t}.$$

Next we substitute into the differential equation and form the auxiliary equation

$$\lambda^2 A e^{\lambda t} - A e^{\lambda t} = 0.$$

Canceling $Ae^{\lambda t}$ we have

$$\lambda^2 - 1 = 0, \text{ so } \lambda = \pm 1.$$

Hence we can write down the general solution, which has two arbitrary constants.

$$y_{GS} = Ae^t + Be^{-t}.$$

But we have been asked to write down the general solution such that $y \to 0$ as $t \to +\infty$. For y_{CS} above, we can see that $y \to \pm\infty$ as $t \to +\infty$ unless $A = 0$. If $A = 0$ the condition is satisfied, as $e^{-t} \to 0$ as $t \to +\infty$. Thus,

$$y_{CS} = Be^{-t}.$$

Now we can apply the boundary condition $y(0) = 1$. At $t = 0$, $y(0) = Be^0 = B = 1$. Hence

$$y = e^{-t}.$$

Exercise 8.5 Find the solution of $y'' + 2y' - 3y = e^{-x}$ satisfying $y(0) = 0$ that tends to zero as $x \to +\infty$.

8.2. AUXILIARY EQUATIONS WITH COMPLEX ROOTS

The second part of this chapter is going to address solving differential equations which involve complex numbers in one way or another. Here you will need to use the knowledge gained from working through Chapter 7. We are going to begin by continuing the work started in Section 6.7 on homogeneous linear equations, and have a look at what happens if the roots of the auxiliary equation are complex numbers.

Example 8.7 Find the general solution of $y'' + y = 0$.

This is a second order, linear equation with constant coefficients; hence we try

$$y = Ae^{\lambda x},$$
$$y' = \lambda Ae^{\lambda x}, \text{ and } y'' = \lambda^2 Ae^{\lambda x}.$$

Substituting into the differential equation, we form the auxiliary equation

$$\lambda^2 A e^{\lambda x} + A e^{\lambda x} = 0,$$
$$\lambda^2 + 1 = 0,$$
$$\lambda^2 = -1 \text{ thus } \lambda = \pm i.$$

The general solution contains two arbitrary constants

$$y_{GS} = A e^{ix} + B e^{-ix}.$$

Recall from Euler's equation (see Section 7.6) that

$$e^{i\theta} = \cos(\theta) + i \sin(\theta) \text{ and,}$$
$$e^{-i\theta} = \cos(\theta) - i \sin(\theta).$$

Hence we can rewrite our general solution in terms of sin and cos as follows:

$$y_{GS} = A(\cos(x) + i \sin(x)) + B(\cos(x) - i \sin(x)),$$
$$= (A + B) \cos(x) + i(A - B) \sin(x),$$
$$= a \cos(x) + b \sin(x).$$

Here we have written $a = A + B$ and $b = i(A - B)$. If y is required to be real (which it often is) then a and b will be real. Solving for A and B we get

$$A = \frac{1}{2}(a - ib), \ B = \frac{1}{2}(a + ib).$$

There is no need to go through all of the details every time we use this technique. The following example is similar but shows you how to set out the solution correctly and succinctly.

Example 8.8 Find the general solution of $y'' + 4y = 0$.

The equation is linear, homogeneous and with constant coefficients so we try $y = A e^{\lambda x}$ (and hence $y'' = \lambda^2 A e^{\lambda x}$ as in previous examples). Substituting in the results for y and y'', then canceling factors of $A e^{\lambda x}$, we have the auxiliary equation

$$\lambda^2 A e^{\lambda x} + 4 A e^{\lambda x} = 0,$$
$$\Rightarrow \lambda^2 = -4,$$
$$\text{So } \lambda = \pm 2i.$$

We can write the general solution in explicitly real form with a and b as the arbitrary real constants, as in Example 8.7 above.

$$y_{GS} = A e^{2ix} + B e^{-2ix},$$
$$y_{GS} = a \cos(2x) + b \sin(2x).$$

Note that in the examples above we have found that the complex roots of our auxiliary equations are both of the form $\pm \beta i$. This has allowed us to write the general solution y_{GS} directly as a real function of cos and sin, instead of a function of complex exponentials. In the next example we will see what happens when the complex roots are of the form $a \pm \beta i$.

Example 8.9 Find the general solution of $y'' + 8y' + 20y = 0$.

The equation is linear, homogeneous and with constant coefficients so we try $y = A e^{\lambda x}$ (and hence $y' = \lambda A e^{\lambda x}$, and $y'' = \lambda^2 A e^{\lambda x}$ as in previous examples).

$$\lambda^2 A e^{\lambda x} + 8\lambda A e^{\lambda x} + 20 A e^{\lambda x} = 0,$$
$$\lambda^2 + 8\lambda + 20 = 0,$$
$$\lambda = \left(-8 \pm \sqrt{64 - 80}\right)/2 = -4 \pm 2i.$$

We can then write the general solution, with A and B as two arbitrary constants

$$y_{GS} = A e^{(-4+2i)x} + B e^{(-4-2i)x}.$$

This is the same as

$$y_{GS} = A e^{-4x} e^{2ix} + B e^{-4x} e^{-2ix},$$
$$= e^{-4x}[A e^{2ix} + B e^{-2ix}].$$

We can again write the complex exponential terms in real form using Euler's equation, with arbitrary constants a and b

$$y_{GS} = e^{-4x}[a \cos(2x) + b \sin(2x)].$$

Key point: If the auxiliary equation has real coefficients but complex roots, the roots will be of the form $\lambda = a \pm i\beta$ with a and β real. We can then directly write the general solution of the form $y_{GS} = Ae^{(a+\beta i)x} + Be^{(a-\beta i)x}$ in terms of real functions as $y_{GS} = e^{ax}[a \cos(\beta x) + b \sin(\beta x)]$. Here a and b are related to A and B as $a = A + B$ and $b = i(A - B)$.

Here is an example from electric circuit theory.

Example 8.10 The charge $q(t)$ on the capacitor in a circuit with a resistor, capacitor and inductor in series is governed by

$$L\frac{d^2q}{dt^2} + R\frac{dq}{dt} + q/C = 0. \tag{8.1}$$

where R, C and L are positive constants and $R^2 < 4L/C$. Find the charge as a function of time.

For a trial solution $q = Ae^{\lambda t}$ the auxilliary equation is

$$L\lambda^2 + R\lambda + 1/C = 0.$$

This is a quadratic equation that can be solved to give

$$\lambda = -\frac{R}{2L} \pm \frac{1}{2L}\left(R^2 - \frac{4L}{C}\right)^{1/2}. \tag{8.2}$$

If $(R^2 - 4L/C) < 0$, then the roots are complex numbers, and we can write this as

$$\lambda = -\frac{R}{2L} \pm i\omega_c, \qquad \text{where } \omega_c = \frac{1}{2L}\left(\frac{4L}{C} - R^2\right)^{1/2}$$

$$= \left(\frac{1}{LC} - \frac{R^2}{4L^2}\right)^{1/2}. \tag{8.3}$$

Then the solutions have the form $e^{(-R/2L\pm i\omega_c)t}$, and the general solution is

$$q(t) = e^{-Rt/2L}\left(Fe^{i\omega_c t} + Ge^{-i\omega_c t}\right).$$

Or, using the relation that $e^{i\omega_c t} = \cos(\omega_c t) + i\sin(\omega_c t)$, we can write the equivalent form

$$q(t) = e^{-Rt/2L}\left(A\cos(\omega_c t) + B\sin(\omega_c t)\right), \qquad (8.4)$$

where the constants A and B (or F and G) are determined by the initial values of the charge and the current in the circuit.

Exercise 8.6 Show that the general solution of $y'' + 2y' + 5y = 0$ is $y = e^{-x}(a\cos(2x) + b\sin(2x))$.

Exercise 8.7 Find the general solution of $y'' + y' + \dfrac{5}{2}y = 0$.

Exercise 8.8 Find the general solution of $y''' - 2y'' + 3y' - 2y = 0$ given that one solution is $y = e^x$.

8.3. EQUATIONS WITH COMPLEX COEFFICIENTS

We can use the same approach to solve a linear, homogeneous equation with constant complex coefficients. In this case the solution will turn out to be complex. When we are dealing with a complex function we use z instead of y.

Example 8.11 Find the general solution of $z'' - iz = 0$.

The equation is linear, homogeneous and with constant coefficients so in the usual way we try $z = Ae^{\lambda x}$ (and hence $z' = \lambda Ae^{\lambda x}$, and $z'' = \lambda^2 Ae^{\lambda x}$ as in previous examples).

$$\lambda^2 Ae^{\lambda x} - iAe^{\lambda x} = 0,$$

$$\text{or } \lambda^2 - i = 0,$$

$$\text{so } \lambda = \pm\sqrt{i}.$$

Recall that the square root of i is another complex number, and was found in Example 7.7:

$$\sqrt{i} = \pm \frac{1+i}{\sqrt{2}};$$

so $\quad z_{GS} = Ae^{(1+i)x/\sqrt{2}} + Be^{-(1+i)x/\sqrt{2}}.$

Here the constants A and B are complex. This is as far as we need to go since we do not require that z be real, so we are not required to write this in terms of real functions. The complex constants A and B would be fixed by the boundary conditions in an application.

Exercise 8.9 Find the general solution of $z'' + iz' + 2z = 0.$

8.4. COMPLEX INHOMOGENEOUS TERM

Examples of the type shown in Section 8.3 do not commonly occur in physical sciences, and so we will not take this any further. However, the following type of differential equation is very common—that is, a differential equation with real coefficients and a complex inhomogeneous term. Let us look at some examples of how to solve these types of equation.

Example 8.12 Find a particular integral of $\ddot{z} + z = 3e^{2it}$.

Since the right hand side of this second order, inhomogeneous complex equation contains an exponential we will try a particular integral solution of a similar form $z_{PI} = Ae^{2it}$ (see Section 6.9). It then follows that

$$\dot{z} = 2iAe^{2it},$$
$$\ddot{z} = (2i)^2 Ae^{2it},$$

where the coefficient A is to be determined. We substitute these results into the differential equation to obtain

$$(2i)^2 A e^{2it} + A e^{2it} = 3 e^{2it}.$$

Next we divide through by e^{2it} and replace i^2 by -1.

$$-4A + A = 3,$$

$$\text{hence } -3A = 3, \text{ and } A = -1.$$

Finally we can write the PI as

$$z_{PI} = -e^{2it}.$$

Example 8.13 Find a particular integral of $z'' + 2z' + 3z = 6 e^{3ix}$.

Again, the right hand side of this second order, inhomogeneous complex equation contains an exponential. We will therefore try a particular integral solution of a similar form for $z_{PI} = A e^{3ix}$.

$$z' = 3i A e^{3ix} \text{ and } z'' = (3i)^2 A e^{3ix}.$$

To determine the coefficient, A, we substitute into the differential equation,

$$(3i)^2 A e^{3ix} + 6i A e^{3ix} + 3A e^{3ix} = 6 e^{3ix}.$$

Next we divide through by e^{3ix} and replace i^2 by -1.

$$-9A + 6iA + 3A = 6,$$

$$-6A + 6iA = 6,$$

$$\text{hence } A = \frac{6}{-6 + 6i} = \frac{1}{-1 + i}.$$

Before we write down the solution for the PI, we can tidy up A by multiplying through by the complex conjugate of the denominator top and bottom (see Section 7.2)

$$A = \frac{(-1 - i)}{(-1 - i)(-1 + i)}, \quad \text{or} \quad A = -\frac{1}{2}(1 + i).$$

Finally we can write the PI as

$$z_{PI} = -\frac{1}{2}(1 + i) e^{3ix}.$$

Exercise 8.10 Find a PI of $\ddot{z} + 2\dot{z} - 3z = 2e^{tt}$.

Now let us consider an example where we are being asked to find not just a particular integral, but the general solution.

Example 8.14 Find the general solution of $\ddot{z} + 2\dot{z} + 3z = 6e^{3it}$.

This equation is a second order, linear, inhomogeneous equation with constant coefficients. Therefore we know that the general solution will be a combination of the complementary function, CF, and a particular integral, PI. First, let us solve the homogeneous equation for the CF. The homogeneous part is

$$\ddot{z} + 2\dot{z} + 3z = 0$$

We will try $z_{CF} = Ae^{\lambda t}$ hence $\dot{z}_{CF} = \lambda Ae^{\lambda t}$, and $\ddot{z}_{CF} = \lambda^2 Ae^{\lambda t}$. We substitute into the differential equations and divide through by $Ae^{\lambda t}$ to obtain the auxiliary equation

$$\lambda^2 + 2\lambda + 3 = 0,$$

$$\lambda = \frac{-2 \pm \sqrt{4 - 12}}{2},$$

$$= -1 \pm \frac{\sqrt{-8}}{2} = -1 \pm \frac{\sqrt{4}\sqrt{2}}{2}i = -1 \pm i\sqrt{2}.$$

Then,

$$z_{CF} = Ae^{(-1+i\sqrt{2})t} + Be^{(-1-i\sqrt{2})t},$$

$$z_{CF} = e^{-t}\left(Ae^{i\sqrt{2}t} + Be^{-i\sqrt{2}t}\right).$$

The general solution is

$$z_{GS} = z_{CF} + z_{PI}.$$

We found z_{PI} in example 8.13 above, so

$$z_{GS} = e^{-t}\left(Ae^{i\sqrt{2}t} + Be^{-i\sqrt{2}t}\right) - \frac{1}{2}(1+i)e^{3it}.$$

Note that since the differential equation is complex, so is the solution (thus A and B are two arbitrary complex numbers).

Exercise 8.11 Find the general solution of $z'' + 2z' - 3z = 20e^{ix}$.

We now come to one of the most useful applications of complex numbers to solving differential equations. This is where we use our knowledge of complex numbers to solve *real equations* with *real coefficients* and a *harmonic driving term* on the right hand side, i.e. equations of the form

$$\ddot{y} + a\dot{y} + by = c\cos(\omega t) + d\sin(\omega t). \tag{8.5}$$

Example 8.15 Find a PI for $y'' + 2y' + 3y = 6\cos(3x)$.

Compare this with the complex equation that we solved in Examples 8.13 and 8.14:

$$z'' + 2z' + 3z = 6e^{3ix}$$

If we use Euler's equation, $e^{i\theta} = \cos\theta + i\sin\theta$, we can rewrite the inhomogeneous term (i.e. the RHS) as

$$z'' + 2z' + 3z = 6\cos(3x) + 6i\sin(3x).$$

Now we can see that the equation that we have been asked to find a PI for (in y) is just the real part of the complex equation above (in z).

So the solution to the equation, $y'' + 2y' + 3y = 6\cos(3x)$, is just the real part of the solution to $z'' + 2z' + 3z = 6\cos(3x) + 6i\sin(3x)$.

Similarly, a PI solution for the real equation in y is just the real part of the PI solution for the complex equation in z, found in Example 8.13:

$$y_{PI} = \text{Re}\left\{-\frac{1}{2}(1+i)e^{3ix}\right\},$$

or, using $e^{i\theta} = \cos(\theta) + i\sin(\theta)$, we have

$$y_{PI} = \mathrm{Re}\left\{-\frac{1}{2}(1+i)\big(\cos(3x)+i\sin(3x)\big)\right\},$$

$$= \mathrm{Re}\left\{-\frac{1}{2}\big(\cos(3x)+i\sin(3x)+i\cos(3x)+i^2\sin(3x)\big)\right\}.$$

But $i^2 = -1$, so

$$y_{PI} = -\frac{1}{2}\cos(3x)+\frac{1}{2}\sin(3x).$$

This solution can be checked by substituting back into the equation that we have been asked to solve.

Similarly, if we wanted to solve the equation in Example 8.15 with $6\sin(3x)$ instead of $6\cos(3x)$ as the inhomogeneous term, we should take the *imaginary part* of a PI to the complex equation $z'' + 2z' + 3z = 6\cos(3x) + 6i\sin(3x)$. Let us look at another example.

Example 8.16 Find the general solution of

$$y'' + 8y' + 20y = 8\cos(4x)$$

To find the general solution we need to find the CF and a PI for this equation, and add them together. We have previously found the CF for this equation in Example 8.9, so we can use that directly here:

$$y_{CF} = e^{-4x}(a\,\cos(2x) + b\,\sin(2x)).$$

Next we need to find a PI. To do this we are going to first of all write down a complex equation whose PI has a real part which is equal to $8\cos(4x)$. We do this because it will be quicker to solve the complex equation and take the real part than it is to solve for the PI in real form (i.e. in terms of cos and sin). We write the following complex equation:

$$z'' + 8z' + 20z = 8e^{4ix}.$$

Now we try a PI of the form $z_{PI} = Ae^{4ix}$.

$$z = Ae^{4ix},$$
$$z' = 4iAe^{4ix}, \text{ and } z'' = -16Ae^{4ix}.$$

We substitute this into the complex differential equation, and divide through by e^{4ix}.

$$-16A + 32iA + 20A = 8,$$

$$A = \frac{2}{(1+8i)}.$$

Then, multiplying the top and bottom by the complex conjugate of the denominator, $(1 - 8i)$, we have

$$A = \frac{2}{65}(1 - 8i).$$

$$\text{Hence } z_{PI} = \frac{2}{65}(1 - 8i)e^{4ix}.$$

To find the real PI we take the real part of the complex PI, $y_{PI} = \text{Re}(z_{PI})$:

$$y_{PI} = \text{Re}\left\{\frac{2}{65}(1 - 8i)e^{4ix}\right\},$$

$$\text{hence } y_{PI} = \frac{2}{65}\text{Re}\left\{(1 - 8i)(\cos(4x) + i\sin(4x))\right\}$$

$$= \frac{2}{65}\left(\cos(4x) - 8i^2\sin(4x)\right)$$

$$= \frac{2}{65}\left(\cos(4x) + 8\sin(4x)\right),$$

$$\text{hence } y_{PI} = \frac{2}{65}\cos(4x) + \frac{16}{65}\sin(4x).$$

Finally we can write the general solution as the sum of the CF and PI:

$$y_{GS} = CF + PI$$

$$= e^{-4x}\left(a\cos(2x)+b\sin(2x)\right)+\frac{2}{65}\cos(4x)+\frac{16}{65}\sin(4x).$$

Note that the solution contains both $\cos(4x)$ and $\sin(4x)$ terms even though the inhomogeneous term in the differential equation only has $\cos(4x)$. Physically this means that the response of the system is out of phase with the driving force, $8\cos(4x)$. As discussed above we could find a PI by guessing $y_{PI} = c\cos(4x) + d\sin(4x)$, with c and d to be found, but this is usually much more long-winded and the chances of making an error are higher.

Exercise 8.12 Use the method above to find a PI for each of the following equations:

(i) $y'' + 2y' + 9y = 2\cos(3x)$;

(ii) $y'' + 7y' + 12y = -150\sin(3x)$.

Exercise 8.13 A series LCR circuit driven with voltage $V = V_0\cos(\omega t)$ satisfies the differential equation

$$L\frac{d^2q}{dt^2}+R\frac{dq}{dt}+q/C=V_0\cos(\omega t)$$

where L, R and C are positive constants with $(R^2 - 4L/C) < 0$. If the steady state response (i.e. the PI) is $q(t) = q_0\cos(\omega t - \delta)$ show that

$$q_0 = \frac{V_0}{\omega\sqrt{(1/\omega C - \omega L)^2 + R^2}}$$

and the phase difference δ is given by

$$\delta = \tan^{-1}\left(\frac{R}{1/\omega C - \omega L}\right).$$

Show that the steady state current dq/dt has its maximum amplitude when $\omega^2 = 1/LC$. (This value of ω is called the resonant frequency of the circuit.)

Exercise 8.14 Find the general solution (CF + PI) of

$$y'' + 2y' + 10y = 13\cos(2x).$$

8.5. BOUNDARY (OR INITIAL) CONDITIONS

The following example puts together all we have learned in this chapter. We can now use our knowledge of complex numbers to solve complex differential equations, and we can apply boundary or initial conditions to those solutions in exactly the same way that we did at the start of the chapter.

Example 8.17 Find the solution of $\ddot{y} + 2\dot{y} + 3y = 6\cos(3t)$ subject to $y(0) = 0$, $\dot{y}(0) = 1$.

We will begin by finding the general solution. First we find the CF, by guessing $y_{CF} = Ae^{\lambda t}$, and hence we can form the auxiliary equation and solve for λ,

$$\lambda^2 + 2\lambda + 3 = 0,$$

$$\lambda = \frac{-2 \pm \sqrt{(4-12)}}{2},$$

hence $\lambda = -1 \pm \sqrt{2}i$

So $\alpha = -1$, $\beta = \sqrt{2}$. Having solved the auxiliary equation we can write down y_{CF} directly in real form as

$$y_{CF} = e^{-t}\left(a\cos\left(\sqrt{2}t\right) + b\sin\left(\sqrt{2}t\right)\right).$$

We can use the result of example 8.15,

$$y_{PI} = -\frac{1}{2}\cos(3t) + \frac{1}{2}\sin(3t).$$

We add the CF and the PI to form the general solution

$$y_{CS} = e^{-t}\left(a\cos\left(\sqrt{2}t\right) + b\sin\left(\sqrt{2}t\right)\right) - \frac{1}{2}\cos(3t) + \frac{1}{2}\sin(3t).$$

Then, differentiating, we can also find $\dot{y}_{CS}$,

$$\dot{y}_{CS} = -e^{-t}\left(a\cos\left(\sqrt{2}t\right) + b\sin\left(\sqrt{2}t\right)\right)$$

$$+e^{-t}\left(-\sqrt{2}a\sin\left(\sqrt{2}t\right) + \sqrt{2}b\cos\left(\sqrt{2}t\right)\right)$$

$$+\frac{3}{2}\sin(3t) + \frac{3}{2}\cos(3t).$$

Now we impose the initial conditions. At $t = 0$,

$$y_{CS}(0) = e^{0}\left(a\cos(0) + b\sin(0)\right) - \frac{1}{2}\cos(0) + \frac{1}{2}\sin(0) = a - \frac{1}{2} = 0.$$

And so $a = 1/2$. We also have $\dot{y}_{CS}(0) = 1$, so

$$\dot{y}_{CS}(0) \quad = -e^{0}\left(a\cos(0) + b\sin(0)\right) + e^{0}\left(-\sqrt{2}a\sin(0) + \sqrt{2}b\cos(0)\right)$$

$$+\frac{3}{2}\sin(0) + \frac{3}{2}\cos(0) = 1,$$

$$\Rightarrow -a + \sqrt{2}b + \frac{3}{2} = 1.$$

And using $a = 1/2$ we find $b = 0$. Therefore, the solution is

$$y = \frac{1}{2}e^{-t}\cos\left(\sqrt{2}t\right) - \frac{1}{2}\cos(3t) + \frac{1}{2}\sin(3t).$$

Exercise 8.15 Find the solution of $y'' + 2y' + 10y = 13\cos(2x)$ subject to the initial conditions $y(0) = 0$, $y'(0) = 2$.

8.6.   SYSTEMS OF FIRST ORDER EQUATIONS

Just as we can have simultaneous linear equations, we can have simultaneous differential equations. The simplest case is a set of homogeneous first order linear equations with constant coefficients. The following example looks at two such equations and illustrates how these can sometimes be solved. (The example can also be solved more simply using addition and subtraction to derive two independent equations, but we use it here to illustrate the more general method.)

Example 8.18 Find the general solution of the system for $u(t)$, $v(t)$:

$$\dot{u} + v = 0,$$
$$\dot{v} - u = 0.$$

The system is linear with constant coefficients so we guess exponential forms for both u and v with $A \neq 0$, $B \neq 0$.

$$\text{Try} \quad u = Ae^{\lambda t} \quad \text{and} \quad v = Be^{\lambda t},$$
$$\text{then} \quad \dot{u} = \lambda Ae^{\lambda t} \quad \text{and} \quad \dot{v} = \lambda Be^{\lambda t}.$$

Substitute and divide through by $e^{\lambda t}$. This is possible since we have chosen u and v to have the same exponential dependence.

$$\lambda A + B = 0,$$
$$\text{and } \lambda B - A = 0.$$

Eliminate A and B to get the auxiliary equation for λ, and solve for λ.

$$B = A/\lambda,$$
$$\lambda^2 A + A = 0.$$

Since $A \neq 0$

$$\lambda^2 + 1 = 0,$$
$$\text{so } \lambda = \pm i.$$

The general solution is a sum of the two possible solutions with arbitrary coefficients.

$$u_{GS} = Ce^{it} + De^{-it}.$$

C and D are arbitrary complex constants. The general solution can be rewritten in explicitly real form, with $\alpha = 0$ and $\beta = 1$ (see Example 8.8)

$$u_{GS} = a\cos(t) + b\sin(t).$$

a and b are arbitrary real constants. The easiest way to obtain v is to return to the original system

$$v_{GS} = -\dot{u},$$

$$v_{GS} = a\sin(t) - b\cos(t).$$

Notice how this works: we add together the possible solutions for one function (u here), choosing the constants to make the solution real, and the general solution for the other function (v here) then follows.

Exercise 8.16 Find the general solution of the following system of first order equations.

$$\dot{u} - v - u = 0,$$

$$\dot{v} + 4v + 6u = 0.$$

What is the solution satisfying $u(0) = 1$, $v(0) = 0$?

8.7. COMPLEX IMPEDANCE

We have seen that a PI of a linear differential equation with constant coefficients subject to a harmonic forcing term (the right hand side) is a harmonic at the same frequency as the forcing term but a different amplitude and phase. Since the equation is linear the amplitude of the force and the response must be linearly related. We represent the driving term as the real part of a complex exponential and we write the constant of proportionality as Z, called the impedance of the system. For a resistive mechanical system the impedance relates the amplitude of the steady state velocity v to the amplitude of the applied harmonic force by $v = F/Z$; for electrical circuits the impedance relates the current and the applied voltage: $I = V/Z$. Note that this gives only the steady state behavior and only for a harmonic driving term. Note also that to get the physically measured real quantities we take the real parts: for example, the applied voltage is $\text{Re}\{V\,e^{i\omega t}\}$ and the response is $\text{Re}\{(V/Z)e^{i\omega t}\}$. The impedance Z is constant in time but a function of the frequency ω (and the parameters of the system). Once we know the impedance Z, solving for the steady state of a resistive system subject to a harmonic force becomes an exercise in complex algebra.

Exercise 8.17 The complex impedance Z of a circuit (an inductor, capacitor and resistor in series) is given as

$$Z = R + \frac{1}{i\omega C} + i\omega L,$$

where R, C, and L are real positive constants. Show that Z can be written in the form

$$Z = \sqrt{R^2 + \left(\omega L - \frac{1}{\omega C}\right)^2}\;e^{i\delta},$$

where

$$\delta = \tan^{-1}\left(\frac{\omega L - \dfrac{1}{\omega C}}{R}\right).$$

Hence show that an applied voltage $\text{Re}(V_0 e^{i\omega t})$, with V_0 real, gives rise to a current

$$I = \frac{V_0}{\sqrt{R^2 + \left(\omega L - \dfrac{1}{\omega C}\right)^2}}\cos\left(\omega t - \delta\right).$$

8.8. SOME EXAMPLES FROM PHYSICS

We have studied numerous examples of solutions of equations for some abstract functions $y(x)$ or $y(t)$. In real physical problems the variables represent physical quantities and the notation will reflect this. The major difference between the problems studied so far and the examples in the following set is just the notation. The procedures are the same as those you have applied in this chapter and in Chapter 6.

Exercise 8.18 By seeking a solution of the form $x = \text{constant} \times e^{mt}$ (and not otherwise), find the general solutions of the equation

$$\ddot{x} + r\dot{x} + \lambda x = 0, \quad r > 0$$

(where $\dot{x} = dx / dt$) in the cases

(i) $\lambda = r^2$

(ii) $\lambda = r^2/4$

(iii) $\lambda = -r^2$.

Exercise 8.19 By seeking a solution of the form $y = \text{constant} \times e^{mx}$ (for suitable m), find a particular integral of

$$y'' + 2y' + 4y = 3 \cos(\omega x)$$

and hence find the general solution. At what driving frequency, ω, is the amplitude of the forced oscillation, y_{PI}, a maximum? (The amplitude of an oscillation $a \cos(\lambda x) + b \sin(\lambda x)$, where a, b and λ are real constants, is $(a^2 + b^2)^{1/2}$. Equivalently, the amplitude of an oscillation $\text{Re}(Ae^{i\lambda x})$ is $|A|$.)

Exercise 8.20 The equation

$$\frac{dI}{ds} + \kappa I = j$$

determines the specific intensity $I(s)$ of radiation propagating in a medium. (The positive functions $\kappa(s)$ and $j(s)$ characterize the absorption and emission properties of the medium and s measures distance along the ray path.) Find the solution $I(s)$ satisfying $I(0) = 0$ in the cases that (i) j, κ are constants, (ii) $j = as$, $\kappa = \beta s$, with a, β constants. Show that in both cases $I(s)$ can never exceed $j(s)/\kappa(s)$ for $s > 0$.

Exercise 8.21 The angular velocity Ω of a rapidly rotating shaft can be shown to obey an equation of the form

$$\frac{d^3\Omega}{dt^3} - \frac{d^2\Omega}{dt^2} + 2\Omega = 10$$

under certain conditions (t is time).

1. Find a particular integral of this equation, and hence the general solution. (Hint: $m^3 - m^2 + 2 = (m + 1)(m^2 - 2m + 2)$.)

2. Show that, except for certain special initial conditions, the angular velocity will vary periodically, with an amplitude increasing with time. (This type of unstable motion is called "hunting.")

Exercise 8.22 A "critically damped" oscillator obeys the equation

$$\frac{d^2y}{dt^2} + r\frac{dy}{dt} + \frac{r^2}{4}y = 0,$$

where y is the displacement, t is the time and r is a non-zero constant. Find the solution obeying the initial conditions $y = 0$ at $t = 0$ and $dy/dt = v > 0$ at $t = 0$. Show that the displacement is a maximum at a time t_{max} which is independent of v and show that the maximum displacement is $y_{max} = 2v/re$ where e is the base of natural logarithms.

8.9. EXTENSION: GREEN FUNCTIONS

So far we have used guesswork to obtain a particular integral of a differential equation. In fact, if we know two independent solutions of a homogeneous equation, then we can construct a formula for a particular integral. In this section, we will outline how this is done, leaving the details to more advanced texts.

Suppose we seek a PI of the equation $y'' + y = f(x)$ where $f(x)$ is given. We know that $\cos(x)$ and $\sin(x)$ are two linearly independent solutions of the homogeneous equation. (By linearly independent we mean that there are no constants a and b such that $a\cos(x) + b\sin(x) = 0$.) We construct the function

$$G(x,x') = \begin{cases} (\cos(x)\sin(x') - \sin(x)\cos(x'))/W(x') & x' \le x \\ 0 & x' > x \end{cases}$$

where $W(x) = \sin(x)(\cos(x))' - (\sin(x))'\cos(x) = -1$. We now assert that a particular integral of $y'' + y = f(x)$ is

$$y_{PI}(x) = \int G(x, x') f(x') dx'$$
$$= \int_{x_0}^{x} \left[(\cos(x)\sin(x') - \sin(x)\cos(x'))/W(x')\right] f(x') dx'$$

with $x_0 < x$ an arbitrary constant. It is an exercise in differentiating an integral with respect to a parameter (x here) to verify this result (see section 10.8). Different choices of the lower limit x_0 and different linear combinations of $\cos(x)$ and $\sin(x)$ as the independent solutions correspond to different PIs satisfying different boundary conditions.

The function $G(x, x')$ is called a Green's function (sometimes a Green function). If we think of the differential equation as a differential operator on y giving us f, then the Green's function is the inverse integral operator on f giving us back y.

In general, if we have two independent solutions $u(x)$ and $v(x)$ of the equation $y'' + a(x)y' + b(x)y = 0$, then a PI of the equation $y'' + a(x)y' + b(x)y = f(x)$ is

$$y(x) = \int_{x_0}^{x} G(x, x') f(x') dx'$$

Where

$$G(x, x') = \begin{cases} (u(x)v(x') - u(x')v(x))/(W(x') & x' \le x \\ 0 & x' > x \end{cases}$$

and where $W(x) = u'(x)v(x) - u(x)v'(x)$. This again can be verified by back-substitution. In practice, there are only a limited number of choices of the coefficients $a(x)$ and $b(x)$ for which we can find explicit solutions of the homogeneous equation. Nevertheless, Green functions have found widespread applications in theoretical physics.

Revision Notes

After completing this chapter you should be able to

- Apply boundary and initial conditions to the general solution of a differential equation

- Solve differential equations involving complex roots in the auxiliary equation, and write the solution in explicitly real terms

- Solve differential equations with complex coefficients

- Solve differential equations with a complex inhomogeneous term, and use this to simplify solving real equations with harmonic inhomogeneous terms (i.e. a function of cos and/or sin)

- Solve a simple system of first order differential equations

8.10. EXERCISES

1. Solve the differential equation

$$\frac{dy}{dx} - 3x^2 y = 0$$

subject to the condition $y(0) = 1$.

2. Solve the differential equation

$$\frac{dy}{dx} + 3x^2 y^2 = 0$$

subject to the condition $y(0) = 1$.

3. Solve the differential equation

$$\frac{dy}{dx} + xy^2 = 0$$

subject to the condition $y(0) = 1$.

4. Integrate the differential equation $\frac{dy}{dx} = x^2 \sec(y)$.

5. Solve the differential equation

$$e^y \frac{dy}{dx} = x^2$$

by separation of variables for y as a function of x.

6. Find the solution of $y' + \frac{2}{x} y = 1$ that approaches a finite value as $x \to 0$.

7. Solve the equation

$$\frac{dy}{dx} + \frac{2}{x} y = \frac{1}{x^3}$$

subject to the condition $y(1) = 0$.

8. Solve the differential equation

$$\frac{dy}{dx} + \frac{y}{x} = 1$$

subject to the condition $y(1) = 1$.

9. Solve the differential equation

$$\frac{d^2y}{dx^2} - 2\frac{dy}{dx} - 3y = 0$$

subject to the conditions $y \to 0$ as $x \to +\infty$, $y(0) = 1$.

10. Solve the differential equation

$$\frac{d^2y}{dx^2} + \frac{dy}{dx} - 2y = 0$$

subject to the conditions $y \to 0$ as $x \to \infty$, $y(0) = 3$.

11. Solve the differential equation

$$\frac{d^2y}{dx^2} - 3\frac{dy}{dx} - 4y = 0$$

subject to the conditions $y \to 0$ as $x \to +\infty$ and $y(0) = 3$.

12. By direct substitution show that $y = \sin(\omega t + \pi/6)$ is a solution of the differential equation

$$\frac{d^2y}{dx^2} + \omega^2 y = 0.$$

Write down one other independent solution.

13. Find the general solution of $y'' + 10y' + 34y = 0$.

14. Find a particular integral of the equation $y'' + 2y' + 4y = \sin(3x)$.

15. Find a particular integral of the differential equation

$$\frac{d^2y}{dx^2} + 2\frac{dy}{dx} - 4y = -4x^2.$$

16. Find a particular integral of the equation

$$\frac{d^3y}{dx^3} + 4\frac{d^2y}{dx^2} + 3\frac{dy}{dx} + y = -2e^{-x}.$$

17. Find a particular integral of the equation $\dfrac{d^2y}{dx^2} + \dfrac{dy}{dx} + y = e^{2x}.$

18. Find a particular integral of the differential equation
$y'' + 7y' + 4y = 8x^2$.

19. Find the general solution of the differential equation
$y'' - y' + 3y = 0$.

20. The general solution of the equation $\ddot{y} - 2\dot{y} + 5y = 0$ is

$$y = e^t(A\cos(2t) + B\sin(2t)).$$

What is the solution satisfying $y = 0$ and $\dot{y} = 1$ at $t = 0$?

21. Find the most general solution of the differential equation
$y'' - y' - 2y = 0$ such that $y(x) \to 0$ as $x \to \infty$.

8.11. PROBLEMS

1. Find the solution of the differential equation

$$\frac{d^2y}{dx^2} + 3\frac{dy}{dx} + 2y = \cos(x)$$

subject to the conditions $y = 11/10$, $dy/dx = -7/10$ at $x = 0$.

2. Find the solution of the differential equation

$$\frac{d^2y}{dx^2} + 2\frac{dy}{dx} + 5y = 5x$$

subject to the conditions $y = -2/5$ and $dy/dx = 0$ at $x = 0$.

3. Solve the differential equation

$$\frac{d^2y}{dx^2} - \frac{dy}{dx} - 2y = 0$$

subject to the conditions $y \to 0$ as $x \to +\infty$, and $y(0) = 1$.

4. Find a particular integral of the equation

$$\frac{d^2y}{dx^2} - \frac{dy}{dx} - 6y = 13.$$

Hence (or otherwise) find the solution of the equation
satisfying $y(0) = 0$ and $dy/dx = 0$ at $x = 0$.

5. The complex function $z(t)$ satisfies the differential equation

$$\ddot{z} - 2\dot{z} + 2z = 2(1 - 5t)e^{2it}.$$

Find the general solution of this equation.
What is the solution which satisfies $z(0) = 0$ and $\dot{z}(0) = 0$?

6. Solve the differential equation

$$\frac{d^2y}{dx^2} + 2\frac{dy}{dx} + 2y = 5\cos(3x)$$

subject to the conditions $y = dy/dx = 0$ for $x = 0$.

7. Solve the differential equation

$$\frac{d^2y}{dx^2} + 2\frac{dy}{dx} + 2y = \sin(2x)$$

subject to the conditions $y(0) = y(\pi/2) = 0$.

8. Solve the differential equation

$$\frac{d^2y}{dx^2} - 4\frac{dy}{dx} + 8y = \cos(x)$$

subject to the conditions $y = dy/dx = 0$ for $x = 0$.

9. Find the general solution of the equation

$$\frac{d^2y}{dt^2} + \frac{dy}{dt} - 12y = e^{-t}.$$

For what values of the arbitrary constants is the solution bounded (i.e. remains finite) as $t \to +\infty$? Determine the relation that must be satisfied between $y(0)$ and $y'(0)$ in order that the solution be bounded as $t \to +\infty$. (Such a system is physically unstable since for the smallest change in the initial conditions $y \to \pm\infty$.)

10. Find the solution of $y'' - 4y' + 5y = 65\cos(x) - 5$ subject to the conditions $y(0) = 0$, $y'(0) = 1$.

11. Find the general solutions of the following differential equations:
  (i) $y' = (1 - y^2)^{1/2}x^3$
  (ii) $y' + y\cot(x) = 2\cos(x)$

     (iii) $y'' - 4y' + 3y = 0$

     (iv) $y'' - 4y' + 4y = 0.$

12. Find the general solutions of the following differential equations:

     (i) $y^2 y' - 1 = \cos(x)$

     (ii) $x^3 y' + x^2 y = 1$

     (iii) $y'' - 3y' - 4y = 0$

     (iv) $y'' - 6y' + 9y = 0.$

13. Find the general solutions of the following differential equations:

     (i) $y' = 3x^2 y$

     (ii) $\cot(x)y' - y = \dfrac{2x}{\sin(x)}$

     (iii) $y'' - 10y' + 25y = 0$

     (iv) $y'' - 7y' + 12y = 6e^x.$

14. (i) Find the general solution of the differential equation

$$y'' + 6y' + 13y = 0.$$

Express your solution in terms of *real* functions of the variable x.

     (ii) The function y satisfies the differential equation

$$y'' - 5y' + 4y = 8x - 6.$$

Find the general solution of this equation. What is the solution which satisfies $y(0) = 1$ and $y'(0) = -1$?

CHAPTER 9

MULTIPLE INTEGRALS

In this chapter, we learn how to integrate functions of two or three variables over two- or three-dimensional regions with simple shapes. As well as the more obvious applications – for example calculating the volume of a given body or the rate of energy emission from a given surface – there are important but less obvious ones – for example finding the mean speed of stars in a galaxy. By using suitably chosen coordinates such integrals can often be evaluated relatively simply. One very important application is to the definition of solid angle. As we shall see, the concept of solid angle extends the notion of an angle subtended at a point by an arc to two-dimensional extended objects.

9.1. REPEATED INTEGRALS

Given a function $f(x, y)$ of two variables (x, y), we can obtain the integral

$$F(x) = \int_{y_1}^{y_2} f(x, y)\, dy.$$

by holding x temporarily constant. Given some limits y_1 and y_2 the result is a function $F(x)$ of just one variable (x). We can then integrate the result with respect to x:

$$I = \int_{x_1}^{x_2} F(x)\, dx = \int_{x_1}^{x_2} \left(\int_{y_1}^{y_2} f(x, y)\, dy \right) dx. \tag{9.1}$$

We can write this with fewer parentheticals as

$$I = \int_{x_1}^{x_2} \int_{y_1}^{y_2} f(x, y)\, dy dx \qquad (9.2)$$

with the understanding that equation (9.2) means "do the integrals in the order indicated in equation (9.1)," i.e. "from the inside, working out." We call this a *repeated integral* (also an *iterated integral* or a *double integral*). Note that the order indicated by the bracketing in equation (9.1) is one convention. Some authors use an alternative convention in which the first integral sign is associated with the first differential and so on in sequence.

Example 9.1 Evaluate $\int_{-1}^{1} \int_{0}^{1} x^2 y\, dy dx$.

The inner integral is $\int_{0}^{1} x^2 y\, dy$. For this inner integral with respect to y, x is a constant, thus

$$\int_{0}^{1} x^2 y\, dy = x^2 \int_{0}^{1} y\, dy = x^2 \left[\frac{y^2}{2}\right]_0^1 = \frac{x^2}{2}.$$

So

$$\int_{-1}^{1} \int_{0}^{1} x^2 y\, dy dx = \int_{-1}^{1} \frac{x^2}{2}\, dx = \left[\frac{x^3}{6}\right]_{-1}^{1} = \frac{1}{3}.$$

Example 9.2 Evaluate $\int_{0}^{1} \int_{a}^{\infty} xe^{-xy}\, dy dx$ (where $a > 0$ is a given constant).

The inner integral is $\int_{a}^{\infty} xe^{-xy}\, dy$. Here x is treated as a constant parameter for the integration with respect to y:

$$\int_{a}^{\infty} xe^{-xy}\, dy = x\int_{a}^{\infty} e^{-xy}\, dy = x\left[-\frac{e^{-xy}}{x}\right]_{y=a}^{y=\infty} = e^{-ax}.$$

So

$$\int_0^1 \int_a^\infty xe^{-xy}\,dydx = \int_0^1 e^{-ax}\,dx = \left[-\frac{e^{-ax}}{a}\right]_{x=0}^{x=1} = -\frac{1}{a}\left(e^{-a}-1\right).$$

Exercise 9.1 Evaluate $\displaystyle\int_0^2 \int_0^1 xy^3\,dydx$.

The extension to more than two integrals is straightforward, as in the following example.

Example 9.3 Evaluate $\displaystyle\int_0^1 \int_0^{\pi/2} \int_0^2 zx^2 \sin(y)\,dzdydx$.

The innermost integral is

$$\int_0^2 zx^2 \sin(y)\,dz = x^2 \sin(y) \int_0^2 z\,dz$$
$$= 2x^2 \sin(y).$$

So now

$$\int_0^{\pi/2} \int_0^2 zx^2 \sin(y)\,dzdy = \int_0^{\pi/2} 2x^2 \sin(y)\,dy$$
$$= 2x^2 \int_0^{\pi/2} \sin(y)\,dy$$
$$= 2x^2.$$

And finally

$$\int_0^1 \int_0^{\pi/2} \int_0^2 zx^2 \sin(y)\,dzdydx = \int_0^1 2x^2\,dx = \frac{2}{3}.$$

Exercise 9.2 Find $\displaystyle\int_0^1 \int_0^{\pi/2} \int_0^1 xy\sin(zy)\,dzdydx$.

If, as in all these examples, the limits of integration are *constants*, then (for the integrands that are likely to arise in practice in physical sciences) the integrals can be evaluated in any order.

This is obvious for cases like Example 9.1 because the double integral is a product of single integrals:

$$\int_{-1}^{1}\left(\int_{0}^{1}x^2 y\, dy\right)dx = \int_{-1}^{1}x^2\left(\int_{0}^{1}y\, dy\right)dx$$

$$= \left(\int_{-1}^{1}x^2\, dx\right)\left(\int_{0}^{1}y\, dy\right)$$

$$= \begin{pmatrix}\text{product} \\ \text{of single} \\ \text{integrals}\end{pmatrix}.$$

which, reversing the sequence of steps,

$$= \left(\int_{0}^{1}y\, dy\right)\left(\int_{-1}^{1}x^2\, dx\right)$$

$$= \int_{0}^{1}y\left(\int_{-1}^{1}x^2\, dx\right)dy$$

$$= \int_{0}^{1}\left(\int_{-1}^{1}yx^2\, dx\right)dy.$$

In fact, it is generally true that when the integrand can be factored into the product of two functions, $f(x,y) = g(x)h(y)$, then the repeated integral over x and y can be factored into the product of two single integrals

$$\int_{y_0}^{y_1}\int_{x_0}^{x_1}f(x,y)\, dx dy = \int_{y_0}^{y_1}\int_{x_0}^{x_1}g(x)h(y)\, dx dy$$

$$= \left(\int_{x_0}^{x_1}g(x)\, dx\right)\left(\int_{y_0}^{y_1}h(y)\, dy\right).$$

We may also change the order of integration in cases, such as Example 9.2, where the double integral is not equal to a product of single integrals. (In these cases one choice of order may result in an easier calculation than another, but the final result will be the same.)

But if the limits of integration are not constants (e.g. if a in Example 9.2 were a function of x), then the order cannot simply be reversed. This is dealt with in Section 9.3.

Exercise 9.3 Evaluate the integral

$$\int_a^\infty \left(\int_0^1 x e^{-xy} \, dx \right) dy,$$

and check that the result is the same as obtained by integrating in the reverse order in Example 9.2. (Hint: the inner integral can be evaluated by integration by parts; also

$$\int \left(\frac{e^{-y}}{y} + \frac{e^{-y}}{y^2} \right) dy = -\frac{e^{-y}}{y}.$$

9.2. INTEGRALS OVER RECTANGLES IN THE PLANE

The area under the graph of a function $y = f(x)$ is given by the definite integral of f. We therefore expect that the volume under a surface, $z = f(x,y)$, is the double integral of $f(x,y)$. To demonstrate this we shall adapt the proof from one dimension to two.

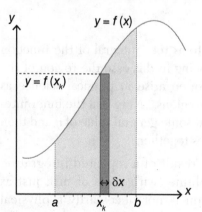

First, consider a curve in the plane. The area A_k of the rectangle (shaded) of width δx at x_k

$$A_k \approx \text{height of curve} \times \text{base}$$

$$\approx f(x_k)\delta x$$

while we have that the area under the curve between a and b is

$$A \approx \sum_k f(x_k) \delta x.$$

In the limit $\delta x \to 0$, $k \to \infty$ we get the exact result:

$$\begin{pmatrix} \text{Area under the curve} \\ \text{in the interval} \\ \text{between } a \text{ and } b \end{pmatrix} = \int_a^b f(x)dx.$$

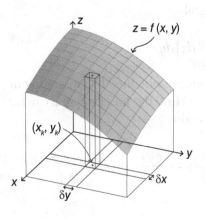

Now, for the analogous volume under a surface. Looking at the figure we have the volume V_k of the box above the point (x_k, y_k) with base $\delta x \times \delta y$:

$V_k \approx$ height of surface $\times$ area of base

$$\approx f(x_k, y_k)\, \delta x \delta y$$

Meanwhile, the volume under the surface is

$$V \approx \sum_k f(x_k, y_k)\, \delta x \delta y.$$

In the limit $\delta x, \delta y \to 0$, $k \to \infty$ we get the exact result

$$\begin{pmatrix} \text{Volume under the surface} \\ \text{above the plane rectangle} \\ x_1 \le x \le x_2, y_1 \le y \le y_2 \end{pmatrix} = \int_{y_1}^{y_2}\left(\int_{x_1}^{x_2} f(x, y)dx \right)dy.$$

This *volume* is also referred to as the "integral of the function $f(x, y)$ over an area," the "area" being in this case the region of the plane rectangle. Such integrals often arise in physical science as "integrals of functions over plane regions." Note that the inner integration is over a slab of width dy at some general value of y and then the outer integral adds all the slabs together.

As in the next example, the result of a repeated integration is not necessarily a physical volume (with units of m³), just as the result of a single integration is not necessarily a physical area (with units of m²). When we talk about the "volume" of a double integral, we mean the volume under the surface in the graph of $f(x, y)$. If $f(x, y)$ and x and y all have units of length, then the double integral does indeed give a physical volume. But if $f(x, y)$ is a charge density over a surface then its integral with respect to its two spatial dimensions (x, y) will be a charge, not a volume.

Example 9.4 A square plate with sides of unit length has a surface mass density

$$\sigma(x, y) = 1 + \cos(\pi x)\cos(\pi y),$$

where x and y are measured from the center of the plate parallel to its sides. Find the mass M of the plate.

The mass of an element $\delta x \delta y$ of surface is the mass per unit area (i.e. the surface density) × area, hence $\sigma \delta x \delta y$. The mass M of the plate is the integral of this, i.e. of the density per unit area over the region of the plate, thus:

$$M = \int_{-1/2}^{1/2} \int_{-1/2}^{1/2} \sigma(x, y)\, dx\, dy.$$

Notice the limits of each integration. We are told in the question that the plate is square, has sides of length 1, and x and y are measured from the center. So x runs from $-1/2$ to $+1/2$, and so does y.

Evaluate the inner integral:

$$\int_{1/2}^{1/2} \left(1 + \cos(\pi x)\cos(\pi y)\right) dx = \left[x + \frac{1}{\pi}\sin(\pi x)\cos(\pi y) \right]_{-1/2}^{1/2}$$

$$= 1 + \frac{2}{\pi}\cos(\pi y).$$

Now evaluate the outer integral:

$$M = \int_{1/2}^{1/2} \left(1 + \frac{2}{\pi}\cos(\pi y)\right) dy$$

$$= \left[y + \frac{2}{\pi^2}\sin(\pi y) \right]_{-1/2}^{1/2}$$

$$= 1 + \frac{4}{\pi^2}.$$

Exercise 9.4 The electromagnetic radiation emitted between frequencies v and $v + dv$ from the annular region of a disc between radii r and $r + dr$ has a power

$$Cr^{-a} \exp(-\beta v r^{1/4}) d v dr$$

where C, a and β are positive constants and $a > 3/4$. If the inner radius of the disc is r_0 and the outer radius is taken to be infinite, write down and evaluate the double integral giving the total power radiated by the disc over all (positive) frequencies.

9.3. MULTIPLE INTEGRALS OVER IRREGULAR REGIONS

In all the examples above, we were given the coordinates with which to integrate (dx, dy) and the region we integrated over was rectangular in these coordinates, defined by lines of constant x and y. Suppose we have some function $f(x, y)$ and we wish to integrate over some non-rectangular region – for example to calculate the charge on a disc if $f(x, y)$ is the surface density of the charge. To integrate a function over an irregular region in the plane we divide the region into tiny area elements, each of area δA_k (each given a number, k). Then we sum $\delta I_k = f(x_k, y_k)\delta A_k$ over all the area elements that fall entirely within the region.

$$I = \sum_{k \, in \, R} \delta I_k = \sum_{k \, in \, R} f\left(x_k, y_k \right) \delta A_k \qquad (9.3)$$

In the limit of $\delta A \to 0$ and $k \to \infty$ we get the double integral over the region

$$I = \iint_R f\left(x, y \right) dA \qquad (9.4)$$

This is subtly different from the repeated integrals we examined previously. Here we evaluate over area elements dA – we did not specify which coordinates to use to integrate over. In practice, double or multiple integrals can be evaluated as repeated integrals.

(This result is known as *Fubini's theorem* and holds for almost all integrals you are likely to encounter.) So we choose coordinates that make the integrand or the region simple to deal with, and evaluate the repeated integral.

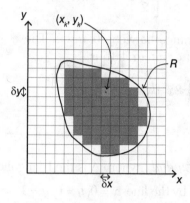

Left: A bounded, irregular region in the (x, y) plane, partitioned into rectangular cells. In order to define the double integral of a function $f(x, y)$ over a non-rectangular region R, we imagine the region partitioned into rectangular cells of area $\delta A = \delta x \times \delta y$. And for each cell (numbered k) we compute the volume over that cell, $V_k = f(x_k, y_k)\delta x\, \delta y$, just as in the case of the rectangular region. We then sum these small volumes over only those cells that fall entirely within the region R:

$$V \approx \sum_{k\, in\, R} f\left(x_k, y_k \right) \delta x \delta y.$$

As before, as the grid becomes finer ($\delta x, \delta y \to 0$) the difference between the grid and the integration region tends to zero and we have

$$I = \iint_R f\left(x, y \right) dA = \iint_R f\left(x, y \right) dx\, dy. \qquad (9.5)$$

So we can write the double integral over an area as a repeated integral over x and y. But we must be careful with the limits as the following example shows.

Example 9.5 Evaluate $I = \iint_\Delta x^2 y \; dy dx$ where Δ is the triangle (Figure 9.1) bounded by the lines $x = 1, y = 0, y = 2x$.

We begin by putting in the limits of integration. In the inner integral with respect to y, x has a general value somewhere in Δ. Draw the dotted line (Figure 9.1) parallel to the y-axis at some general value of x in Δ. From this we see that, within

Δ, y runs from 0 to $2x$. These are the limits for the inner (y) integral. Thus, the inner integral is

$$\int_0^{2x} x^2 y \, dy = x^2 \int_0^{2x} y \, dy = x^2 \left[\frac{y^2}{2} \right]_{y=0}^{y=2x} = 2x^4.$$

Now, to cover Δ we must let x range between 0 and 1. These are the required limits for the outer integral, thus

$$I = \int_0^1 \left(\int_0^{2x} x^2 y \, dy \right) dx = \int_0^1 2x^4 \, dx = 2 \left[\frac{x^5}{5} \right]_0^1 = \frac{2}{5}.$$

Exercise 9.5 Evaluate $I = \iint_\Delta x^2 y \, dy dx$ where Δ is the triangular region in Figure 9.2(a), bounded by the lines $x = 0$, $y = 0$, $y = 1 - x$.

Exercise 9.6 Evaluate $I = \iint_D x^2 y \, dy dx$ over the quarter of the unit disc in Figure 9.2(b).

Note that in these examples we have drawn the diagram for you. In general you will have to begin by sketching your own figure (see Exercise 9.7).

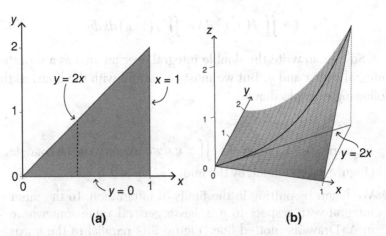

FIGURE 9.1: (a) The region bounded by $y = 2x$, $x = 1$ and $y = 0$. (b) The function $z = x^2 y$ with the triangular region indicated on the (x, y) plane.

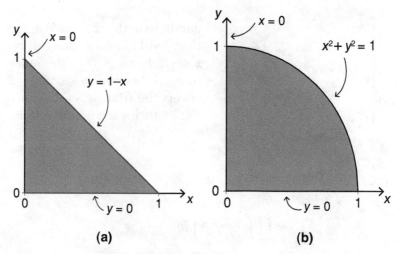

FIGURE 9.2: (a) The region for Exercise 9.5 bounded by $y = 1 - x$, $x = 0$ and $y = 0$. (b) The region for Exercise 9.6 bounded by $x^2 + y^2 = 1$, $x = 0$ and $y = 0$.

9.4. CHANGE OF ORDER OF INTEGRATION

As stated, for most functions order of integration does not matter if the limits of integration are simple constants, e.g. integrating over a rectangular region (with $x_1 \le x \le x_2$ and $y_1 \le y \le y_2$). When the limits of the integrals are themselves functions, we can still change the order of integration if we are careful with the limits, as in Example 9.6. (There are some cases, with functions that are not of constant sign or involving infinities or discontinuities, where we cannot change the order of integration. But these are seldom encountered in undergraduate physics.)

Example 9.6 Evaluate $I = \iint_{\Delta} x^2 y \, dx dy$ where Δ is the region outlined in the figure (below). This is the same as in example 9.5 but with the order of the integrations reversed.

For the inner integration with respect to x, y has a general value in Δ. Draw the dotted line at a general y within Δ

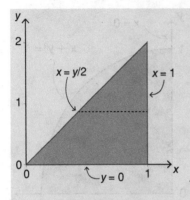

parallel to the x-axis. We see that, within Δ, x runs from $x = y/2$ to $x = 1$. For the outer integral the range $y = 0$ to $y = 2$ covers the triangle so these are the limits for the y integration.

$$I = \int_0^2\left(\int_{y/2}^1 x^2 y\,dx\right)dy$$

$$= \int_0^2 y\left[\frac{x^3}{3}\right]_{x=y/2}^{x=1}dy = \int_0^2 \frac{y}{3}\left(1-\frac{y^3}{8}\right)dy$$

$$= \frac{1}{3}\left[\frac{y^2}{2}-\frac{y^5}{8\times 5}\right]_0^2 = \frac{1}{3}\left(\frac{4}{2}-\frac{32}{8\times 5}\right)$$

$$= \frac{1}{3}\left(2-\frac{4}{5}\right) = \frac{2}{5}.$$

The function to be integrated and the region of integration are the same as in the example of Section 9.3, so the answer is, of course, the same. But note that you have to work out the limits (e.g. from a diagram) each time. This requires thought and care.

Sometimes integrals that arise in one order are best evaluated in another. The following is an artificial example.

Example 9.7 Evaluate $\displaystyle\int_0^1\left(\int_{\sin^{-1}(y)}^{\pi/2}\operatorname{cosec}(x)\,dx\right)dy$ by changing the order of integration (assuming that $\sin^{-1}(y)$ is in the range 0 to $\pi/2$).

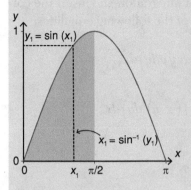

First we draw a sketch. The inner integral is an integral with respect to x, where the limits are $x = \sin^{-1}(y)$ to $x = \pi/2$. So we plot $y = \sin(x)$ and examine the horizontal extent of the region between this curve and $x = \pi/2$. Looking at the outer integral tells us that y ranges between $y = 0$ and $y = 1$. This gives us the shaded area in the figure.

Now we can safely change the order of integration. Now the inner integration is with respect to y. For a general value of x (draw a vertical line at some x), y ranges between 0 and $\sin(x)$. So these are now the limits for the inner y integral.

Within the shaded region x ranges over 0 to $\pi/2$, so these are the limits for the outer x integral, so we now have that

$$
\begin{aligned}
\int_0^1 \left(\int_{\sin^{-1}(y)}^{\pi/2} \csc(x)\, dx \right) dy &= \int_0^{\pi/2} \left(\int_0^{\sin(x)} \csc(x)\, dy \right) dx \\
&= \int_0^{\pi/2} \left(\csc(x) \int_0^{\sin(x)} dy \right) dx \\
&= \int_0^{\pi/2} \csc(x) \left[y \right]_0^{\sin(x)} dx \\
&= \int_0^{\pi/2} \csc(x) \sin(x)\, dx \\
&= \int_0^{\pi/2} dx = \frac{\pi}{2}.
\end{aligned}
$$

Exercise 9.7 Sketch the regions of integration and insert the correct limits on the right hand sides of the following equalities:

(i) $\int_0^1 \left(\int_0^{y} f(x,y)\, dx \right) dy = \int \left(\int f(x,y)\, dy \right) dx$,

(ii) $\int_0^{\infty} \left(\int_{\sqrt{y}}^{\infty} f(x,y)\, dx \right) dy = \int \left(\int f(x,y)\, dy \right) dx$.

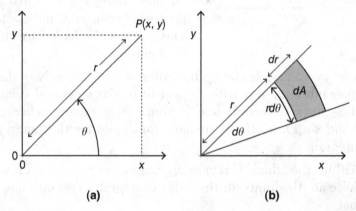

FIGURE 9.3: (a) Polar coordinates (r, θ) and Cartesian coordinates (x, y). (b) The area element in terms of polar coordinates: $dA = r\, dr\, d\theta$.

9.5. POLAR COORDINATES IN TWO DIMENSIONS

Instead of the Cartesian labels (x, y) for points in the plane, we can use a distance r from the origin and an angle θ (from the x-axis) as in Figure 9.3(a). From the figure we deduce the relation between the (x, y) and (r, θ) labels of the point P:

$$x = r \cos(\theta),$$
$$y = r \sin(\theta). \tag{9.6}$$

The inverse relations are

$$r = (x^2 + y^2)^{1/2},$$
$$\theta = \tan^{-1}(y/x). \tag{9.7}$$

Functions in the plane may take on a simpler form in polar coordinates.

Example 9.8 What is the function $f(x, y) = (x^2 + y^2)^{3/2}$ in polar coordinates?

We substitute $x = r\cos(\theta), y = r\sin(\theta)$ in $f(x, y)$ to get

$$f(r, \theta) = (r^2\cos^2(\theta) + r^2\sin^2(\theta))^{3/2} = r^3$$

Exercise 9.8 What is the function $f(x, y) = x\left(1 + \dfrac{y^2}{x^2}\right)^{-1/2}$ in polar coordinates (r, θ)?

Exercise 9.9 What are the curves

(i) $r = $ constant

(ii) $\theta = $ constant?

9.6. INTEGRALS OVER REGIONS IN THE PLANE

In polar coordinates the element of area is $dA = dr \times r d\theta$ (see Figure 9.3b). We can use this to integrate a function over a plane region. Instead of

$$\int f\, dA = \iint f(x, y)\, dxdy$$

we can write

$$\int f\, dA = \iint f(r, \theta)\, rdrd\theta.$$

This is useful if the limits of integration are more readily expressed in (r, θ) coordinates. The temptation to write $drd\theta$ (instead of $rdrd\theta$) must be resisted; $dxdy$ is an area, but $drd\theta$ is not – they cannot be equal.

Example 9.9 Find the integral of the function $f(x, y) = xy$ over the region $x^2 + y^2 \leq 1, x \geq 0, y \geq 0$.

Let

$$I = \iint_R xy \, dxdy$$

where R is the given region. Because the area is a quadrant of a circular disc, we transform the problem to polar coordinates. So, first transform the integrand:

$$
\begin{aligned}
f(x, y) &= xy \\
&= (r\cos(\theta))(r\sin(\theta)) \\
&= r^2 \cos(\theta)\sin(\theta)
\end{aligned}
$$

Next transform the element of area:

$$dxdy = rd\theta dr$$

Finally, we transform the limits: the area of integration is the first quadrant for which $0 \leq r \leq 1, 0 \leq \theta \leq \pi/2$, and so now the integrand is a product of a function of r and a function of θ with constant limits:

$$
\begin{aligned}
I &= \int_0^{\pi/2} \int_0^1 r^2 \cos(\theta)\sin(\theta) \, rd\theta dr \\
&= \int_0^{\pi/2} \sin(\theta) d(\sin\theta) \int_0^1 r^3 dr \\
&= \left[\frac{1}{2}\sin^2(\theta)\right]_0^{\pi/2} \left[\frac{1}{4}r^4\right]_0^1 \\
&= \frac{1}{2} \times \frac{1}{4} = \frac{1}{8}.
\end{aligned}
$$

Note how the integral in θ is evaluated using $d(\sin(\theta)) = \cos(\theta)d\theta$. This is equivalent to changing the variable to $u = \sin(\theta)$. It is not necessary to change the limits for $\sin(\theta)$ explicitly.

Exercise 9.10

(i) Find the integral of the function $f(x, y) = x^2 + y^2$ over the disc $x^2 + y^2 \leq 4$.

(ii) A disc of radius 2m centered on the origin has a variable surface density given by $\sigma(r) = 0.1r^2$ kg m^{-1}. What is the mass of the disc?

9.7. AN IMPORTANT DEFINITE INTEGRAL

As a further example of integration over a region in polar coordinates we look at the following trick to evaluate $\int_0^\infty e^{-x^2} dx$.

Example 9.10 Evaluate the integral of the function $f(x, y) = e^{-(x^2+y^2)}$ over the whole plane. In Cartesian coordinates this is

$$I = \int_{-\infty}^{\infty} \int_{-\infty}^{\infty} e^{-x^2} e^{-y^2} dx dy$$

$$= \left(\int_{-\infty}^{\infty} e^{-x^2} dx \right) \left(\int_{-\infty}^{\infty} e^{-y^2} dy \right)$$

$$= \left(\int_{-\infty}^{\infty} e^{-x^2} dx \right)^2 .$$

But there are no elementary methods to evaluate $\int_0^\infty e^{-x^2} dx$. We can, however, transform the problem to polar coordinates. The problem becomes: evaluate the integral of the function $f(r, \theta) = e^{-r^2}$ over the whole plane. This is

$$I = \int_0^{2\pi} \int_0^\infty e^{-r^2} r dr d\theta$$

$$= \left(\int_0^{2\pi} d\theta \right) \left(\int_0^\infty e^{-r^2} r dr \right)$$

$$= 2\pi \times \left[-\frac{1}{2} e^{-r^2} \right]_0^\infty = \pi$$

Exercise 9.11 Deduce from the example that

$$\int_{-\infty}^{\infty} e^{-x^2} dx = \sqrt{\pi}$$

and hence that

$$\int_0^{\infty} e^{-x^2} dx = \frac{1}{2}\sqrt{\pi}$$

The example is important because (i) this integral appears in various branches of physics and mathematics, and (ii) it cannot be evaluated by elementary methods (do not waste time trying!).

9.8. CYLINDRICAL POLAR COORDINATES

We can extend two-dimensional polar coordinates to three dimensions by adding the z-coordinate. Thus we label points in space by two distances and an angle as in Figure 9.4(a). Note that the usual convention is to label the radial coordinate ρ here (in place of r in section 9.5) and the angular coordinate ϕ (in place of θ) to avoid confusion with the spherical polar coordinates introduced in Section 9.11 below.[1]

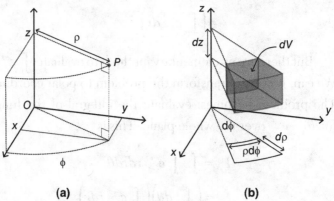

(a)   **(b)**

FIGURE 9.4: (a) Cylindrical polar coordinates (ρ, ϕ, z) and Cartesian coordinates (x, y, z). (b) The volume element in terms of cylindrical polar coordinates $dV = \rho\,d\rho\,d\phi\,dz$.

---

[1]This notation is common in physics, but not universal. When working on a particular problem, always check the definitions.

From Figure 9.4(a) we deduce the relation between the (x, y, z) and (ρ, ϕ, z) labels of the point P:

$$x = \rho \cos(\phi),$$
$$y = \rho \sin(\phi), \qquad (9.8)$$
$$z = z.$$

The inverse relation is

$$\rho = \left(x^2 + y^2\right)^{1/2},$$
$$\phi = \tan^{-1}\left(y / x\right), \qquad (9.9)$$
$$z = z.$$

Exercise 9.12 What are the surfaces

  (i) ρ = constant,

 (ii) z = constant,

(iii) ϕ = constant?

Exercise 9.13 What does the function

$$f(x, y, z) = z^n \ln(1 + x^2 + y^2)$$

become as a function of cylindrical polar coordinates (ρ, ϕ, z)?

Note that in Exercise 9.13 the function is independent of ϕ; we say that it is *axially symmetric*. If $n = 0$ the function is independent of both ϕ and z; we say that the function is *cylindrically symmetric*.

Example 9.11 Show that a function that is cylindrically symmetric is constant on cylinders centered on the z-axis.

If f is cylindrically symmetric it is independent of both ϕ and z, i.e. it is a function of ρ only, $f(\rho)$. Thus f is constant on surfaces ρ = constant, i.e. on cylinders.

9.9. VOLUME INTEGRALS (CYLINDRICAL POLAR)

In cylindrical polar coordinates (ρ, ϕ, z) the element of volume is (from Figure 9.4b)

$$dV = d\rho \times \rho d\phi \times dz = \rho d\rho d\phi dz.$$

We can use this to integrate a function over a region of three-dimensional space (solid body). Instead of

$$\int f\, dV = \iiint f(x,y,z)\, dx\, dy\, dz$$

we can write

$$\int f\, dV = \iiint F(\rho,\phi,z)\, \rho d\rho\, d\phi dz,$$

where $F(\rho, \phi, z)$ is the function $f(x, y, z)$ expressed in terms of ρ, ϕ and z. This is useful if the limits of integration are readily expressed in (ρ, ϕ, z) coordinates.

Example 9.12 Find the integral of the function $x^2 + y^2$ over the cylinder $0 \le \rho \le 1, 0 \le z \le 1$.

First we transform the function to cylindrical coordinates

$$x^2 + y^2 = (\rho \cos(\phi))^2 + (\rho \sin(\phi))^2 = \rho^2.$$

The correct element of volume is

$$dV = \rho d\rho d\phi dz,$$

while the limits of integration are $0 \le \rho \le 1, 0 \le z \le 1, 0 \le \theta \le 2\pi$. Put these together and evaluate the integral and we have that

$$\int_V \left(x^2 + y^2\right) dV = \int_0^1 \int_0^{2\pi} \int_0^1 \rho^2 \times \rho\, d\rho\, d\phi\, dz = \left[\frac{\rho^4}{4}\right]_0^1 \left[\phi\right]_0^{2\pi} \left[z\right]_0^1$$

$$= \frac{1}{4} \times 2\pi \times 1 = \frac{\pi}{2}.$$

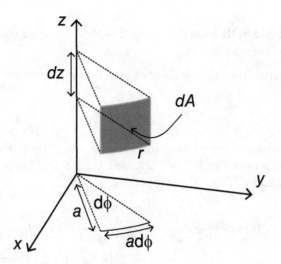

FIGURE 9.5: The cylindrical element of area for a cylinder of radius $\rho = a$: $dA = ad\phi dz$.

Exercise 9.14

(i) Find the integral of the function $f(x, y, z) = xyz$ over the wedge defined by $0 \le \rho \le 1$, $0 \le \phi \le \pi/2$, $0 \le z \le 2$.

(ii) The number per unit volume of bacterial fossils in a cylindrical wedge of rock, $0 \le \phi \le \pi/2$, of radius 1m and height 2m is $10^5 \, xyz$ m^{-3}. How many fossils are in the rock?

9.10. INTEGRALS OVER CYLINDRICAL SURFACES

Suppose we are given a function $f(\phi, z)$ at each point on some part of the surface of a cylinder $\rho = a =$ constant. The function might represent the surface mass density of the cylinder, or the surface density of electric charge, or the flux of energy through the surface, and so on. We might then wish to evaluate the integral of f over the surface (to get the total mass, charge etc.). From Figure 9.5 we have $dA = ad\phi dz$.

So

$$\int f \, dA = \iint f(\phi, z) \, ad\phi dz$$

the surface integral of f over the cylindrical surface.

Example 9.13 Find the integral of the function $z \sin(\phi/2)$ over the cylindrical surface $\rho = 4$, $1 \le z \le 2$.

We write down the element of area on the surface:

$$dA = 4d\phi dz.$$

The limits of integration required to cover the portion of the cylinder once are $0 \le \phi \le 2\pi$ and $1 \le z \le 2$. Finally, we write down and evaluate the integral:

$$\int_1^2 \int_0^{2\pi} z \sin(\phi/2) 4 d\phi dz = 4\left[\frac{z^2}{2}\right]_1^2 \left[-2\cos(\phi/2)\right]_0^{2\pi}$$

$$= (8-2)(-2\times -1 + 2\times 1) = 6\times 4 = 24.$$

Exercise 9.15 Find the integral of the function $\dfrac{\cos(\phi)}{(1+z)^2}$ over the part of the surface of the cylinder $\rho = 2$, $-\pi/2 \le \phi \le \pi/2$, $z \ge 0$.

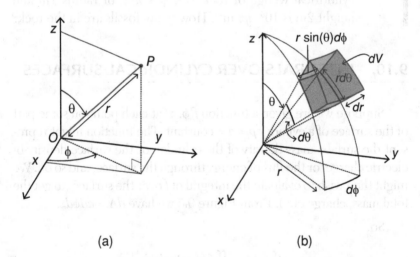

(a)   (b)

FIGURE 9.6: (a) Spherical polar coordinates (r, ϕ, θ) and Cartesian coordinates (x, y, z). (b) The volume element in terms of spherical polar coordinates $dV = r^2 \sin(\theta) dr d\theta d\phi$.

9.11. SPHERICAL POLAR COORDINATES

We can label points in space by a distance and two angles as in Figure 9.6(a). The coordinates (r, θ, ϕ) are the spherical polar coordinates of the point P. From the figure we deduce the relation between the (x, y, z) and (r, θ, ϕ) labels of P:

$$x = r\sin(\theta)\cos(\phi),$$
$$y = r\sin(\theta)\sin(\phi),$$
$$z = r\cos(\theta). \tag{9.10}$$

The inverse relation is

$$r = \left(x^2 + y^2 + z^2\right)^{1/2},$$
$$\theta = \cos^{-1}\left(z / \left(x^2 + y^2 + z^2\right)^{1/2}\right),$$
$$\phi = \tan^{-1}\left(y / x\right). \tag{9.11}$$

Note carefully (and remember) how the angles θ and ϕ are defined: this is the standard convention and you need to know it – θ is the co-latitude of P (not the latitude; i.e. it is measured down from the z-axis, not up from the equator). The angle ϕ is measured in the (x, y) plane from the x-axis.

Exercise 9.16 What are the surfaces

(i) ρ = constant,

(ii) θ = constant,

(iii) ϕ = constant?

Exercise 9.17 What does the function $f(x, y, z) = z^n \ln(1 + x^2 + y^2)$ become in spherical polar coordinates?

Note that the function in Exercise 9.17 is independent of ϕ: it is axially symmetric. A function that is independent of both θ and ϕ is said to be spherically symmetric.

Exercise 9.18

(i) Show that $f(x, y, z) = \ln(1 + x^2 + y^2 + z^2)$ is spherically symmetric.

(ii) Show that a spherically symmetric function is constant on spheres centered on the origin.

9.12. VOLUME INTEGRALS (SPHERICAL POLAR)

In spherical polar coordinates (r, θ, ϕ) the element of volume is (Figure 9.6b):

$$dV = dr \times rd\theta \times r\sin(\theta)d\phi = r^2 \sin(\theta)\, drd\theta d\phi.$$

We can use this to integrate a function over the volume of a solid body. Instead of

$$\int f\, dV = \iiint f(x, y, z)\, dxdydz.$$

we can write

$$\int f\, dV = \iiint f(r, \theta, \phi)\, r^2 \sin(\theta)\, drd\theta d\phi.$$

This is useful if the limits of integration are readily expressed in (r, θ, ϕ) coordinates.

Example 9.14 Find the integral of the function $f(x, y, z) = z$ over the volume of the hemisphere $0 \le r \le 1$, $0 \le \theta \le \pi/2$, $0 \le \phi \le 2\pi$.

For a volume that is part of a sphere we choose to work in spherical polar coordinates. Write the function in spherical polars:

$$f(x, y, z) = z = r\cos(\theta).$$

Next, write down the limits for the integration: these are given as

$$0 \le r \le 1, 0 \le \theta \le \pi/2, 0 \le \phi < 2\pi.$$

We can now write down the integral

$$\int z\, dV = \int_0^{2\pi} \int_0^{\pi/2} \int_0^1 (r\cos(\theta))(r^2 \sin(\theta)\, dr\, d\theta\, d\phi)$$

$$= \int_0^1 r^3\, dr \int_{\theta=0}^{\theta=\pi/2} \sin(\theta)\, d(\sin(\theta)) \int_0^{2\pi} d\phi.$$

$$= \left[\frac{r^4}{4}\right]_0^1 \left[\frac{1}{2}\sin^2(\theta)\right]_0^{\pi/2} [\phi]_0^{2\pi}$$

$$= \frac{1}{4} \times \frac{1}{2} \times 2\pi = \frac{\pi}{4}.$$

Note the range of θ required to cover the hemisphere. If we choose points in the quadrant of the $(x$-$z)$ plane, $\phi = 0$, $0 \le \theta \le \pi/2$ and swing this around through 2π we cover the *complete hemisphere*. Rotating the arc $\phi = 0$, $0 \le \theta \le \pi$ gives the complete sphere.

Exercise 9.19 What range of coordinates covers the whole sphere of radius a?

Exercise 9.20 Find the integral of the function $f(x, y, z) = z^2$ over the sphere $r \le 1$.

9.13. INTEGRALS OVER SPHERICAL SURFACES

Suppose we are given a function $f(\theta, \phi)$ at each point on some part of the surface of a sphere, $r = a = $ constant. We might need to evaluate the integral of f over the surface. From Figure 9.7 the element of area is $dA = a\sin(\theta)d\phi \times ad\theta = a^2 \sin(\theta)\, d\theta d\phi$. So

$$\int f\, dA = \iint f(\theta, \phi) a^2 \sin(\theta)\, d\theta\, d\phi,$$

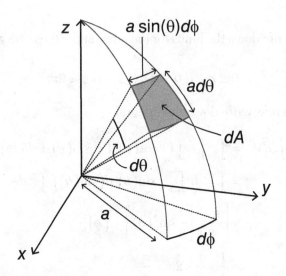

FIGURE 9.7: The spherical element of area for a sphere of radius $r = a$.

the surface integral of f over the spherical surface. Note the factor of $\sin(\theta)$ in the integrand. The area is not $a^2 d\theta d\phi$. You can see from the figure that the same $d\theta d\phi$ near the poles has a smaller area than near the equator. The factor of $\sin(\theta)$ takes this into account.

Example 9.15 Find the integral of the function $f(x, y, z) = z$ over the surface of the hemisphere $r = a$, $0 \le \theta \le \pi/2$, $0 \le \phi \le 2\pi$.

Because the surface is part of a sphere it is convenient to use spherical polar coordinates. Write the function in spherical polars:

$$z = a \cos(\theta).$$

Next, write down the element of area

$$dA = a^2 \sin(\theta) \, d\theta d\phi.$$

Write down the limits of integration for each of the variables:

$$0 \le \theta \le \pi/2, \; 0 \le \phi \le 2\pi.$$

Now we write down and evaluate the integral:

$$\int z \, dA = \int_0^{2\pi} \int_0^{\pi/2} \left(a \cos(\theta) \right) a^2 \sin(\theta) \, d\theta d\phi$$

$$= a^3 \int_{\theta=0}^{\theta=\pi/2} \sin(\theta) \, d(\sin(\theta)) \int_0^{2\pi} d\phi$$

$$= a^3 \left[\frac{1}{2} \sin^2(\theta) \right]_0^{\pi/2} \left[\phi \right]_0^{2\pi} = a^3 \times \frac{1}{2} \times 2\pi = \pi a^3.$$

Exercise 9.21

(i) Find the integral of $f(x, y, z) = 1 - z^2$ over the surface of the sphere $r = 1$.

(ii) The moment of inertia of a uniform sphere of radius a and unit surface density about an axis through its center at the origin is given by the integral of $a^2 - z^2$ over the surface of the sphere. Find the moment of inertia of such a sphere.

9.14. SOLID ANGLE

The angle, measured in radians, subtended by an arc of the unit circle at its center is the length of the arc (Figure 9.8a). The angle, in radians, subtended at a point O by an element of length δl, distance r from O, is

$$\delta\theta = \frac{\delta l \cos(i)}{r}$$

where i is the angle between the normal to the element and the line from O to the element (Figure 9.8b).

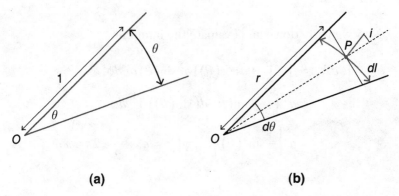

FIGURE 9.8: (a) The length of an arc of the unit circle about point O is equal to the angle (in radians) at O subtended by the arc. (b) A tiny length element dl viewed at a distance r from point O is inclined by an angle i from the circle of radius r centered on O. The angle is $d\theta = (dl \cos i)/r$.

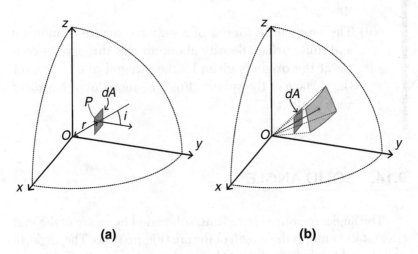

FIGURE 9.9: (a) The area element dA at point P, a distance r from point O, with angle i between the normal to the area element and the line $\overrightarrow{OP}$. (b) The solid angle subtended by dA from O can be thought of as the area of the image of dA projected onto a unit sphere from point O.

The solid angle, measured in steradians (sr), subtended by a cap of the unit sphere at its center is the area of the cap (Figure 9.10). For the purpose of visualization the solid angle subtended at O by some surface S can be thought of as the area of the shadow that S casts (when lit from O) on the surface of a large sphere, center O, regarded as of unit radius. Equivalently, for the purpose of calculation: the solid angle, in steradians, subtended at a point O by an element of area, A, distance r from O is

$$\delta\Omega = \frac{\delta A \cos(i)}{r^2} \qquad (9.12)$$

where i is the angle between the normal to the element and the line from O to the element (Figure 9.9a,b). The solid angle, Ω, subtended by a surface is the integral of $\cos(i)/r^2$ over the surface, i.e.

$$\Omega = \int d\Omega = \int \frac{\cos(i)}{r^2} dA. \qquad (9.13)$$

The concept of solid angle is important in many areas of physical science. To evaluate the integral for solid angle in any particular case you will need i and r as functions of θ and ϕ. This will either be given or deducible for the given surface.

Example 9.16 Find the solid angle subtended by a sphere at its center.

We use the formula (9.13). We write down the element of area at $P = (r, \theta, \phi)$ on the sphere assuming a radius r:

$$dA = r^2 \sin(\theta)\, d\theta d\phi.$$

Now work out the angle i at P, the angle between the radius vector and the normal to the surface: here $i = 0$, because the normal is along the radius vector, and so $\cos(i) = 1$. Write down the limits of integration, chosen here to cover the whole sphere:

$$0 \le \theta \le \pi,\ 0 \le \phi \le 2\pi.$$

Substitute in equation (9.13):

$$\Omega = \int d\Omega = \int_0^{2\pi} \int_0^{\pi} \frac{r^2 \sin(\theta)}{r^2} d\theta\, d\phi$$

$$= \int_{\theta=0}^{\theta=\pi} \sin(\theta)\, d\theta \int_0^{2\pi} d\phi = 2\pi \left[-\cos(\theta) \right]_{\theta=0}^{\theta=\pi} = 4\pi.$$

Note that the result is independent of r.

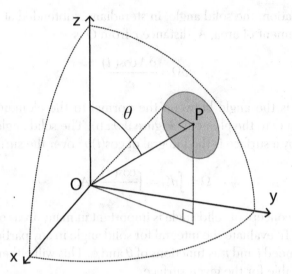

FIGURE 9.10: The solid angle on a sphere.

In fact the sphere subtends solid angle 4π steradians at every interior point, not just the center. (You need to know this.)

Exercise 9.22 What is the solid angle subtended by the sector of a sphere $0 < \phi < \pi/4$, $0 < \theta < \pi/4$ at its center? See Figure 9.11(a).

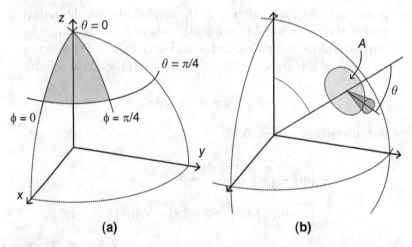

FIGURE 9.11: (a) Exercise 9.22. (b) Exercise 9.24.

Exercise 9.23 What is the solid angle subtended by a portion of a cylindrical surface of radius a, $0 \le \phi \le \pi/2$, $-a \le z \le a$ at the origin?

Exercise 9.24 A small hole of area A is cut in a sphere containing a uniform radiation field which carries an energy per unit time per unit normal area per unit solid angle I. Therefore the rate at which energy is emitted into the surrounding space is

$$\int_{\text{hemisphere}} IA\cos(\theta)\, d\Omega$$

integrated over the hemisphere $0 \le \theta \le \pi/2$. Show that equation (24) gives πAI. See Figure 9.11(b).

Exercise 9.25 The pressure exerted by a gas of kinetic energy density u is

$$P = \frac{1}{\pi} \int_{\text{hemisphere}} u\cos^2(\theta)\, d\Omega;$$

assuming that u is independent of angle, perform the integration to find P.

Exercise 9.26 Show that the solid angle subtended by a plane rectangle of sides $2a$, $2b$ at a point distance l along the normal through its center can be written

$$\Omega = \int_{-b}^{b} \int_{-a}^{a} \frac{l\, dx dy}{\left(x^2 + y^2 + l^2\right)^{3/2}}.$$

9.15. SKETCHING SURFACES

Sketching three-dimensional surfaces from a given formula is hard: it can be simplified if the surface has some symmetry (e.g. axial, cylindrical, spherical). In such cases the formula will simplify in the appropriate coordinate system (found by trial and error).

Example 9.17 Sketch the surface $z = x^2 + y^2$.

In cylindrical polar coordinates this simplifies to $z = \rho^2$, i.e. z is independent of ϕ, so the surface is axially symmetric (i.e. a surface of revolution about the z-axis). We need therefore only plot the curve of z against ρ for fixed ϕ and rotate the result. In Figure 9.12(a), z has been plotted against ρ in the plane $\phi = 0$ (i.e. the $(z - x)$ plane, so $\rho = x$ there) giving a parabola. This parabola has then been rotated about the z-axis to generate the paraboloid represented by the given equation.

Example 9.18 Sketch the surface $z = +(1 - x^2 - y^2)^{1/2} + 2$.

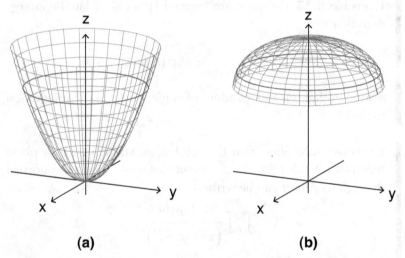

(a)   **(b)**

FIGURE 9.12: (a) Example 9.17: $z = x^2 + y^2$. (b) Example 9.18: $z = (1 - x^2 - y^2)^{1/2} + 2$. In both cases the functions have axial symmetry (because they depend on x and y only through $r^2 = x^2 + y^2$), so we can draw the graph of the function $z = f(0, y)$ on the (y, z) plane, and then rotate about the z-axis.

Write $Z = (z - 2)$; then the surface becomes

$$Z^2 = 1 - x^2 - y^2, Z > 0.$$

In spherical polar coordinates based on the Z-axis (i.e. $Z^2 = r^2 \cos^2(\theta)$ with $Z > 0$), this becomes $r^2 = 1$; $Z > 0$. The surface $r^2 = 1$ is a sphere and the condition $Z > 0$ gives the hemisphere in Figure 9.12(b).

Exercise 9.27 Sketch the surfaces

(i) $z^2 = x^2 + y^2$,

(ii) $z^2 = 1 + x^2 + y^2$.

Revision Notes

After completing this chapter you should be able to

- Compute a given *multiple integral*
- Insert limits for an integral over a given plane region
- Insert the limits corresponding to a *change in the order of integration* over a plane region
- Define *cylindrical* and *spherical polar* coordinates
- Write down the *transformations* between Cartesian coordinates and cylindrical coordinates and spherical polar coordinates and their inverses
- Write down the *elements of area* and *volume* in these coordinate systems
- Use the appropriate elements of area and volume to integrate over parts of *spheres* and *cylinders* and their surfaces
- Define and compute a *solid angle*
- Recognize axial, cylindrical, and spherical *symmetry*

9.16. EXERCISES

1. Evaluate $\int_0^1 \left[\int_0^y x^2 y\, dx \right] dy$.

2. Write down an expression for the integral

$$\int_1^2 \left[\int_1^y f(x,y)\, dx \right] dy$$

   with the order of integration reversed, after sketching the area of integration in the (x, y) plane.

3. Draw a diagram to show the area of integration in

$$\int_0^1 \left[\int_x^{\sqrt{x}} f(x,y)\, dy \right] dx,$$

   and hence rewrite the integral in the form $\int \left[\int f(x,y)\, dx \right] dy$ with the correct limits of integration inserted.

4. Write down the following integral with the order of integration reversed (including the correct limits):

$$\int_0^1 \int_{y/2}^{1/2} f(x,y)\, dx\, dy .$$

5. Show that the integral of the function $f(x, y) = xy^2$ over the region of the (x, y) plane bounded by the lines $y = 0$, $x = 1$ and the curve $y^2 = x$ is 2/21.

6. Let R be the tetrahedron given by $x \geq 0$, $y \geq 0$, $z \geq 0$ and $x + y + z \leq 1$. Show that

$$\iiint_R y\, dx\, dy\, dz = \frac{1}{24}$$

   where the integration is over the volume of the tetrahedron.

7. Draw a diagram to show the relation between Cartesian coordinates (x, y, z) and standard spherical polar coordinates (r, θ, ϕ). What does the function $f(x, y, z) = (x^2 + y^2)^{1/2}/z$ become in spherical polar coordinates?

8. Draw a diagram showing the relation of standard spherical polar coordinates (r, θ, ϕ) to Cartesian coordinates (x, y, z). Indicate on your diagram the standard cylindrical polar coordinates (ρ, ϕ, z) and obtain the relation between (ρ, ϕ, z) and (r, θ, ϕ).

9. Find the integral of the function $\cos^2(\theta)\cos(\phi/4)$ over the surface of the unit sphere centered at the origin.

10. Find the integral of $\cos(2\theta)$ over the surface of the unit sphere. (θ is the angle between the radius vector and a fixed direction.)

11. If $f = (x^2+y^2)z$ calculate the integral of f over the solid cylindrical wedge of unit radius, unit height in $x \geq 0$, $y \geq 0$, $1 \geq z \geq 0$.

12. A spherical cap of angle a is the surface of a sphere, center O, cut off by a right circular cone of semi-angle a with vertex at O. Find the solid angle subtended at O by such a spherical cap. Hence show that a circular disc of angular diameter $2a$ at a very large distance (compared with its radius) subtends a solid angle at a point on its axis of approximately πa^2.

13. Find the solid angle subtended by a circular disc of radius 3 cm at a point a distance 4 cm along the axis perpendicular to the disc through its center.

14. Write down an expression for the solid angle subtended at the origin by an element of surface dS with normal $\mathbf{n}$ at a point with position vector $\mathbf{r}$. Use this expression to estimate the solid angle subtended by a coin of radius 1.295 cm (or approximately 5/4 cm) at a distance of 1/2 m.

9.17. PROBLEMS

1. Write down the element of volume dV in spherical polar coordinates (r, θ, ϕ). Find

$$I = \int_V \frac{\exp\left(ir\cos(\theta) - \epsilon r\right)}{r}\,dV,$$

where ϵ is a constant, $i = \sqrt{-1}$, and the integral is taken over all space. Hence show that $\lim_{\epsilon \to 0} I = 4\pi$.

2. A hemisphere of radius $\sqrt{2}$ is drilled from the pole to the center to remove material within the surface $z = x^2 + y^2$. Show that the volume remaining is that of a unit sphere.

3. Show that the solid angle subtended at the origin by the portion of the surface of a right circular cylinder of radius a with axis along the z-axis between the planes $x = z$ and $x = -z$, $x > 0$ can be written

$$\Omega = \int_{-\pi/2}^{\pi/2}\left[\int_{-a\cos(\phi)}^{a\cos(\phi)} \frac{a^2}{\left(a^2 + z^2\right)^{3/2}}\,dz\right]d\phi.$$

4. (a) Show that the solid angle subtended by a circular disc of unit area at a point on its axis unit distance from its center is

$$\Omega_D = 2\pi \int_0^{1/\sqrt{\pi}} \frac{\rho\,d\rho}{\left(1 + \rho^2\right)^{3/2}}.$$

   (b) Show that the solid angle subtended at the origin by a triangle with vertices at $(0, 0, 1), (a, 0, 1), (a, aa, 1), (a > 0, a > 0)$ is

$$\Omega_\Delta = \int_0^a \int_0^{aa} \frac{dy\,dx}{\left(1 + x^2 + y^2\right)^{3/2}}.$$

(c) From (b) deduce an expression for the solid angle, Ω_n, subtended by an n-sided regular polygon of unit area at a point on its axis unit distance from its center, and show that this can be written as

$$\Omega_n = 2n \int_0^{1/\sqrt{\pi}} \int_0^{\pi x/n} \frac{dy}{\left(1+x^2\right)^{3/2}\left(1+\dfrac{y^2}{1+x^2}\right)^{3/2}}\, dx.$$

By approximating the inner integral in the expression for Ω_n show that $\Omega_n \le \Omega_D$, and hence that a regular polygon subtends a smaller solid angle than a disc of the same area.

10

PARTIAL DERIVATIVES

In this unit we extend the rules for differentiating functions of a single variable to functions of several variables. We shall then be able to deal with functions of the three space coordinates and functions of space and time. This step is essential if we are to be able to formulate the laws of the physical sciences in more than one space dimension in a general way. As an example, you are familiar with how to find a maximum or minimum of a function of one variable, but how do you find these for a function of two variables? Pictorially this is equivalent to asking how to find the summit of a hill; you might also need to find the steepest way down. We also consider the multi-dimensional analogues of the chain rule for differentiation and Taylor series for the expansion of a function about a given point, and we introduce the notion of a differential.

10.1. FUNCTIONS OF TWO VARIABLES

Let $f(x, y) = x^2 + xy^2$. This means that for each x and y (chosen independently) there corresponds a value for f given by this expression.

Example 10.1 At $(x, y) = (1, 0)$, $f(x, y) = x^2 + xy^2$ has the value $1^2 + 1 \times 0^2 = 1$; at $(x, y) = (1, 2)$, f has the value $1^2 + 1 \times 2^2 = 5$.

Exercise 10.1

(i) What value does the function $f(x, y) = x^3 y + y^2$ have at $(x, y) = (-1, 3)$?

(ii) What value does $g(u, v) = u^3 v + v^2$ have at $(u, v) = (-1, 3)$?

(iii) And at $(u, v) = (2, 4)$?

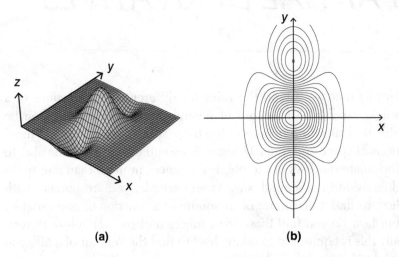

(a)   **(b)**

FIGURE 10.1: The function $z = \left(2 + x^2 - y^2\right)e^{1 - x^2 - y^2/4}$. (a) A surface plot. (b) A contour(level curves) plot. The axes have been offset from the origin in (a) to avoid cluttering the plot.

The graph of a function of two variables $f(x, y)$ is the set of all points (x, y, z) for which $z = f(x, y)$. We can think of this as a surface, with z being the height of the surface that varies with x and y. Using a computer we can render such a surface, as in Figures 10.1(a) and 10.2(a). Another way to visualize a function of two variables is to plot the *level curves*, the curves of constant $z = f(x, y)$. These are just *contour lines* of the surface, as shown in Figures 10.1(b) and 10.2(b). It is usually best to plot contours for values of z that are evenly spaced. Where the contour lines are close together, the function must be changing rapidly (increasing or decreasing over short distances in the (x, y) plane in a direction perpendicular to the contours); where the contours are widely separated the function must be varying slowly (staying nearly constant).

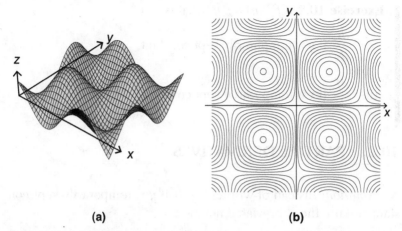

(a)　　　　　　　　　　　　　　　　**(b)**

FIGURE 10.2: The function $z = sin(x)\ sin(y)$. (a) A surface plot. (b) A contour (level curves) plot. The axes have been offset from the origin in (a) to avoid cluttering the plot.

Let $f(x, y) = x^2 y + y^2$. Keeping y constant temporarily we find $\dfrac{df}{dx}$ by the standard rules of differentiation:

$$\frac{df}{dx} = 2xy + 0 \quad \left(\text{treating } y \text{ as constant temporarily}\right).$$

If instead we keep x constant temporarily we can find $\dfrac{df}{dy}$ by the standard rules:

$$\frac{df}{dy} = x^2 + 2y \quad \left(\text{treating } x \text{ as constant temporarily}\right).$$

Example 10.2 If $f(x, y) = x^3/y$ what is (i) $\dfrac{df}{dx}$ if y is temporarily kept constant and (ii) $\dfrac{df}{dy}$ if x is temporarily kept constant?

(i) $\dfrac{df}{dx} = \dfrac{3x^2}{y}$ (keeping y constant).

(ii) $\dfrac{df}{dy} = -\dfrac{x^3}{y^2}$ (keeping x constant).

Exercise 10.2 If $f(x, y) = x^2 y$, what is

(i) $\dfrac{df}{dx}$ if y is temporarily kept constant,

(ii) $\dfrac{df}{dy}$ if x is temporarily kept constant?

10.2. PARTIAL DERIVATIVES

Notation: Instead of writing "df/dx if y is temporarily kept constant" we use the abbreviated notation

$$\left(\frac{\partial f}{\partial x}\right)_y.$$

We call this the partial derivative of f with respect to x (with y held constant). The expression is read as "partial df by dx." Similarly "df/dy (keeping x temporarily constant)" becomes

$$\left(\frac{\partial f}{\partial y}\right)_x,$$

the partial derivative of f with respect to y, and is read as "partial df by dy." If f is already understood to be a function of x and y we can abbreviate these further to

$$\frac{\partial f}{\partial x} \quad \text{and} \quad \frac{\partial f}{\partial y}.$$

The "curly" $\partial/\partial x$ signifies that everything other than x is to be kept constant temporarily (i.e. during the process of differentiation). There are a number of alternative notations; the following mean exactly the same thing:

$$\left(\frac{\partial f}{\partial x}\right)_y, \quad \frac{\partial f}{\partial x}, \quad f_{,x}, \quad f_x, \quad \partial_x f.$$

Interpretation: The function $f(x, y)$ can be represented by plotting its value at (x, y) as the height of a surface above the point (x, y) in the (x, y) plane, as in Figure 10.3. If we keep y constant, f is then

Exercise 10.3 If $f(x, y) = x \sin(y)$, what is

(i) $\dfrac{\partial f}{\partial x}$

(ii) $\dfrac{\partial f}{\partial y}$

(iii) $\dfrac{\partial f}{\partial y}(1,0)$ –i.e. $\dfrac{\partial f}{\partial y}$ evaluated at $(x, y) = (1, 0)$?

Exercise 10.4 If $z(u, V) = u^2 + V^2$, what is z_u?

a function of x only (for that value of y); e.g. if $y = 1$ and $f(x, y) = x^2 + y^2$ then $f = x^2 + 1$. This is represented as the cross-section of f in the plane $y = 1$, i.e. the curve $z = x^2 + 1$. (Figure 10.3 shows a similar function.)

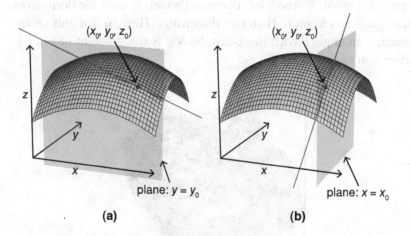

(a)   (b)

FIGURE 10.3: The partial derivative as the slope of a cross-section. (a) A function $z = f(x, y)$ plotted as a surface. A plane of constant y ($= y_0$) is shown. The curve $z = f(x, y_0)$ is the intersection of the surface with this plane. The gradient of the tangent of this curve at some point $P = (x_0, y_0)$ is the partial derivative of z with respect to x at P. (b) The same surface, plotted with a plane of constant x ($= x_0$). The curve $z = f(x_0, y)$ is the intersection of the surface with this plane. The gradient of the tangent of this curve at P is the partial derivative of z with respect to y at P.

The function obtained from $f(x, y)$ by keeping $y = y_0$ constant is the cross-section of the surface $z = f(x, y)$ in the plane $y = y_0$. The partial derivative $f_x(x_0, y_0)$ is the slope of this curve at the point (x_0, y_0).

10.3. DIRECTIONAL DERIVATIVES

We have seen that $(\partial f/\partial x)_y$ is the slope of the surface $z = f(x, y)$ in the x-direction, i.e. parallel to the x-axis (see Figure 10.3). Similarly, $(\partial f/\partial x)_x$ is the slope of the surface in the y-direction. The slope of the surface in the direction making angle θ (counterclockwise) with the x-axis is

$$\nabla_{\hat{u}} f = \cos(\theta)\frac{\partial f}{\partial x} + \sin(\theta)\frac{\partial f}{\partial y}. \tag{10.1}$$

We call this expression the *directional derivative* of f. It uses a special symbol, ∇ (the "del" operator) which is used for derivatives and gradients in more than one dimension. Here, $\hat{u}$ is a unit vector making an angle θ with the x-axis. So $\nabla_{\hat{u}} f$ is the gradient of f in the direction of the vector $\hat{u}$.

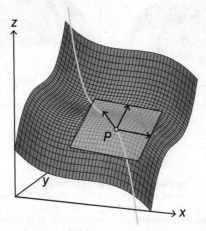

FIGURE 10.4: Illustration of the directional derivative. A function $z = f(x, y)$ is plotted as a surface. At a point P the tangent plane along with tangents parallel to the x- and y-axes are marked. Also marked is the tangent of the function z in some arbitrary direction following a curve along the surface. The gradient of this is the directional derivative, which is a linear combination of the gradients along each of the axes.

Example 10.3 Find the derivative of $f = 1 - x^2 - y^2$ in the direction making angle $\pi/4$ (counterclockwise) with the x-axis at the point $(1, 2)$.

The derivative in a particular direction is given by equation (10.1) so we first need the partial derivatives of f:

$$\frac{\partial f}{\partial x} = -2x, \quad \frac{\partial f}{\partial y} = -2y.$$

Inserting these values into (10.1) we get

directional derivative $= \cos(\pi/4)(-2x) + \sin(\pi/4)(-2y)$.

Since we want the directional derivative at the point $(1, 2)$ we let $x = 1$ and $y = 2$:

$$\text{directional derivative} = \frac{\sqrt{2}}{2}(-2) + \frac{\sqrt{2}}{2}(-4) = -3\sqrt{2}.$$

Exercise 10.5 Find the derivative of $f = x^3 + y^2$ in the direction making an angle $60°$ (counterclockwise) with the x-axis at the point $\left(-1, \sqrt{3}\right)$.

There is a connection here to vectors that we will mention in passing, but we will return to this idea when we study vector calculus. If $\hat{\mathbf{u}}$ is a unit vector at an angle θ with the x-axis then $\hat{\mathbf{u}} = \cos(\theta)\mathbf{i} + \sin(\theta)\mathbf{j}$, and its components are $(\cos \theta, \sin \theta)$. We can also define a vector whose components are the partial derivatives of f. We give this the special symbol ∇f, so

$$\nabla f = \frac{\partial f}{\partial x}\mathbf{i} + \frac{\partial f}{\partial y}\mathbf{j} = \left(\frac{\partial f}{\partial x}, \frac{\partial f}{\partial y}\right).$$

Now, the dot product of $\hat{\mathbf{u}}$ and ∇f gives the directional derivative

$$\nabla_{\hat{u}} f = \hat{\mathbf{u}} \cdot \nabla f = \cos(\theta)\frac{\partial f}{\partial x} + \sin(\theta)\frac{\partial f}{\partial y},$$

which is the same as we had before.

10.4. FUNCTIONS OF MANY VARIABLES

The notion of a partial derivative extends to functions of more than two variables (except that we cannot now draw a picture).

Example 10.4 Let (x, y, z) be independent variables and let $r(x, y, z) = (x^2 + y^2 + z^2)^{1/2}$. Find $(\partial r/\partial x)_{y,z}$, the derivative of r w.r.t. x keeping y and z constant. (We usually write this as just $\partial r/\partial x$ if it is understood that r is a function of x, y, z.) Note that, by assertion, z is here an independent variable, on the same footing as x and y, whereas in Section 10.2, z was used for a function of x and y. Care is needed to distinguish the dependent and independent variables in a given problem.

We are to differentiate r with respect to the variable x. Thus we shall treat the other variables as constants for now. If this is not clear consider the following: y^2 and z^2 are to be treated as constant so use a notation that makes this clear, say $y^2 + z^2 = a^2$. Then $r = (x^2 + a^2)^{1/2}$ which can be differentiated with respect to x, as a function of a function, in the usual way (i.e. by the chain rule):

$$\frac{dr}{dx} = \frac{1}{2}\left(x^2 + a^2\right)^{-1/2} \cdot 2x.$$

Now replace a^2 by $y^2 + z^2$ and we have that

$$\frac{\partial r}{\partial x} = \frac{1}{2}\frac{2x}{\left(x^2 + y^2 + z^2\right)^{1/2}} = \frac{x}{r}.$$

With more practice you can simply imagine y and z to be constants and write the calculation directly.

Exercise 10.6 If $f(x, y, z) = (x^2 + y^2 + z^2)^{-1/2}$ write down
(i) $\partial f/\partial x$,   (ii) $\partial f/\partial y$,   (iii) $\partial f/\partial z$.

Exercise 10.7 If $g(t, u, v, w) = t + u^2 v^3 w^4$, what is $\dfrac{\partial g}{\partial v}$?

10.5. HIGHER DERIVATIVES

Let $f(x, y) = 3x^3y^2$. We have $\partial f/\partial x = 9x^2y^2$ and $\partial f/\partial y = 6x^3y$. We can now differentiate again, apparently in four ways. We have

$$\frac{\partial}{\partial x}\left(\frac{\partial f}{\partial x}\right) = \frac{\partial}{\partial x}\left(9x^2y^2\right) = 18xy^2.$$

Similarly

$$\frac{\partial}{\partial y}\left(\frac{\partial f}{\partial x}\right) = \frac{\partial}{\partial y}\left(9x^2y^2\right) = 18x^2y.$$

We also have that

$$\frac{\partial}{\partial x}\left(\frac{\partial f}{\partial y}\right) = \frac{\partial}{\partial x}\left(6x^3y\right) = 18x^2y,$$

and

$$\frac{\partial}{\partial y}\left(\frac{\partial f}{\partial y}\right) = \frac{\partial}{\partial y}\left(6x^3y\right) = 6x^3.$$

We write these second derivatives more compactly as

$$\frac{\partial^2 f}{\partial x^2} = 18xy^2, \quad \frac{\partial^2 f}{\partial x \partial y} = 18x^2y, \quad \frac{\partial^2 f}{\partial y \partial x} = 18x^2y, \quad \text{and} \quad \frac{\partial^2 f}{\partial y^2} = 6x^3.$$

If we define the mixed second derivative f_{xy} to mean $(f_x)_y$ (i.e. $\partial/\partial y(\partial f/\partial x)$) we can write even more succinctly $f_{xx} = 18xy^2$, $f_{xy} = 18x^2y$ etc.

Exercise 10.8 If $f(x, y, z) = (x^2 + y^2 + z^2)^{1/2}$, find
(i) f_x, (ii) f_y, (iii) f_{xx}, (iv) f_{yx}, (v) f_{xy}, (vi) f_{yy}.

Note in the above example and in Exercise 10.8 we have that

$$\frac{\partial^2 f}{\partial x \partial y} = \frac{\partial^2 f}{\partial y \partial x} \quad \text{equivalently} \quad f_{yx} = f_{xy}. \tag{10.2}$$

This can be shown to be the case for reasonably smooth functions (those normally occurring in physics): *the order in which partial derivatives are taken does not matter.*

Exercise 10.9 Let $f = e^x \log(x + y)$. Find f_x and f_y and verify that $f_{xy} = f_{yx}$.

10.6. FUNCTION OF A FUNCTION

We now examine the rule for differentiating a "function of a function" (recall Section 1.3) for functions of many variables. If we have some function $f(u)$ where $u(x, y)$ is itself a function of (x, y) we wish to find the rule that gives us $\partial f / \partial x$ etc. Here, f can be written either as a function of just u, or as a function of both x and y together. If u were a function only of x the chain rule would give

$$\frac{df}{dx} = \frac{df}{du}\frac{du}{dx}.$$

But $u(x, y)$ is a function of x and y. However, we hold y constant when we compute the partial derivative with respect to x. So

$$\frac{\partial f}{\partial x} = \frac{df}{du}\frac{\partial u}{\partial x}. \tag{10.3}$$

where the curly ∂'s remind us that y is being held constant. Otherwise there is nothing new here. (Notice we do not need the curly ∂'s for df/du since f can be written as a function of u only.)

Example 10.5 Let $r(x, y, z) = (x^2 + y^2 + z^2)^{1/2}$, and let $f = 1/r$. Find $\partial f / \partial x$.

To obtain $\partial f / \partial x$, we treat y and z as constants. We apply

$$\frac{\partial f}{\partial x} = \frac{df}{dr}\frac{\partial r}{\partial x}.$$

We have from Example 10.4 that

$$\frac{\partial r}{\partial x} = \frac{x}{r},$$

and we can see that

$$\frac{df}{dr} = -\frac{1}{r^2}.$$

So, these give

$$\frac{\partial f}{\partial x} = -\frac{1}{r^2} \cdot \frac{x}{r} = -\frac{x}{r^3}.$$

This example is important in calculations that involve the inverse square law.

Exercise 10.10

(i) Let $r(x, y, z) = (x^2 + y^2 + z^2)^{1/2}$ and let $f = 1/r^n$. Find $\dfrac{\partial r}{\partial x}$.

(ii) Let $f = 1/r$; find $\dfrac{\partial^2 f}{\partial x^2}$ and write down expressions for $\dfrac{\partial^2 f}{\partial y^2}$ and $\dfrac{\partial^2 f}{\partial z^2}$.

Let $f(u)$ be a given function of u and let $u(x, y, z)$ be a given function of (x, y, z). If $u = u(x, y, z)$ is substituted for the variable in $f(u)$, then we have seen that

$$\frac{\partial f}{\partial x} = \frac{df}{du}\frac{\partial u}{\partial x},$$

but in general

$$\frac{\partial^2 f}{\partial x^2} \neq \frac{d^2 f}{du^2} \times \frac{\partial^2 u}{\partial x^2}; \tag{10.4}$$

i.e. we cannot simply change the first derivatives for second derivatives and expect the equation to be valid.

Exercise 10.11 Demonstrate in equation (10.4) by means of an example (of your choice) for the functions $f(u)$ and $u(x, y, z)$. Establish the correct formula

$$\frac{\partial^2 f}{\partial x^2} = \frac{d^2 f}{du^2} \times \left(\frac{\partial u}{\partial x}\right)^2 + \frac{df}{du} \times \frac{\partial^2 u}{\partial x^2},$$

and verify that this is correct for your example.

10.7. IMPLICIT DIFFERENTIATION

We can partially differentiate both sides of an equation using the "function of a function" rule, and then rearrange to find the partial derivative that we seek (recall Section 1.5). Apart from the use of curly ∂ there is nothing new here.

Example 10.6 Let $r(x, y, z)$ be given by $r^2 = x^2 + y^2 + z^2$. Find $\dfrac{\partial r}{\partial x}$.

First, differentiate the left side, which is a function (r^2) of a function $r(x, y, z)$, with respect to x:

$$\frac{\partial}{\partial x}\left(r^2\right) = \frac{d\left(r^2\right)}{dr}\frac{\partial r}{\partial x} = 2r\frac{\partial r}{\partial x}.$$

On the right side y and z are constants when differentiating partially with respect to x; thus

$$\frac{\partial}{\partial x}\left(r^2\right) = 2x + 0 + 0.$$

Comparing these two we see that

$$\frac{\partial r}{\partial x} = \frac{x}{r}.$$

Exercise 10.12 Let $\sin^2(w) = x^2 + y^2 + z^2$. Show that

$$\frac{\partial w}{\partial x} = \frac{2x}{\sin(2w)}.$$

10.8. DERIVATIVE WITH RESPECT TO A PARAMETER

The following is a simple but important application of partial differentiation. We have a function of two variables, $f(x, t)$, and we wish to know the derivative with respect to x of its integral with respect to t. To differentiate an integral with respect to a parameter the rule is

$$\frac{d}{dx}\left(\int_a^b f(x,t)\,dt\right) = \int_a^b \frac{\partial f(x,t)}{\partial x}\,dt \qquad (10.5)$$

On the left side we imagine performing the integral to obtain a function of x which we then differentiate (giving another function of x).On the right side we imagine differentiating the integrand first with respect to x and then integrating with respect to t to obtain a function of x. Equation (10.5) says that the order in which we do the differentiation does not matter: we get the same result either way. (The partial derivative is required on the right of equation (10.5) to indicate that t is held temporarily constant for the differentiation. On the left the integral is a function only of x anyway, so we write this as an ordinary derivative.)

Given two ways of evaluating the same thing we are free to choose the easier. This enables us to find quickly some useful integrals that would otherwise be difficult.

Example 10.7 Find $\dfrac{d}{dx}\displaystyle\int_0^1 \dfrac{\sin(tx)}{t}\,dt$.

Using equation (10.5) we have that

$$\frac{d}{dx}\int_0^1 \frac{\sin(tx)}{t}\,dt = \int_0^1 \frac{\partial}{\partial x}\left(\frac{\sin(tx)}{t}\right)dt$$

$$= \int_0^1 \cos(tx)\,dt$$

$$= \frac{1}{x}\Big[\sin(tx)\Big]_{t=0}^{t=1}$$

$$= \frac{\sin(x)}{x}.$$

354 • Mathematical Physics

Example 10.8 From the equation

$$\int_0^1 t^x dt = \frac{1}{x+1}, \quad \text{for} \quad x > -1,$$

deduce that

$$\int_0^1 t^x \left(\log(t)\right)^m dt = \frac{(-1)^m m!}{(x+1)^{m+1}}, \quad \text{for} \quad m = 1, 2, \dots.$$

First, differentiate the right side of the given equation:

$$\frac{d}{dx}\int_0^1 t^x dt = \frac{d}{dx}\left(\frac{1}{x+1}\right) = -\frac{1}{(x+1)^2}.$$

Next we differentiate the left side (with respect to x) using equation (10.5). (Recall that you learned how to differentiate a^x in Section 2.7.)

$$\frac{d}{dx}\int_0^1 t^x dt = \int_0^1 \frac{\partial}{\partial x}(t^x) dt$$

$$= \int_0^1 t^x \left(\log(t)\right) dt.$$

Therefore

$$\int_0^1 t^x \left(\log(x)\right) dt = -\frac{1}{(x+1)^2}.$$

If we differentiate both sides again we get another factor of $\log(t)$ from the differentiation of t^x with respect to x:

$$\int_0^1 t^x \left(\log(t)\right)^2 dt = -\frac{(-1)(-2)}{(x+1)^3},$$

and hence repeating this for m applications gives the desired result

$$\int_0^1 t^x \left(\log(t)\right)^m dt = \frac{(-1)^m m!}{(x+1)^{m+1}}, \quad \text{for} \quad m = 1, 2, \dots.$$

Exercise 10.13 Show that $\int_0^{\pi/2} t\sin(tx)dt = -\dfrac{d}{dx}\int_0^{\pi/2}\cos(tx)dt$.

Given that $\int_0^{\pi/2}\cos(tx)dt = \dfrac{\sin(\pi x/2)}{x}$, find $\int_0^{\pi/2} t\sin(t)dt$.

(This is an alternative to integration by parts for evaluating this integral.)

Exercise 10.14 Given that $\int_0^\infty e^{-\lambda t}dt = \dfrac{1}{\lambda}$, use the method of

Example 10.8 to find $\int_0^\infty t^n e^{-\lambda t}dt$, for $n = 1,2,\ldots$.

Hence show that $\int_0^\infty x^n e^{-x}dx = n!$ and show that $0! = 1$ using this formula. (Thus, this formula agrees with the usual definition of $n!$ for n a positive integer, and gives a meaning to cases when n is not a positive integer.)

10.9. TAYLOR EXPANSION ABOUT THE ORIGIN

One of the most important applications of partial derivatives in physics is the approximation of a function by a simple formula near a given point. Suppose we are given the function $f(x,y)$. We can evaluate f and its partial derivatives at the point $P = (0, 0)$. Let

$$f(P) = f(0,0),\quad f_x(P) = \left(\frac{\partial f}{\partial x}\right)(0,0),\quad f_y(P) = \left(\frac{\partial f}{\partial y}\right)(0,0),$$

$$f_{xx}(P) = \left(\frac{\partial^2 f}{\partial x^2}\right)(0,0),$$

(10.6)

and so on. The multiple variable form of Taylor's theorem says that we can approximate the function $f(x, y)$ near the point $(0, 0)$ by a finite number of terms from the following series:

$$f(x,y) = f(P) + xf_x(P) + yf_y(P)$$
$$+ \frac{1}{2!}\left[x^2 f_{xx}(P) + 2xy f_{xy}(P) + y^2 f_{yy}(P)\right]$$
$$+ \frac{1}{3!}\left[x^3 f_{xxx}(P) + 3x^2 y f_{xxy}(P) + 3xy^2 f_{xyy}(P) + y^3 f_{yyy}(P)\right] + \ldots.$$

(10.7)

Equation (10.7) is the *Taylor expansion of f(x, y)* about (0, 0). Taylor series centered on the origin, like these, are often called *Maclaurin series*. We discussed both Maclaurin and Taylor series for functions of a single variable in Sections 2.2 and 2.3.

The coefficients in each group of terms follow the binomial pattern (Section 2.1), although in practice we almost never need the higher order ones. Note that however complicated our initial function f, the Taylor series involves powers of x and y only (in practice usually up to quadratic or cubic at most). This is the point of the calculation. If you end up with anything more complicated you have made a mistake.

Example 10.9 Find the Taylor series expansion up to second order of the function $f(x,y) = \sin(xy)$ about the point $P = (0, 0)$.

We evaluate each of the required quantities at the given point $(0, 0)$ in turn, starting with the function itself:

$$f(P) = f(0, 0) = \sin(0) = 0.$$

Next, each of the first derivatives at P:

$$f_x = y \cos(xy), \text{ so } f_x(P) = 0 \times \cos(0) = 0,$$
$$f_y = x \cos(xy), \text{ so } f_y(P) = 0 \times \cos(0) = 0.$$

Then each of the second derivatives. Note that we compute all the derivatives and only then substitute values for x and y at P:

$$f_{xx} = -y^2 \sin(xy), \text{ so } f_{xx}(P) = 0 \times \sin(0) = 0,$$
$$f_{yy} = -x^2 \sin(xy), \text{ so } f_{yy}(P) = 0 \times \sin(0) = 0.$$

We can compute either of f_{xy} or f_{yx} (i.e. differentiate either f_x or f_y) since these are equal:

$$f_{xy} = \cos(xy) - xy \sin(xy), \text{ so } f_{xy}(P) = \cos(0) - 0 \times \sin(0) = 1.$$

Finally, using the definition of the Taylor series (10.7)

$$\sin(xy) \approx 0 + x \cdot 0 + y \cdot 0 + \frac{1}{2!}\left[x^2 \cdot 0 + 2xy \cdot 1 + y^2 \cdot 0\right] + \dots$$

and so, $\sin(xy) \approx xy + \dots$ to second order near $(0,0)$.

The result agrees with the first term of the series expansion for $\sin(u)$ with $u = xy$, as, of course, it must. You will find it extremely helpful if you set out your work systematically like this.

> **Exercise 10.15** Let $f(x,y) = 1 + e^x e^y$ and $P = (0, 0)$. Find $f_x(P)$, $f_y(P), f_{xx}(P), f_{xy}(P), f_{yy}(P)$, and hence obtain the Taylor expansion of f about the origin to second order.

10.10. EXPANSION ABOUT AN ARBITRARY POINT

Suppose we are given the function $f(x, y)$. We can evaluate f and its partial derivatives at the point $P = (x_0, y_0)$. Let

$$f(P) = f(x_0, y_0), \quad f_x(P) = \frac{\partial f}{\partial x}(x_0, y_0),$$

and so on. Taylor's theorem says we can approximate the function $f(x,y)$ near the point (x_0, y_0) by a finite number of terms from the following series:

$$f(x,y) = f(P) + (x - x_0) f_x(P) + (y - y_0) f_y(P)$$
$$+ \frac{1}{2!} \Big[(x - x_0)^2 f_{xx}(P) + 2(x - x_0)(y - y_0) f_{xy}(P) + (y - y_0)^2 f_{yy}(P) \Big]$$
$$+ \dots \tag{10.8}$$

Equivalently, we can write this as

$$f(x_0 + h, y_0 + k) = f(x_0, y_0) + h f_x(x_0, y_0) + k f_y(x_0, y_0)$$
$$+ \frac{1}{2!} \Big[h^2 f_{xx}(x_0, y_0) + 2hk f_{xy}(x_0, y_0) + k^2 f_{yy}(x_0, y_0) \Big] + \dots \tag{10.9}$$

where $h = x - x_0$ and $k = y - y_0$ are the displacements from P. Equation (10.8) (or (10.9)) is the Taylor expansion of $f(x, y)$ about the point $P = (x_0, y_0)$. It is easier to see the form of the series if we leave out explicit reference to the point P.

$$f(x_0 + h, y_0 + k) = f + h f_x + k f_y$$
$$+ \frac{1}{2!} \Big[h^2 f_{xx} + 2hk f_{xy} + k^2 f_{yy} \Big] \tag{10.10}$$
$$+ \frac{1}{3!} \Big[h^3 f_{xxx} + 3h^2 k f_{xxy} + 3hk^2 f_{xyy} + k^3 f_{yyy} \Big] + \dots.$$

Example 10.10 Find the Taylor expansion of $f = \sin(xy)$ about the point $P = (\pi/2, 1)$.

Evaluate each of the coefficients in the Taylor series in turn at the point P:

$$f(P) = \sin(\pi/2) = 1,$$

$$f_x = y\cos(xy), \text{ so } f_x(P) = \cos(\pi/2) = 0,$$

$$f_y = x\cos(xy), \text{ so } f_y(P) = (\pi/2)\cos(\pi/2) = 0,$$

$$f_{xx} = -y^2\sin(xy), \text{ so } f_{xx}(P) = -\sin(\pi/2) = -1,$$

$$f_{xy} = \cos(xy) - xy\sin(xy),$$

so $f_{xy}(P) = \cos(\pi/2) - \dfrac{\pi}{2}\cdot 1\cdot\sin(\pi/2) = -\pi/2,$

$$f_{yy} = -x^2\sin(xy), \text{ so } f_{yy}(P) = -(\pi/2)^2\sin(\pi/2) = -\pi^2/4.$$

Substitute into the general form of the Taylor series (10.8):

$$\sin(xy) = 1 + (x - \pi/2)\times 0 + (y-1)\times 0$$

$$+\frac{1}{2!}\Big[(x-\pi/2)^2(-1) + 2(x-\pi/2)(y-1)(-\pi/2)$$

$$+(y-1)^2(-\pi^2/4)\Big] + ...,$$

and so

$$\sin(xy) = 1 - \frac{1}{2}\Big[(x-\pi/2)^2 + \pi(x-\pi/2)(y-1) + \frac{\pi^2}{4}(y-1)^2\Big] + ...$$

The result is a power series (or a quadratic if we stop after the second order terms). Near the point $(\pi/2, 1)$ we can put $X = x - \pi/2$ and $Y = y - 1$ and write the series more succinctly as

$$\sin(xy) = 1 - \frac{1}{2}X^2 - \frac{\pi}{2}XY - \frac{\pi^2}{8}Y^2 + ...$$

Exercise 10.16 Let $f(x,y) = x^3y^4 + x^4y^3$. Find the partial derivatives of f up to second order at $(1,1)$ and hence obtain the Taylor expansion of f about the point $(1, 1)$ to second order.

10.11.   FUNCTION OF MORE THAN TWO VARIABLES

Suppose we are given the function $f(x, y, z)$. We can evaluate f and its partial derivatives at the point $P = (x_0, y_0, z_0)$. Write

$$f(P) = f(x_0, y_0, z_0), \quad f_x(P) = \frac{\partial f}{\partial x}(x_0, y_0, z_0) \quad \text{etc.}$$

Taylor's theorem says that we can approximate the function $f(x, y, z)$ near the point (x_0, y_0, z_0) by a finite number of terms from the following series:

$$f(x, y, z) = f(P) + (x - x_0) f_x(P) + (y - y_0) f_y(P) + (z - z_0) f_z(P)$$
$$+ \frac{1}{2!} \Big[(x - x_0)^2 f_{xx}(P) + 2(x - x_0)(y - y_0) f_{xy}(P)$$
$$+ 2(x - x_0)(z - z_0) f_{xz}(P) + (y - y_0)^2 f_{yy}(P)$$
$$+ 2(y - y_0)(z - z_0) f_{yz}(P) + (z - z_0)^2 f_{zz}(P) \Big] + \dots.$$
$$(10.11)$$

Extensions to more variables follow a similar pattern, but you are unlikely to meet them.

To sum up then, if we need to find the behavior of a function near a given point we can approximate the function by a polynomial, usually a quadratic, which is much simpler than the original function.

10.12.   STATIONARY POINTS

We return to the problem of finding stationary points (from Section 1.8), now considering functions of more than one variable.

In two dimensions a stationary point can be a maximum, minimum or saddle point. At a stationary point, by definition, $f(x, y)$ is approximately constant (to first order) (Figure 10.5). A necessary condition for $P = (x_0, y_0)$ to be a stationary point is that

$$\frac{\partial f}{\partial x}(x_0, y_0) = 0, \quad \text{and} \quad \frac{\partial f}{\partial y}(x_0, y_0) = 0, \qquad (10.12)$$

or, in more compact notation, $f_x(P) = f_y(P) = 0$. If the derivatives along the two directions are both zero, then the derivative in any direction is zero (by Equation 10.1); i.e. the tangent plane at P is horizontal.

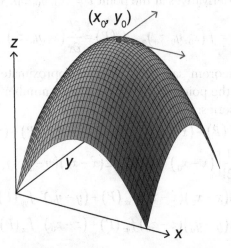

FIGURE 10.5: Near a stationary point the function $f(x, y)$ is constant to first order (in all directions).

Example 10.11 Find the stationary points of the function $f(x, y) = x^2y + 2xy^2 + 6xy$.

First, calculate f_x and f_y and set them equal to 0:

$$f_x = 2xy + 2y^2 + 6y = y(2x + 2y + 6) = 0 \text{ and}$$
$$f_y = x^2 + 4xy + 6x = x(x + 4y + 6) = 0$$

for a stationary point. Now solve the two equations for x and y. One solution is $(x, y) = (0, 0)$. If this is not the case then

(i) $y \neq 0$ and so $2x + 2y + 6 = 0$, or

(ii) $x \neq 0$ and so $x + 4y + 6 = 0$.

If $y = 0$ then (ii) gives $x = -6$, while if $x = 0$ then (i) gives $y = -3$. If $x \neq 0$, and $y \neq 0$ then (i) and (ii) are two simultaneous equations for x and y which can be solved to give $x = -2$, $y = -1$. So we have found four stationary points

$$(x, y) = (0, 0), (-6, 0), (0, -3), \text{ and } (-2, -1).$$

These are the stationary points of f.

Exercise 10.17 Find the stationary points of the functions

(i) $f(x, y) = x^3 - 2x^2 - 4x - y^2 + 6y$,

(ii) $f(x, y) = xy^2 - 2y - x$.

In order to identify the type of stationary point we use a second derivative test (compare with the test used for functions of a single variable in Section 1.8). We calculate the three second derivatives at P:

$$a = f_{xx} = \frac{\partial^2 f}{\partial x^2}, \quad b = f_{xy} = \frac{\partial^2 f}{\partial x \partial y}, \quad c = f_{yy} = \frac{\partial^2 f}{\partial y^2}$$

(There is no need to compute f_{yx} because it is the same as f_{xy}.) We then compute

$$D = ac - b^2 = f_{xx}f_{yy} - (f_{xy})^2. \qquad (10.13)$$

If $D > 0$ then the point is a maximum or minimum; if not then it is a saddle point or requires more detailed investigation to identify its nature (see Figure 10.6). The second derivative test works as follows:

- If $D > 0$ and $f_{xx} > 0$ then P is a minimum (f_{yy} will also be > 0).
- If $D > 0$ and $f_{xx} < 0$ then P is a maximum (f_{yy} will also be < 0).
- If $D < 0$ then P is a saddle point (a maximum in one direction and a minimum in another, like the top of a mountain pass).
- If $D = 0$ then further investigation is needed.

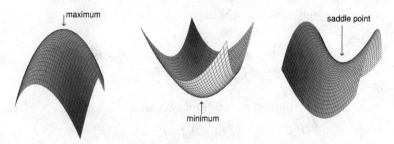

FIGURE 10.6: Three different types of stationary point for a function of two variables. Maximum (left), minimum (center) and saddle point (right). Near a stationary point the function $f(x, y)$ is constant to first order. As you can see from the figures, stationary points are important for the analysis of stability.

Example 10.12 Find the stationary point of the function $f(x, y) = 3x^2 - 2xy + y^2$ and determine its nature.

First we compute f_x and f_y to find the stationary point(s)

$$f_x = 6x - 2y \text{ and } f_y = -2x + 2y.$$

At a stationary point $f_x = 0$ and $f_y = 0$ so from the second equation $x = y$ and from the first $6x - 2x = 4x = 0$ gives $x = y = 0$; hence $P = (0, 0)$ is the only stationary point. To determine whether this is a maximum or minimum or saddle point we compute the second derivatives and D.

$$f_{xx} = 6; f_{xy} = -2; f_{yy} = 2;$$
$$f_{xx}(P) = 6 > 0; \text{ and } f_{yy}(P) = 2 > 0;$$
$$D = f_{xx}(P)f_{yy}(P) - f_{xy}^2(P) = 12 - 4 > 0.$$

So $P = (0, 0)$ is a minimum.

Example 10.13 Find and determine the nature of the stationary points of the function

$$f(x, y) = x^2y + 2y^2x + 6xy.$$

The stationary points have been found in Example 10.11; they were $(0, 0)$, $(-6, 0)$, $(0, -3)$, $(-2, -1)$. So next we compute the second partial derivatives:

$$f_x = 2xy + 2y^2 + 6y, f_y = x^2 + 4xy + 6x,$$
$$f_{xx} = 2y, \quad f_{xy} = 2x + 4y + 6, \quad f_{yy} = 4x.$$

Now evaluate them at the stationary points:

$$P = (0, 0): f_{xx}(P) = 0, f_{xy}(P) = 6, f_{yy}(P) = 0.$$

Therefore $D = f_{xx}(P)f_{yy}(P) - f_{xy}(P)^2 < 0$ and $(0, 0)$ is a saddle point.

$$P = (-6, 0): f_{xx}(P) = 0, f_{xy}(P) = -6, f_{yy}(P) = -24.$$

Therefore $D = f_{xx}(P)f_{yy}(P) - f_{xy}(P)^2 < 0$ and $(-6, 0)$ is a saddle point.

$$P = (0, -3): f_{xx}(P) = -6, f_{xy}(P) = -6, f_{yy}(P) = 0.$$

Therefore $f_{xx}(P)f_{yy}(P) - f_{xy}(P)^2 < 0$ and $(0, -3)$ is a saddle point.

$$P = (-2, -1): f_{xx}(P) = -2, f_{xy}(P) = -2, f_{yy}(P) = -8.$$

Therefore $D = f_{xx}(P)f_{yy}(P) - f_{xy}(P)^2 > 0$ and so P is either a maximum or a minimum. Since $f_{xx}(P)$ and $f_{yy}(P)$ are negative, $(-2, -1)$ is a maximum.

Exercise 10.18

(i) Find and determine the nature of the stationary points of the function

$$f(x,y) = x^2 - 2x - y^3 + y^2 + 8y.$$

(ii) Show that the function

$$f(x,y) = -[x^2y - (x + 1)]^2 - (x^2 - 1)^2$$

has maxima at $(1, 2)$ and $(-1, 0)$.

(The function in Exercise 10.18(ii) has two peaks and no passes. This is a counter-example to the intuitive but incorrect idea that two mountain peaks must be separated by a pass. In effect the pass here has been shifted to infinity.)

Examination Using Taylor Series

Now we have seen how to find and test stationary points we can examine how this works using Taylor's theorem. Suppose we are given a function $f(x, y)$. We can use Taylor's theorem to find the stationary points (maxima and minima) of f as follows. Suppose $P = (x_0, y_0)$ is a stationary point and expand f as a Taylor series about (x_0, y_0):

$$f(x,y) \approx f(P) + (x - x_0)f_x(P) + (y - y_0)f_y(P) + \dots$$

At a stationary point f is constant to first order; hence $(x - x_0)$ $f_x(P) + (y - y_0)f_y(P) = 0$ for all (x, y) near (x_0, y_0). Therefore P is a stationary point if $f_x = 0$ and $f_y = 0$ at P.

Just as was the case for functions of a single variable (Section 1.8) we can classify stationary points of a function of two variables as follows:

- (x_0, y_0) is a maximum if $f(x, y) < f(x_0, y_0)$ for all (x, y) near (x_0, y_0).

- (x_0, y_0) is a minimum if $f(x, y) > f(x_0, y_0)$ for all (x, y) near (x_0, y_0).

If neither condition holds, i.e. if for some (x, y), $f(x, y) > f(x_0, y_0)$ and for some (x, y), $f(x, y) < f(x_0, y_0)$, then

- (x_0, y_0) is a saddle point (see Figure 10.6).

To translate these conditions into a usable test we use Taylor's theorem to tell us what f looks like near $P = (x_0, y_0)$. This is clearer if we use $h = x - x_0$, $k = y - y_0$ to represent small displacements from x_0, y_0 and let $a = f_{xx}(P)$, $b = f_{xy}(P)$, $c = f_{yy}(P)$. Then Taylor's theorem up to second order is

$$f(x,y) = f(P) + \frac{1}{2}\left[h^2 f_{xx}(P) + 2hk f_{xy}(P) + k^2 f_{yy}(P)\right] + \cdots$$

$$= f(P) + \frac{1}{2}\left[ah^2 + 2bhk + ck^2\right] \qquad (10.14)$$

The linear terms vanish by equation (10.12). We can rewrite this as

$$f(x,y) - f(P) = \frac{1}{2}\left[ah^2 + 2bhk + ck^2\right] = \frac{1}{2}Q \qquad (10.15)$$

with $Q = ah^2 + 2bhk + ck^2$.

Now, if P is a maximum this means $f(x, y) < f(P)$ near P, and so $Q < 0$; or if P is a minimum then $f(x, y) > f(P)$ near P and so $Q > 0$, for all values of the displacements h and k. In both these cases $Q \neq 0$ which means the quadratic equation

$$Q = ah^2 + 2bhk + ck^2 = 0$$

can have no real solution except $h = k = 0$. This is the case if $b^2 - ac < 0$, or equivalently, $D = ac - b^2 > 0$. This is therefore the condition for a maximum or a minimum.

Hence the stationary point $P = (x_0, y_0)$ is

1. A minimum if

$$f_{xx}(P) > 0, \text{ and } f_{yy}(P) > 0, \text{ and } f_{xx}(P) f_{yy}(P) - f_{xy}^2(P) > 0$$

2. A maximum if

$$f_{xx}(P) < 0, \text{ and } f_{yy}(P) < 0, \text{ and } f_{xx}(P) f_{yy}(P) - f_{xy}^2(P) > 0$$

3. A saddle point if

$$f_{xx}(P) f_{yy}(P) - f_{xy}^2(P) < 0 \text{ (if } f_{xx}(P), f_{yy}(P) \text{ have opposite signs)}$$

If the second derivatives, f_{xx}, f_{yy} and f_{xy}, are all zero one would need to investigate the behavior of the function in more detail to correctly identify the type of stationary point.

10.13.   DEFINITION OF THE TOTAL DIFFERENTIAL

Given a function $f(x, y)$, its *total differential*, df, is defined by

$$df = \frac{\partial f}{\partial x} dx + \frac{\partial f}{\partial y} dy \quad \text{equivalently} \quad df = f_x dx + f_y dy. \quad (10.16)$$

Since this is a definition it is not necessary to ask what it means we shall simply show that it is useful.

Example 10.14 Find the total differential of the function $f(x, y) = xy^3$.

We calculate $f_x = y^3$ and $f_y = 3xy^2$. So from the definition (10.16) we have

$$df = y^3 dx + 3xy^2 dy.$$

> **Exercise 10.19** Find the total differential of
>
> (i) $f(x, y) = x^3y + x/y$,
>
> (ii) $f(u, v) = u^2/v$.

10.14. EXACT DIFFERENTIALS

Now consider the converse: given $a(x, y)$ and $b(x, y)$ can we write $adx + bdy = df$ for some f? If such an f exists we say that $adx + bdy$ is an *exact differential* (it is also called a *perfect differential*). It is often useful to know if $adx + bdy$ is an exact differential, because, if it is, then there exists a curve $f(x, y) =$ constant, such that (a, b) is normal to the curve at each point. As an example, let $a = 3x^2y + 1/y$ and $b = x^3 - x/y^2$; does there exist an f such that

$$df = (3x^2y + 1/y)dx + (x^3 - x/y^2)dy?$$

If you are good at this sort of thing you can guess (with the help of Exercise 10.19) that $f = x^3y + x/y$ will do. Can we find a more mechanical method?

Suppose we can find an f such that

$$df = adx + bdy.$$

But then, given this f, we have

$$df = f_x dx + f_y dy,$$

by equation (10.16). Thus, if both are to be valid, $a = f_x$, $b = f_y$. Hence, $a_y = f_{xy}$, $b_x = f_{yx}$. Since $f_{xy} = f_{yx}$, this means that if f exists then

$$a_y = b_x.$$

Although we shall not prove it, this is in fact a necessary and sufficient condition for the differential to be exact.

Example 10.15 Suppose $a = (3x^2y + 1/y)$ and $b = (x^3 - x/y^2)$. Show that $adx + bdy$ is indeed an exact differential.

We check if $a_y = b_x$. We have $a_y = 3x^2 - 1/y^2$, $b_x = 3x^2 - 1/y^2 = a_y$, so $adx + bdy$ is an exact differential.

A differential $adx + bdy$ is said to be *inexact* if this is not true, i.e. there exists no function $f(x, y)$ with this differential. For example, $2ydx + 3xdy$ is an inexact differential. If it were an exact differential there would be some function $f(x, y)$ whose partial derivatives are $\partial f/\partial x = 2y$ and $\partial f/\partial y = 3x$. But we can integrate each of these to find $f(x, y)$. If we integrate the first with respect to x we get $f(x, y) = 2xy + g(y)$, where $g(y)$ is some function of y only; if we integrate the second with respect to y we get $f(x, y) = 3xy + h(x)$, where $h(x)$ is some function of x only. These two are inconsistent (for any choice of $g(y)$ and $h(x)$); therefore no such $f(x, y)$ exists. You can prove to yourself that this differential does not meet the criterion above (i.e. that $a_y \neq b_x$). The difference between exact and inexact differentials is important in several areas of physics, most notably thermodynamics.

Exercise 10.20 If $a = x$, $b = xy$, show that $adx + bdy$ is not an exact differential.

Exercise 10.21 If $a = \ln(yx^{3/2})$ and $b = x/y$, show that $adx + bdy$ is an exact differential. (This arises in thermodynamics, with, in the standard thermodynamic notation, $x = T$, $y = V$, $a = S$ and $b = P$.)

10.15. THE CHAIN RULE

Recall (Section 1.3) the rule for differentiating a "function of a function" of a single variable: given $f(u)$ and $u = g(x)$ we find df/dx by the chain rule:

$$\frac{df}{dx} = \frac{df}{du}\frac{du}{dx}. \tag{10.17}$$

Now we consider the case for a function of many variables. Given $f(u, v)$ and $u = u(x, y)$, $v = v(x, y)$ we find $\partial f/\partial x$ and $\partial f/\partial y$ by the chain rule:

$$\frac{\partial f}{\partial x} = \frac{\partial f}{\partial u}\frac{\partial u}{\partial x} + \frac{\partial f}{\partial v}\frac{\partial v}{\partial x}, \tag{10.18}$$

$$\frac{\partial f}{\partial y} = \frac{\partial f}{\partial u}\frac{\partial u}{\partial y} + \frac{\partial f}{\partial v}\frac{\partial v}{\partial y}. \tag{10.19}$$

(You can remember this by thinking of it as applying equation (10.17) to each variable in turn and summing the results.) Note that, for example, in equation (10.18), $\partial f/\partial x$ is found keeping y constant (so f is considered as a function of x and y) and $\partial f/\partial u$ is found keeping v constant (so f is considered as a function of u and v).

Example 10.16 Given $f(u, v)$ with $u = xy$ and $v = (x - y)$ find f_x and f_y in terms of f_u and f_v.

By the chain rule we have

$$\frac{\partial f}{\partial x} = \frac{\partial f}{\partial u}\frac{\partial u}{\partial x} + \frac{\partial f}{\partial v}\frac{\partial v}{\partial x}.$$

However

$$\frac{\partial u}{\partial x} = y \quad \text{and} \quad \frac{\partial v}{\partial x} = 1, \quad \text{so} \quad \frac{\partial f}{\partial x} = y\frac{\partial f}{\partial u} + \frac{\partial f}{\partial v}.$$

Similarly

$$\frac{\partial u}{\partial y} = x, \quad \text{and} \quad \frac{\partial v}{\partial y} = -1, \quad \text{so} \quad \frac{\partial f}{\partial y} = x\frac{\partial f}{\partial u} - \frac{\partial f}{\partial v}.$$

Example 10.17 Given $f(x, y)$ with $x = r\cos(\theta)$ and $y = r\sin(\theta)$ find f_r in terms of f_x and f_y.

By the chain rule we have

$$\frac{\partial f}{\partial r} = \frac{\partial f}{\partial x}\frac{\partial x}{\partial r} + \frac{\partial f}{\partial y}\frac{\partial y}{\partial r}.$$

But

$$\frac{\partial x}{\partial r} = \cos(\theta), \quad \text{and} \quad \frac{\partial y}{\partial r} = \sin(\theta),$$

and so we have that

$$\frac{\partial f}{\partial r} = \cos(\theta)\frac{\partial f}{\partial x} + \sin(\theta)\frac{\partial f}{\partial y}.$$

You will find that the chain rule is important wherever it is necessary to use different coordinate systems (e.g. Cartesians and polars).

Exercise 10.22 Complete Example 10.17 by finding f_θ in terms of f_x and f_y. Hence obtain expressions for f_x and f_y in terms of f_r and f_θ. Deduce that if $f = r\theta$ then

$$f_x = \theta \cos(\theta) - \sin(\theta) = \frac{x}{r}\cos^{-1}\left(\frac{x}{r}\right) - \frac{y}{r}, \quad \text{where} \quad r = \left(x^2 + y^2\right)^{1/2}.$$

10.16. CONSISTENCY WITH THE CHAIN RULE

Given $f(u, v)$ and $u = u(x,y)$, $v = v(x,y)$ we can form the following differentials:

$$df = f_u du + f_v dv \tag{10.20}$$
$$du = u_x dx + u_y dy \tag{10.21}$$
$$dv = v_x dx + v_y dy. \tag{10.22}$$

Substituting equations (10.21) and (10.22) into equation (10.20) we obtain

$$df = f_u(u_x dx + u_y dy) + f_v(v_x dx + v_y dy)$$
$$= (f_u u_x + f_v v_x)dx + (f_u u_y + f_v v_y)\, dy.$$

But equation (10.16) says

$$df = f_x dx + f_y dy.$$

Therefore, comparing coefficients of dx and dy in the last two equations,

$$f_x = f_u u_x + f_v v_x \text{ and } f_y = f_u u_y + f_v v_y \qquad (10.23)$$

which are exactly the chain rule equations (10.18) and (10.19).

Example 10.18 Given $f(x, v)$ and $v = v(x, y)$ find $(\partial f/\partial x)_y$ in terms of $(\partial f/\partial x)_v$ and $(\partial f/\partial v)_x$.

Given $f(x, v)$ we would get $f(x, y)$ by substituting for $v(x, y)$. For differentials, the strategy is the same: compute df in terms of dx and dy by substituting for dv. So, first we compute $df(x, v)$:

$$df = (f_x)_v dx + (f_v)_x dv.$$

Next, compute dv:

$$dv = (v_x)_y dx + (v_y)_x dy.$$

Substituting for dv

$$df = (f_x)_v dx + (f_v)_x[(v_x)_y dx + (v_y)_x dy]$$
$$= [(f_x)_v + (f_v)_x(v_x)_y]dx + (f_v)_x(v_y)_x dy.$$

Write df in the form "$(\cdots)dx + (\cdots)dy$":

$$df = (f_x)_y dx + (f_y)_x dy.$$

Finally, compare coefficients of dx:

$$(f_x)_y = (f_x)_v dx + (f_v)_x(v_x)_y.$$

i.e.

$$\left(\frac{\partial f}{\partial x}\right)_y = \left(\frac{\partial f}{\partial x}\right)_v + \left(\frac{\partial f}{\partial v}\right)_x \left(\frac{\partial v}{\partial x}\right)_y.$$

Exercise 10.23 Suppose we are given a function $f(x, y)$ with $x = x(t)$ and $y = y(t)$. Show that

$$df = \left(f_x \frac{dx}{dt} + f_y \frac{dy}{dt} \right) dt,$$

and hence that

$$\frac{df}{dt} = f_x \frac{dx}{dt} + f_y \frac{dy}{dt}.$$

Verify that this result can be obtained directly from the chain rule.

If we are given that x is some function of a variable u, $x = x(u)$, and also u as some function of x, $u = u(x)$, then we can find dx/du if we know du/dx. In fact

$$\frac{dx}{du} = \left(\frac{du}{dx} \right)^{-1}.$$

Suppose now that we are given $x = x(u, v)$ and $y = y(u, v)$. Then we must have some inverse relation $u = u(x, y)$ and $v = v(x, y)$. How can we find $(\partial x/\partial u)_v$ if we know $(\partial u/\partial x)_y$, $(\partial v/\partial x)_y$, $(\partial u/\partial y)_x$ and $(\partial v/\partial y)_x$? The next exercise shows how. (The answer is *not* $\left(\partial x/\partial u \right)_v = \left(\partial u/\partial x \right)_y^{-1}$ because different variables are being held constant on the two sides.)

Exercise 10.24 If $x = x(u, v)$, $y = y(u, v)$ and the inverse relations $u = u(x, y)$ and $v = v(x, y)$ are given, write down the definitions of the differentials du, dv and dx. Using the fact that also

$$dx = 1 \cdot dx + 0 \cdot dy,$$

show that

$$x_v = -\frac{x_u u_y}{u_y},$$

and that

$$x_u = v_y/(u_x v_y - u_y v_x).$$

This last relation gives us the required expression for $(\partial x/\partial u)_v$.

Revision Notes

After completing this chapter you should be able to

- Calculate the partial derivatives of a given function $f(x, y, \ldots)$

- Calculate the partial derivatives with respect to x, y, z of a function $f(r)$ where $r = (x^2 + y^2 + z^2)^{1/2}$

- Differentiate an integral w.r.t. a parameter

- Calculate a finite number of terms of the Taylor series of a function of several variables

- Calculate the stationary points of a given function and determine whether they are maxima, minima or saddle points

- Calculate the total differential of a given function

- Determine whether a given expression is an exact differential

- Use the chain rule to relate partial derivatives with respect to different coordinates

10.17. EXERCISES

1. By explicit calculation verify that if $f(x, y) = \ln(\cos(x/y))$ then $\dfrac{\partial^2 f}{\partial x \partial y} = \dfrac{\partial^2 f}{\partial y \partial x}$.

2. Write down the chain rule for the partial derivative $\partial f / \partial u$ of a function $f(u, v)$ in terms of $\partial f / \partial x$ and $\partial f / \partial y$ where $x = x(u, v)$ and $y = y(u, v)$. If $f(x, y) = (\ln(x))(\ln(y))$ and $x = ue^{-v}, y = uv$ show that $\partial f / \partial u = (\ln(u^2 v) - v)/u$.

3. If $f(x, y, z) = z/r^2$ where $r = (x^2 + y^2 + z^2)^{1/2}$, what are

    (i) $\dfrac{\partial f}{\partial x}$ and (ii) $\dfrac{\partial f}{\partial z}$?

4. Let $r^2 = x^2 + y^2 + z^2$; find $\dfrac{\partial r}{\partial x}$ and $\dfrac{\partial}{\partial x}(\dfrac{1}{r})$, and show that

$$\frac{\partial^2}{\partial x^2}\left(\frac{1}{r}\right) = -\frac{1}{r^3} + \frac{3x^2}{r^5}.$$

5. Let $r^2 = x^2 + y^2 + z^2$; if $f(r)$ is a function of r only show that

$$\frac{\partial f}{\partial x} = \frac{df}{dr}\frac{x}{r} \quad \text{and} \quad \frac{\partial^2 f}{\partial x^2} = \frac{x^2}{r^2}\frac{d^2 f}{dr^2} + \frac{1}{r}\frac{df}{dr} - \frac{x^3}{r^3}\frac{df}{dr}$$

and $\dfrac{\partial^2 f}{\partial x \partial y} = \dfrac{xy}{r}\dfrac{d}{dr}\left(\dfrac{1}{r}\dfrac{df}{dr}\right)$.

6. Let $f(x, y) = xy^2 \sin(y/x)$; show that $x\dfrac{\partial f}{\partial x} + y\dfrac{\partial f}{\partial y} = 3f$.

7. Let $r = (x^2 + y^2 + z^2)^{1/2}$; show that

$$(x\frac{\partial}{dx} + y\frac{\partial}{dy} + z\frac{\partial}{dz})r^n = nr^n \quad \text{for any } n \neq 0.$$

8. Let $r = (x^2 + y^2 + z^2)^{1/2}$ and $f = 1/r$. Find $\dfrac{\partial f}{\partial x}$, write down expressions for $\dfrac{\partial f}{\partial y}$ and $\dfrac{\partial f}{\partial z}$ and hence show that f satisfies

$$x\frac{\partial f}{\partial x} + y\frac{\partial f}{\partial y} + z\frac{\partial f}{\partial z} = -f.$$

9. Let $u = x^2 - y^2$ and $v = y$. For a function $f(x, y)$ find $\dfrac{\partial f}{\partial x}$ and $\dfrac{\partial f}{\partial y}$ in terms of $\dfrac{\partial f}{\partial u}$ and $\dfrac{\partial f}{\partial v}$. Hence find the general solution of the partial differential equation

$$y\frac{\partial f}{\partial x} + x\frac{\partial f}{\partial y} = 2xy.$$

10. Show that

$$-\frac{d}{d\lambda}\int_0^\infty \exp(-\lambda x^2)\,dx = \int_0^\infty x^2 \exp(-\lambda x^2)\,dx(\lambda > 0);$$

hence (or otherwise) evaluate $\int_0^\infty x^2 e^{-x^2} dx$ given that

$$\int_0^\infty e^{-x^2} dx = \frac{\sqrt{\pi}}{2}.$$

11. Evaluate $\dfrac{d}{dt} \int_0^\infty \dfrac{e^{-x^2 t}}{x^2} dx \ (t > 0)$ given that $\int_0^\infty e^{-x^2} dx = \dfrac{\sqrt{\pi}}{2}$.

12. Find the Taylor series of $f(x, y) = x/y$ about the point $x = 1, y = 1$ up to and including terms of the third order (i.e. as a cubic polynomial).

13. Given that $f(x, y)$ is stationary at the origin and its second derivatives are $f_{xx} = 2, f_{yy} = -2, f_{xy} = 3$, determine the nature of the stationary point.

14. Write down the conditions for $a(x, y)dx + b(x, y)dy$ to be a perfect differential, and deduce that

$$\frac{1}{y} dx - \frac{x}{y^2} dy$$

is a perfect differential.

15. For what value of n is the expression $x^n[(6x^5 + 3y^2)dx - 2xy\,dy]$ a perfect differential?

16. Is $e^{x^2/y}(1 + \dfrac{2x^2}{y})dx - \dfrac{x^3}{y}e^{x^2/y}dy$ a perfect differential? (Show your reasoning.)

17. Write down the condition for $p(x, y)dx + q(x, y)dy$ to be a perfect differential. Is

$$\frac{2x}{y + x^2} dx - \frac{x^2}{y^2 + x^2 y} dy$$

a perfect differential?

10.18. PROBLEMS

1. Show that if $x = r\cos(\theta)$ and $y = r\sin(\theta)$ then $\dfrac{\partial r}{\partial x} = \dfrac{x}{r}$ and $\dfrac{\partial r}{\partial y} = \dfrac{y}{r}$. Suppose now that the function $g(r, \theta)$ has the form $g(r, \theta) = f(r)\sin(\theta)$, where $f(r)$ is a function of r only. By writing $g(r, \theta)$ in the form $yF(r)$, where $F(r) = f(r)/r$, or otherwise, show that

$$(i)\frac{\partial g}{\partial x} = \frac{xy}{r}\frac{dF}{dr}, \quad (ii)\frac{\partial g}{\partial y} = F + \frac{y^2}{r}\frac{dF}{dr}$$

and $(iii)\dfrac{\partial^2 g}{\partial x^2} + \dfrac{\partial^2 g}{\partial y^2} = y\left(\dfrac{d^2 F}{dr^2} + \dfrac{3}{r}\dfrac{dF}{dr}\right)$

Hence prove that if $g(r, \theta)$ satisfies the partial differential equation $\dfrac{\partial^2 g}{\partial x^2} + \dfrac{\partial^2 g}{\partial y^2} = 0$, then $f(r) = Ar + B/r$ where A and B are constants.

2. Given $u(r, \theta)$ with $x = r\cos(\theta), y = r\sin(\theta)$, show that

$$\frac{\partial u}{\partial x} = \cos(\theta)\frac{\partial u}{\partial r} - \frac{\sin(\theta)}{r}\frac{\partial u}{\partial \theta},$$

$$\frac{\partial u}{\partial y} = \sin(\theta)\frac{\partial u}{\partial r} + \frac{\cos(\theta)}{r}\frac{\partial u}{\partial \theta}.$$

Now consider $f(r, \theta)$ with $x = r\cos(\theta), y = r\sin(\theta)$. By setting $u = \dfrac{\partial f}{\partial x}$ use the above results to express $\dfrac{\partial^2 f}{\partial x^2}$ in terms of $\dfrac{\partial^2 f}{\partial r^2}, \dfrac{\partial^2 f}{\partial r\partial \theta}, \dfrac{\partial^2 f}{\partial \theta^2}, \dfrac{\partial f}{\partial r}$ and $\dfrac{\partial f}{\partial \theta}$.

Obtain the corresponding expression for $\dfrac{\partial^2 f}{\partial y^2}$ and show that

$$\frac{\partial^2 f}{\partial x^2} + \frac{\partial^2 f}{\partial y^2} = \frac{\partial^2 f}{\partial r^2} + \frac{1}{r}\frac{\partial f}{\partial r} + \frac{\partial^2 f}{\partial \theta^2}.$$

3. Find the stationary point of the function

$$f(x, y) = x^4 + 2x^2 + 3xy + 3y.$$

Write down the Taylor series about the stationary point up to terms of second order, and hence (or otherwise) show that this point is neither a maximum nor a minimum.

4. (i) Show that the quadratic form $ax^2 + 2bxy + cy^2$ is positive for all $x, y \neq (0, 0)$ if $a > 0$ and $ac - b^2 > 0$.

   (ii) Show that the function

$$f(x, y) = x^4 + 3x^2 + 4xy + 2y^2 - 4x - 4y + 2$$

   has a single stationary point at $(0, 1)$ which is a minimum.

   (iii) Write down the Taylor expansion of $f(x, y)$ about the stationary point and hence show that the minimum is a global minimum, i.e. f nowhere takes a smaller value.

5. Find the stationary points of the function

$$f(x, y) = x^4 - 2x^2 + y^3 - 3y$$

   and for each determine whether it is a minimum, maximum or saddle point. Suppose the true values of x and y are $x = 1$, $y = 1$ and that they are measured each with an error of at most 10%. What is the corresponding maximum percentage error in f? If the true values are $x = 2, y = 2$ what is the maximum percentage error in f corresponding to an error of 1% in x and y? Briefly explain your result.

6. Find the stationary points of the function

$$f(x,y) = x^4 + x^3 y + \frac{5}{8} xy^2$$

and determine whether the stationary point not at the origin is a maximum, minimum or saddle point.

7. Find the stationary point of the function $f(x, y) = xy + 1/x + 1/y$, $(x \neq 0, y \neq 0)$ and determine whether it is a maximum, minimum or saddle point.

8. Obtain the Taylor expansion to second order of the function $f(x, y) = (xy)^{1/2}$ about the point $x = 1, y = 1$. Hence show that $(xy)^{1/2} < \dfrac{1}{2}(x+y)$ near $(1, 1)$.

9. Find the stationary point of the function $\ln(x) - x/y^2 - 2y$ and determine its nature.

10. Show that $f(x,y) = 3x^2 + 2xy + y^2 - 2x + \dfrac{1}{2}$ has one stationary point. Find the position (x_0, y_0) of the stationary point and show that it is a minimum. Deduce that $f(x, y) > 0$ for all x, y. Show that $g(x, y) = \exp(a(x + y))f(x, y)$ has a minimum at (x_0, y_0) for all values of a. Show further that when $a \neq 0$, $g(x, y)$ has a second stationary point and determine its nature.

11

PARTIAL DIFFERENTIAL EQUATIONS

Many situations in physical science are governed by the relations involving partial derivatives of an unknown function, hence by partial differential equations (PDEs). Examples include the gravitational field of a distribution of mass, the electromagnetic field generated by a charge distribution, the diffusion of heat in a reactor, the probability density of an electron in an atom and the flow of a fluid. For example, the rate at which the temperature in a bar is changing at any point (partial derivative of temperature with respect to time) is related to the spatial gradient of temperature at that point (partial derivative with respect to spatial coordinate).

In many real situations the governing partial differential equations are complicated and cannot be solved using analytical methods; they must be solved by numerical methods. But in simple cases there do exist methods of explicit solution, and these provide insight into the behavior of the system and assist the development of physical intuition. In this and the next chapter we concentrate mainly on one particular equation, namely the wave equation in one space dimension that governs, for example, the propagation of disturbances on strings and in tubes of fluid. We consider two methods of solution: the general solution involving arbitrary functions introduced in Section 11.2 and the sum of separable solutions, involving arbitrary coefficients, which is taken up in Section 11.11 and Chapter 12. The former is a special feature of the wave equation, the latter a rather general approach for linear equations.

11.1. EXAMPLES FROM PHYSICS

Let $y(x, t) = \sin(x + ct)$, with c a constant, and let $u = x + ct$. Then, writing y_x for $\partial y/\partial x$ etc., we have, by the chain rule,

$$y_x = \frac{\partial y}{\partial u} \times \frac{\partial u}{\partial x} = \cos(u) \times 1 = \cos(x + ct).$$

Similarly $y_t = c \cos(x + ct)$ and

$$y_{xx} = -\sin(x + ct), \quad y_{tt} = -c^2 \sin(x + ct).$$

Therefore

$$\frac{1}{c^2} y_{tt} = y_{xx}$$

or

$$\frac{1}{c^2} \frac{\partial^2 y}{\partial t^2} = \frac{\partial^2 y}{\partial x^2}. \tag{11.1}$$

We say that the function $y(x, t) = \sin(x + ct)$ satisfies the *partial differential equation* (11.1). But this may not be – in fact, is not – the only solution. To find other solutions of equation (11.1) we must perform a reverse process: start from equation (11.1) and find the possible forms for $y(x, t)$. In practice, solutions are usually required to satisfy further subsidiary conditions (boundary conditions or initial conditions, or a combination), as will be explained later.

Equation (11.1) is called the *wave equation* (in one space dimension). It describes the propagation, for example, of a wave on a uniform string under tension (y is the transverse displacement of the string from its equilibrium position, x is the distance along the string and t the time; compare Figure 11.1) or of a sound wave in a uniform tube (y is the pressure in the tube) etc. The constant c is the speed of the wave. In the case of the string, the physical interpretation of this equation is that the transverse acceleration (perpendicular to the string) of an element of string, $\partial^2 y/\partial t^2$, is a result of the net tension acting on the element arising from the distortion of the string $\partial^2 y/\partial x^2$.

The following are four more examples of partial differential equations in physics.

The "wave equation" for $\phi = \phi(x, y, z, t)$ in three space dimensions:

$$\frac{\partial^2 \phi}{\partial x^2} + \frac{\partial^2 \phi}{\partial y^2} + \frac{\partial^2 \phi}{\partial z^2} = \frac{1}{c^2} \frac{\partial^2 \phi}{\partial t^2}. \tag{11.2}$$

Laplace's equation:

$$\frac{\partial^2 \phi}{\partial x^2} + \frac{\partial^2 \phi}{\partial y^2} + \frac{\partial^2 \phi}{\partial z^2} = 0, \tag{11.3}$$

where $\phi = \phi(x, y, z)$ might be, for example, the electrostatic potential, or the gravitational potential in empty space, or the velocity potential for an inviscid fluid.

The "diffusion equation":

$$\frac{\partial^2 y}{\partial x^2} = \frac{1}{D} \frac{\partial y}{\partial t}, \tag{11.4}$$

where $y = y(x,t)$, D = constant, describing, for example, the diffusion of heat in a bar (y is then the temperature).

Schrödinger's equation:

$$\frac{\partial^2 \psi}{\partial x^2} = -i \frac{\partial \psi}{\partial t} \tag{11.5}$$

for the wave function ψ of a free particle.

These equations are *linear* (i.e. the unknown function and its derivatives occur only linearly) and *homogeneous* (there is no term involving only the independent variables and not the unknown function). They are also of second order (i.e. second derivatives occur but not higher ones). A method of solution will be presented in section 11.11. However, in Sections 11.2–11.9 we explore an exceptional feature of the wave equation (11.1): the existence of an explicit form for the general solution.

Exercise 11.1 For any one of the equations (11.1)–(11.5), show by substitution that if two functions, g and h say, of the independent variables are known to be solutions, then $Ag + Bh$ is also a solution where A, B are arbitrary constants. (This is true because the equations are linear and homogeneous.)

Example 11.1 Find the general solution for $h(x,y)$, a function of x and y, of the equation $\partial h/\partial x = 0$.

The obvious approach is to integrate. If

$$\frac{\partial h}{\partial x} = 0$$

then h is a constant as far as differentiation with respect to x is concerned, i.e. h is an unknown function of y only, say $g(y)$ and so $h(x,y) = g(y)$. It is independent of x. Differentiate $g(y)$ with respect to x to check, remembering that partial differentiation with respect to one variable means that the others are held constant, and so

$$\frac{\partial h}{\partial x} = \frac{\partial g(y)}{\partial x} = 0.$$

This is the general solution. Note that it contains one arbitrary function (of one variable). This is an obvious extension of the corresponding result for a function of a single variable, namely that the solution to a first order ordinary differential equation contains one arbitrary constant. For example, if the function of a single variable $h(x)$ satisfies the ordinary differential equation $dh/dx = 0$, then $f(x)$ is independent of x, i.e. $f(x) = c$, a constant.

Exercise 11.2 Write down an extension of the above example to a function of three variables $h(x, y, z)$.

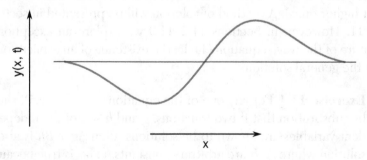

FIGURE 11.1: Wave on a string at some instant of time t.

11.2. GENERAL SOLUTION OF THE WAVE EQUATION

The wave equation in one space dimension for the unknown function $y(x, t)$ is

$$\frac{\partial^2 y}{\partial x^2} = \frac{1}{c^2} \frac{\partial^2 y}{\partial t^2}. \tag{11.6}$$

For a concrete picture we shall think of $y(x, t)$ as the transverse displacement of a uniform stretched string at position x at time t. (The string is undergoing transverse motion of small amplitude in a horizontal plane, so gravitys can be neglected.)

We now come to a fundamental result. Take any functions $f(x)$, $g(x)$ of a single variable (with the property that they vary smoothly enough to have second derivatives at each point). A solution of equation (11.6) is

$$y(x,t) = f(x - ct) + g(x + ct) \tag{11.7}$$

Proof: We calculate f_{xx} and f_{tt} and show they satisfy equation (11.6). Let $u = x - ct$, so $f = f(u(x, t))$, i.e. $f(x - ct)$ is a function of a function $u = (x - ct)$. Differentiate $f(u(x, t))$ by the chain rule:

$$\frac{\partial f}{\partial x} = \frac{df}{du} \frac{\partial u}{\partial x} = \frac{df}{du}$$

since $\partial u/\partial x = 1$. Note that we write df/du (not curly ∂) because f is a function of a single variable, u. Differentiate again to get f_{xx},

$$\frac{\partial^2 f}{\partial x^2} = \frac{d}{du}\left(\frac{df}{du}\right)\frac{\partial u}{\partial x} = \frac{d^2 f}{du^2}.$$

Using the fact that $\partial u/\partial t = -c$,

$$\frac{\partial f}{\partial t} = \frac{df}{du} \frac{\partial u}{\partial t} = \frac{df}{du}(-c),$$

and so

$$\frac{\partial^2 f}{\partial t^2} = \frac{d}{du}\left(-c\frac{df}{du}\right)\frac{\partial u}{\partial t} = c^2 \frac{d^2 f}{du^2}.$$

Hence $f_{xx} = (1/c^2)f_{tt}$, so $y = f(u)$; i.e. $y = f(x - ct)$ satisfies equation (11.6).

Exercise 11.3 Complete the proof by showing also that $g(x + ct)$ satisfies equation (11.6).

In fact, equation (11.7) is the general solution of equation (11.6); i.e. every solution can be written in the form equation (11.7) for some functions f and g. This is an important result. (For a proof, see Section 11.3.)

Example 11.2 Is the function $y(x, t) = x^2 - c^2 t^2$ a solution of the wave equation?

The function can be written as $(x - ct)(x + ct)$ which is not of the form "a function of $(x - ct)$ plus a function of $(x + ct)$." Therefore this is not a solution of the wave equation. (Of course, this can be checked by direct substitution, but this usually requires more effort.)

Exercise 11.4 If $f(u) = u^2$, $g(u) = \cos(u)$, what does equation (11.7) give for $y(x, t)$?

Exercise 11.5 Which of the following are not solutions of the wave equation?

(i) $(x - ct)^2$,

(ii) xt,

(iii) $x^2 + c^2 t^2$,

(iv) $\sin(x)\sin(ct)$.

The function $y = f(x - ct)$, where c is a positive constant, represents a wave. (The shape of this wave will depend on the choice of the function f, which can be anything, not just a sine or cosine

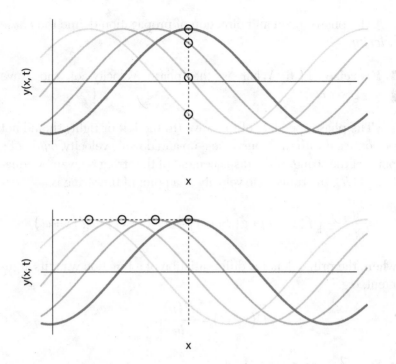

FIGURE 11.2: A wave moving to the right. The circles in the top figure indicate the displacement of a point at $x = 1$ with time t: $y(1, t)$. The circles in the bottom figure indicate the movement with time of a point of constant phase.

(Exercise 11.5).) The quantity $x - ct$ is the *phase* of the wave at position x and time t.

In which direction does the wave $y = f(x - ct)$ move? To answer this consider the point with zero phase at any time. This is the point $x - ct = 0$. So at $t = 0$ the point with zero phase is at the origin $x = 0$. At later times $(t > 0)$ it is at $x = ct$ along the positive x-axis. Therefore the wave moves to the right with wavespeed (or *phase speed*) c. (The choice of zero phase is merely for convenience; any other choice gives the same result.)

Another way to remember the direction of propagation is that for the crest to stay at the constant height $y = f(x_0)$ (i.e. to move with the wave) as t increases, x must increase so as to keep $x - ct$ constant: thus the wave moves to the right with speed c.

The phase speed and direction of propagation define the *phase velocity*.

> **Exercise 11.6** What is the phase velocity of the wave $y = g(x + ct)$?

The phase speed, c, of the wave (to the left or right) should not be confused with the transverse ("up and down") velocity, $\partial y/\partial t$, of a point of the string. If the displacement of the string is given by equation (11.7), the transverse velocity of a point of the string is

$$\frac{\partial y}{\partial t} = \frac{\partial}{\partial t}\left[f(x-ct)+g(x+ct)\right] = -cf'(x-ct)+cg'(x+ct),$$

where the prime denotes differentiation of a function w.r.t. its argument, e.g.

$$f'(x-ct) = \frac{df(u)}{du}\bigg|_{u=x-ct}.$$

> **Exercise 11.7** What is the transverse velocity of the string in exercise 11.4 at the point $x = 5\pi/8$, at time $t = 5\pi/8c$?

11.3. DERIVATION OF THE GENERAL SOLUTION

To derive the general solution we make a transformation from (x, t) to new variables (u, v) defined by $u = x - ct$, $v = x + ct$. We have, by the chain rule,

$$\frac{\partial y}{\partial x} = \frac{\partial y}{\partial u}\frac{\partial u}{\partial x} + \frac{\partial y}{\partial v}\frac{\partial v}{\partial x} = \frac{\partial y}{\partial u}+\frac{\partial y}{\partial v}, \quad \text{since} \frac{\partial u}{\partial x}=1 \text{ and} \frac{\partial v}{\partial x}=1.$$

Hence

$$\frac{\partial y}{\partial x} = \left[\frac{\partial}{\partial u}+\frac{\partial}{\partial v}\right]y$$

and

$$\frac{\partial^2 y}{\partial x^2} = \left[\frac{\partial}{\partial u}+\frac{\partial}{\partial v}\right]\frac{\partial y}{\partial x} = \left[\frac{\partial}{\partial u}+\frac{\partial}{\partial v}\right]\left[\frac{\partial}{\partial u}+\frac{\partial}{\partial v}\right]y = \frac{\partial^2 y}{\partial u^2}+2\frac{\partial^2 y}{\partial u\partial v}+\frac{\partial^2 y}{\partial v^2}.$$

Exercise 11.8 Show similarly that

$$\frac{\partial^2 y}{\partial t^2} = c^2 \left(\frac{\partial^2 y}{\partial u^2} - 2 \frac{\partial^2 y}{\partial u \partial v} + \frac{\partial^2 y}{\partial v^2} \right).$$

Putting these together, the wave equation (11.6) becomes

$$y_{uu} + 2y_{uv} + y_{vv} = y_{uu} - 2y_{uv} + y_{vu}.$$

Now we rearrange to see that

$$4 \frac{\partial^2 y}{\partial u \partial v} = 0, \quad \text{or} \quad \frac{\partial^2 y}{\partial u \partial v} = 0.$$

This is now in a form which can be solved. We can write this as

$$\frac{\partial}{\partial u} \left(\frac{\partial y}{\partial v} \right) = 0.$$

This implies that $\partial y / \partial v$ is independent of u, i.e. an (arbitrary) function of v only:

$$\frac{\partial y}{\partial v} = h(v),$$

where h is an arbitrary function of v. Integrating this equation gives

$$y = \int h(v) \, dv + f(u),$$

where $f(u)$ is independent of v, i.e. a function of u only. Since the integral of an arbitrary function is also an arbitrary function, we can write this as

$$y = g(v) + f(u)$$

where g and f are both arbitrary functions. This is the general solution – it involves two arbitrary functions. (Compare second order ordinary differential equations, where the general solution involves two arbitrary constants.) In terms of x and t we have

$$y(x, t) = f(x - ct) + g(x + ct)$$

which is equation (11.7).

11.4. A STRING INITIALLY AT REST

In applications, the functions f and g in the general solution (Equation 11.7) are determined by the way in which the wave motion is set up, i.e. through given conditions at some specified time, often $t = 0$.

Example 11.3 Find the solution of the wave equation for the displacement of a string $y(x, t)$, given that

$$y(x,0) = \sin(x) \quad \text{and} \quad \frac{\partial y}{\partial t}(x,0) = 0$$

i.e. the string is initially at rest.

Starting from the general solution for the displacement, the general solution is

$$y(x, t) = f(x - ct) + g(x + ct). \tag{11.8}$$

To impose the second initial condition we need $\partial y/\partial t$:

$$\frac{\partial y}{\partial t} = -cf'(x - ct) + cg'(x + ct). \tag{11.9}$$

Now impose the initial conditions on y and $\partial y/\partial t$. We are given $y = \sin(x)$ and $\partial y/\partial t = 0$ initially (the string is at rest), and so

$$y(x, 0) = f(x) + g(x) = \sin(x) \tag{11.10}$$

and

$$\frac{\partial y}{\partial t}(x,0) = -cf'(x) + cg'(x) = 0. \tag{11.11}$$

Thus we have two simultaneous equations for f and g, equations (11.10) and (11.11). If we now integrate equation (11.11) we obtain

$$-f(x) + g(x) = k, k \text{ a constant.} \tag{11.12}$$

If we add this to equation (11.10) we find

$$2g(x) = k + \sin(x),$$

and so

$$g(x) = \frac{1}{2}\left(\sin(x) + k\right).$$

And from equation (11.12),

$$f(x) = \frac{1}{2}\left(\sin(x) - k\right).$$

These are the solutions for f and g. Hence, from $f(x)$ we get $f(x - ct)$ by substitution of $x - ct$ for x. Similarly for $g(x)$. Note that k cancels, and so we arrive at

$$y(x,t) = f(x - ct) + g(x + ct) = \frac{1}{2}\sin(x - ct) + \frac{1}{2}\sin(x + ct).$$

For a string released from rest (and only in this case) the solution can be interpreted graphically as follows: the initial configuration $y = h(x)$ (in this case $\sin(x)$) is made up of two waves: $\frac{1}{2}h(x)$ moving to the right and $\frac{1}{2}h(x)$ moving to the left. These combine to give the solution

$$y = \frac{1}{2}h(x - ct) + \frac{1}{2}h(x + ct)$$

for a string *initially at rest* in the configuration $y = h(x)$.

Exercise 11.9

(i) Repeat the working of Example 11.3 to find the solution of the wave equation for the displacement of a string, $y(x, t)$, released from rest under the initial conditions

$$y(x,0) = e^{-x^2}, \quad \frac{\partial y}{\partial t}(x,0) = 0.$$

(ii) Interpret the solution as was done for Example 11.3 above (see Figure 11.3).

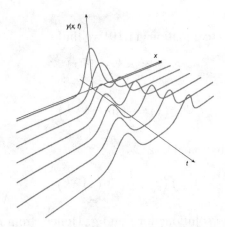

FIGURE 11.3: Motion of a string released from rest, with an initially Gaussian displacement $y(x, 0) = \exp(-x^2)$. As time increases (from back to front in the figure), the displacement splits into two oppositely moving parts (Exercise 11.9).

Example 11.4 An infinite string is released from rest at time $t = 0$ with the displacement

$$y(x,0) = \begin{cases} x+L & -L \le x \le 0 \\ L-x & 0 \le x \le L \\ 0 & \text{otherwise} \end{cases}$$

where L is a given constant. By using the graphical interpretation described above draw a series of figures to show the displacement of the string at times $t < L/2c$, $L/2c < t < L/c$ and $t > L/c$ (where c is the wavespeed).

The string is released from rest so can be considered as two equal parts moving to the left and right. The sum of these parts gives the shape of the wave (Figure 11.4).

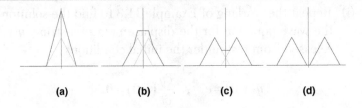

(a) (b) (c) (d)

FIGURE 11.4: Displacement of string released from rest with an intially triangular displacement. The sequence (a)–(d) shows the evolution with increasing time. (a) At $t = 0$ the two waves overlap completely. (b) At $0 < t < L/(2c)$ the two waves are mostly overlapping. (c) At $L/(2c) < t < L/c$ the two waves are only partially overlapping. (d) At $t > L/c$ the two waves have separated completely.

Exercise 11.10 An infinite string is released from rest at time $t = 0$ with the displacement

$$y(x,0) = \begin{cases} a, & -L < x < L \\ 0 & \text{otherwise} \end{cases},$$

where L and a are given constants. *By using the graphical interpretation above* draw a series of figures to illustrate the displacement of the string at times $t < L/2c$, $L/2c < t < L/c$, $t > L/c$, where c is the wavespeed. (Note that this is an idealized example: the string is displaced to a height a between two end-points where it is fixed (so has zero height). But a real string cannot be discontinuous at the points $x = \pm L$. The picture therefore approximates a string that has been given a constant displacement along almost the whole of its length $2L$.)

11.5. A STRING GIVEN AN INITIAL VELOCITY

The graphical method discussed above works only for strings released from rest. In this section and the next (11.6) we give examples of the general method for strings with a non-zero initial velocity.

Example 11.5 An infinite undisplaced string is given an initial transverse velocity

$$\frac{\partial y}{\partial t}(x,0) = \sin(x).$$

Find its subsequent displacement.

For an infinite string we start from the general solution for the displacement and transverse velocity which is

$$y(x, t) = f(x - ct) + g(x + ct) \tag{11.13}$$

and

$$\frac{\partial y}{\partial t} = -cf'(x - ct) + cg'(x + ct). \tag{11.14}$$

Write down the initial conditions: at $t = 0$:

$$y(x, 0) = 0 \tag{11.15}$$

and

$$\frac{\partial y}{\partial t}(x,0) = \sin(x). \tag{11.16}$$

Putting $t = 0$ in equation (11.13) and using equation (11.15)

$$f(x) + g(x) = 0 \text{ or } f = -g \tag{11.17}$$

and putting $t = 0$ in equation (11.14) and using equation (11.16) gives

$$-cf'(x) + cg'(x) = \sin(x).$$

Integrating both sides (with respect to x):

$$- cf(x) + cg(x) = -\cos(x),$$

and since $f = -g$ in equation (11.17) we have

$$2cg(x) = -\cos(x).$$

So

$$f(x) = -g(x) = \frac{1}{2c}\cos(x),$$

and hence

$$y(x,t) = f(x-ct) + g(x+ct) = \frac{1}{2c}\cos(x-ct) - \frac{1}{2c}\cos(x+ct).$$

If we use the rule that

$$\cos(A) - \cos(B) = -2\sin\left(\frac{A+B}{2}\right)\sin\left(\frac{A-B}{2}\right)$$

with $A = x - ct$ and $B = x + ct$ then the cosine terms can be combined in this solution to give

$$y(x,t) = \frac{1}{c}\sin(x)\sin(ct).$$

The initial conditions therefore generate a standing wave pattern.

Exercise 11.11 An infinite string is given an initial displacement $y(x, 0) = \tan^{-1}(x)$ and an initial velocity $y_t(x, 0) = c(1+x^2)^{-1}$ (with c a constant). Find the subsequent displacement $y(x, t)$.

11.6. A FORMULA FOR GENERAL INITIAL CONDITIONS

In this section, we give a general formula for the propagation of a wave on an infinite string from arbitrary starting conditions. It is more important that you understand the method than memorize the general formula.

Example 11.6 Suppose that an infinite string is given the initial displacement $y(x, 0) = q(x)$ and transverse velocity $y_t(x, 0) = p(x)$, with q and p given functions. Find the subsequent displacement of the string.

Since this is an infinite string, start from the general solution of the wave equation:

$$y(x, t) = f(x - ct) + g(x + ct),$$

with

$$\frac{\partial y}{\partial t} = -cf'(x - ct) + cg'(x + ct).$$

Write down the initial conditions. These will determine f and g, and so at $t = 0$

$$y(x, 0) = q(x) \text{ and } y_t(x, 0) = p(x).$$

Put $t = 0$ in the general solution and use the initial conditions to give us two equations for f and g:

$$f(x) + g(x) = q(x) \text{ and } -cf'(x) + cg'(x) = p(x).$$

Integrating the second of these gives

$$-f(x) + g(x) = \frac{1}{c}\int_{x_0}^{x} p(x')\,dx',$$

where x_0 is arbitrary and we have omitted an arbitrary constant of integration (since it will cancel as in Example 11.3). We have relabeled the integration variable on the right to avoid confusing it with the upper limit, which we now want to call x. Solving for g using $f + g = q$:

$$g(x) = \frac{1}{2c}\int_{x_0}^{x} p(x')\,dx' + \frac{1}{2}q(x),$$

and hence

$$f(x) = \frac{1}{2}q(x) - \frac{1}{2c}\int_{x_0}^{x} p(x')\,dx'.$$

Finally, substitute for f and g in the general solution and we have that

$$y(x,t) = \frac{1}{2}q(x-ct) - \frac{1}{2c}\int_{x_0}^{x-ct} p(x')\,dx' + \frac{1}{2}q(x+ct) + \frac{1}{2c}\int_{x_0}^{x+ct} p(x')\,dx'.$$

We can combine these integrals using the fact that

$$\int_a^c f\,dx - \int_a^b f\,dx = \int_b^c f\,dx,$$

and so

$$y(x,t) = \frac{1}{2}\left[q(x-ct) + q(x+ct)\right] + \frac{1}{2c}\int_{x-ct}^{x+ct} p(x')\,dx'. \qquad (11.8)$$

The solution (11.18) for the motion of an infinite string under any initial conditions is sometimes called *d'Alembert's formula*. Note that the arbitrary constant x_0 has canceled out. Any particular problem can now be solved by substituting for the given $p(x)$ and $q(x)$.

Example 11.7 Let $y(x, 0) = \sin(kx)$, with k a constant, and

$$\frac{\partial y}{\partial t}(x,0) = 0.$$

Use equation (11.18) to find $y(x, t)$.

Here $p(x) = 0$, so the integral term in equation (11.18) is equal to 0. Inserting $q(x) = \sin(kx)$ we get

$$y(x,t) = \frac{1}{2}\sin\left[k(x-ct)\right] + \frac{1}{2}\sin\left[k(x+ct)\right].$$

as in example 11.3.

Example 11.8 Let $y(x, 0) = 0$ and

$$\frac{\partial y}{\partial t}(x,0) = \sin(x).$$

Use equation (11.18) to find $y(x, t)$.

Here $q = 0, p = \sin(x)$, so

$$y(x,t) = \frac{1}{2c}\int_{x-ct}^{x+ct}\sin(x')dx' = \frac{1}{2c}\left[\cos(x+ct) - \cos(x-ct)\right].$$

Exercise 11.12 Solve Exercise 11.11 using the general formula (Equation 11.18) and hence check your solution.

11.7. SEMI-INFINITE STRING

Take an infinite string and fix the point at $x = 0$. To obey this condition the general solution must satisfy

$$y(0, t) = f(-ct) + g(ct) = 0$$

for all t. This implies

$$g(u) = -f(-u).$$

Thus the general solution must be of the form

$$y(x, t) = f(x - ct) - f(-(x + ct)).$$

Example 11.9 Sketch this solution for the case

$$f(u) = \begin{cases} -1, & -2 \le u \le -1 \\ 0, & \text{otherwise.} \end{cases}$$

See Figure 11.5.

(a) At $t = 0$, $y(x) = f(x) - f(-x)$. The function $-f(-u)$ is 1 for $2 \ge u \ge 1$ and 0 otherwise (f is reflected in $u = 0$ with a change of sign). Thus at $t = 0$, $y(x, 0) = f(x) - f(-x)$ is the sum of two top hat functions.

(b) At $t = 1/c$, $y(x) = f(x - 1) - f(-(x + 1))$. To the left of the origin, f is non-zero where $x-1$ lies between -2 and -1, i.e. where x lies between -1 and 0. To the right of the origin f is reflected with a change of sign.

(c) At $t = 3/(2c)$, $y(x) = f(x - 3/2) - f(-(x + 3/2))$ On the left f is non-zero where $(x - 3/2)$ lies between -2 and -1, hence for x between $-1/2$ and $+1/2$. Since this is symmetrical about the origin, it must lie on top of the right top hat and the two cancel giving no net disturbance on the string.

(d) At $t = 2/c$, $y(x) = f(x - 2) - f(-(x + 2))$ so $f = -1$ where $x-2$ lies between -2 and -1, hence where x lies between 0 and 1. And y is $+1$ in the reflected region to the left of the origin.

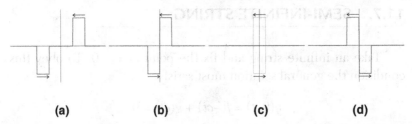

| (a) | (b) | (c) | (d) |

FIGURE 11.5: See Example 11.9. At $t = 0$, a "top hat" displacement $f(u)$ at $x < 0$ and its counterpart $-f(-u)$. The sequence (a)–(d) shows the evolution of the displacement at times $t = 0$, $1/c$, $3/(2c)$ and $2/c$.

Now look at these pictures from the point of view of the semi-infinite string $x > 0$: the left-moving wave is reflected at the origin, with a change of sign, into a right-moving wave. Effectively, a wave coming in from infinity (or at least a long way away) reflects off the fixed point at the origin (e.g. a wall) and changes its phase by 180°.

11.8. SIMPLE HARMONIC WAVES

Any wave progressing in the $+x$ direction with speed c has the form $y = f(x - ct)$. If the function f is such that the shape of the wave repeats over a fixed distance (and fixed time) the wave is called a *harmonic* wave. If f has the special form

$$y = A \cos[k(x - ct) + \phi]$$

(or $y = A \sin[k(x - ct) + \phi]$) the wave is called *simple harmonic*. Let

$$\omega = ck. \tag{11.19}$$

Then $y = A \cos[k(x - ct) + \phi] = A \cos[k(ct - x) - \phi]$ can be written as

$$y = A \cos(\omega t - kx - \phi). \tag{11.20}$$

This gives us the following quantities (Figure 11.6):

- A is the amplitude of the wave,
- $2\pi/k$ is the wavelength,
- k is the wavenumber,
- ω is the frequency,
- $2\pi/\omega$ is the period,
- $c = \omega/k$ is the phase speed,
- $(\omega t - kx - \phi)$ is the phase of the wave at t, x).

Exercise 11.13 What is the change of phase of a simple harmonic wave on reflection at a fixed boundary? (See Example 11.9.)

11.9. BEATS

Suppose we add two simple harmonic waves on a string with slightly different frequencies and equal amplitudes:

$$y(x,t) = A\cos(\omega t - kx) + A \cos(\omega' t - k'x) \tag{11.21}$$

where $\omega' = \omega + \Delta\omega$, $k' = k + \Delta k$ with $\Delta\omega$ and Δk small (i.e. $\Delta\omega/\omega \ll 1$ and $\Delta k/k \ll 1$) and $\omega = ck$, $\omega' = ck'$. Let $\bar{\omega} = (\omega + \omega')/2$ be the mean frequency and $\bar{k} = (k + k')/2$ be the mean wavenumber. Also, we have

$$\frac{\Delta\omega}{\Delta k} = \frac{\omega' - \omega}{k' - k} = \frac{ck' - ck}{k' - k} = c.$$

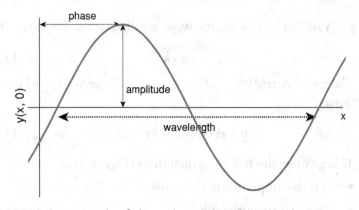

FIGURE 11.6: Some properties of a harmonic wave pictured at a given time.

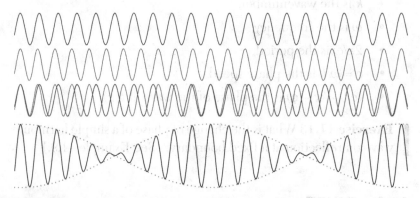

FIGURE 11.7: Beats in the sum of two simple harmonic waves of the same amplitude with slightly different frequencies. The top curve shows $A\cos(\omega t - kx)$. The second curve shows $A\cos(\omega' t - k'x)$. The two curves are overlaid to highlight where they are in phase and out of phase, and the fourth curve shows the sum of the two waves.

Using $\cos(a) + \cos(\beta) = 2\cos\left(\dfrac{a - \beta}{2}\right)\cos\left(\dfrac{a + \beta}{2}\right)$, equation (11.21) can be rewritten as

$$y(x,t) = 2A\cos\left(\frac{\Delta\omega}{2}t - \frac{\Delta k}{2}x\right)\cos\left(\overline{\omega}t - \overline{k}x\right). \qquad (11.22)$$

If we let

$$A(x,t) = 2A\cos\left(\frac{\Delta\omega}{2}t - \frac{\Delta k}{2}x\right),$$

then equation (11.22) is

$$y(x,t) = A(x,t)\cos\left(\overline{\omega}t - \overline{k}x\right). \qquad (11.23)$$

This is therefore a wave (the *carrier wave*) with frequency $\overline{\omega}$ and speed $\overline{\omega}/\overline{k} = c$, with varying amplitude $A(x, t)$. The amplitude $A(x, t)$ is itself a wave of large wavelength, $4\pi/\Delta k$, and low frequency, $\Delta\omega/2$, the *beat* frequency. The beat pattern moves with velocity $\Delta\omega/\Delta k = c$. (See Figures 11.7 and 11.8.)

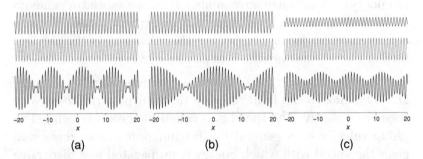

FIGURE 11.8: The beat pattern (at $t = 0$) from the sum of harmonic waves with slightly different wavenumbers: $y(x, t) = A\cos(\omega t - kx) + B\cos(\omega't - k'x)$. (a) Waves with the same amplitudes, $A = B = 1$, and $(k' - k)/k = 0.1$, (b) with same amplitudes, $A = B = 1$ and $(k' - k)/k = 0.05$, (c) with $(k' - k)/k = 0.1$ and $A = 0.4$ and $B = 1$.

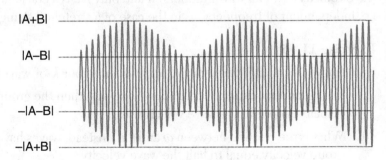

FIGURE 11.9: Beats in the sum of two simple harmonic waves of different amplitudes.

Exercise 11.14 Let $y(x, t) = A\cos(\omega t - kx) + B\cos(\omega' t - k'x)$ (A and B given constants) be the sum of two simple harmonic waves of different amplitudes, $B \neq A$ with $\omega/k = \omega'/k' = c$. By writing $A = (A + B)/2 + (A - B)/2$ and $B = (A + B)/2 - (A - B)/2$, show that this also gives rise to beats, but of a diminished amplitude (compared with the case $A = B$). See Figure 11.9.

11.10. GROUP VELOCITY

Waves on a string move with the same speed c independent of frequency. Waves in other media can move at speeds that differ for different frequencies, so $c = c(\omega)$. Such media are said to be *dispersive*. In a dispersive medium, we have $\omega = c(\omega)k$ or, equivalently, solving for ω, we have $\omega = \omega(k)$. Thus, in a dispersive medium, the frequency is a function of wavenumber. If the wavespeed depends on frequency, we can ask, at what speed does the beat pattern of waves at frequencies ω and $\omega + \Delta\omega$ move? The calculation of the previous section gives the answer: the beat pattern moves with speed $\Delta\omega/\Delta k$.

Consider now a group of waves, with frequencies in the range ω to $\omega + \Delta\omega$. In the limit $\Delta\omega \to 0$ the beat pattern moves with speed $v_g = d\omega/dk$. The speed v_g is called the *group velocity*. The group velocity is in general itself a function of frequency and gives the speed with which energy is propagated in a dispersive medium (Figure 11.10).

Exercise 11.15 Show that the group velocity v_g is equal to the wave velocity c (for all wavenumbers) if and only if c is a constant, i.e. independent of frequency, as in the case of a stretched string.

Exercise 11.16

(i) Show that if the frequency ω and wavenumber k of waves in a certain medium are related by $\omega = \dfrac{1}{2}k^2$, then the group velocity is double the wave velocity.

(ii) What is the relation between ω and k if, instead, waves have group velocity equal to half the wave velocity?

Exercise 11.17 For a wave of frequency ω, wavenumber $k(\omega)$ propagating in a general medium with wavespeed $c(\omega)$, show that

$$\frac{dc}{d\omega} = \frac{1}{k}\left(1 - \frac{c}{v_g}\right),$$

and hence that the group velocity is respectively greater than the wave velocity, equal to it, or smaller, according as the wave velocity increases with the frequency, is constant (as in the case of a stretched string), or decreases as the frequency increases.

Waves for which the group and phase velocities are equal are called non-dispersive. In particular, if a wave is non-dispersive then its phase speed is constant with frequency (and conversely). The group speed provides a useful estimate of the modes (of different frequencies) that transfer the most energy.

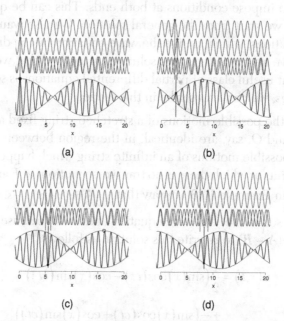

(a) (b)

(c) (d)

FIGURE 11.10: The beat pattern from the sum of harmonic waves with different wavenumbers and different phase velocities: $y_1(x, t) = \cos(\omega t - kx)$ (top curve) and $y_2(x, t) = \cos(\omega' t - k'x)$ (second curve). Here $v_1 = \omega/k = 5/5 = 1$, while $v_2 = \omega'/k' = 6.6/5.5 = 1.2$. (a) At time $t = 0$ the two simple waves are in phase, and they are in phase with their sum. (b)–(d) Times $t = 2.5$, 5 and 7.5 showing the effect of the different phase speeds of the simple waves on the sum. The solid dots represent points of constant phase on each wave (y_1, y_2 and the carrier wave of their sum); these all move with their respective phase velocities. The open circle shows a point of constant phase on the envelope; this moves with the group velocity.

Exercise 11.18 A wave propagates along the x-axis with a frequency given by

$$\omega = -k/(k^2 + 1).$$

Determine the minimum frequency with which the wave can propagate in

   (i) the negative x-direction

  (ii) the positive x-direction.

11.11. SEPARATION OF VARIABLES

So far we have considered infinite strings (i.e. strings so long that we need not consider what is happening at the ends) and semi-infinite strings, where we need consider only one end. For a finite string we have to impose conditions at both ends. This can be quite hard to solve if we start from the general solution. We also want to solve other equations, which, unlike the wave equation in one dimension, do not have an explicit general solution. A method that works for a limited but useful class of partial differential equations is separation of variables, which we describe in this section.

Now, the possible motions of a stretched string fixed at its endpoints P and Q, say, are identical, in the region between P and Q, with the possible motions of an infinite string which happen to have nodes (points at which there is no transverse motion) at P and Q. For example, in Example 11.3, we saw that $y = \frac{1}{2}\sin(x-ct) + \frac{1}{2}\sin(x+ct)$ is a possible solution of the wave equation (11.6). We can use the identity for $\sin(A - B)$ to rewrite this solution as follows:

$$y(x,t) = \frac{1}{2}\big(\sin(x)\cos(ct) - \cos(x)\sin(ct)\big)$$

$$+ \frac{1}{2}\big(\sin(x)\cos(ct) + \cos(x)\sin(ct)\big)$$

$$= \sin(x)\cos(ct).$$

This solution has the property that $y(0, t) = 0$ and $y(2\pi, t) = 0$ for all t. If we restrict our attention to the region $0 \le x \le 2\pi$, this solution also represents one possible wave motion on a string of length 2π fixed at its ends (so the displacement, y, is constrained to be zero at the ends). This leads to two questions:

1. Can other motions of a string fixed at its ends be represented in a form like equation (11.11), namely with $y(x, t) =$ (function of x only) $\times$ (function of t only)? This question will be answered below.

2. Can all motions of the string be derived from such solutions? This question will be answered in Chapter 12.

To find the answer to (1) we take $y(x, t)$ to be of the proposed form, substitute in the wave equation, and determine what conditions must be satisfied if y is to be a solution. So, since we want a function of x times a function of t, let us write $y(x, t) = X(x)T(t)$ as the general form of solution and attempt to determine what functions X and T will give a solution of the wave equation. Note that X is a function of x only and T a function of t only. A solution of this form is said to be separable and the process of seeking a solution in this way is referred to as the method of *separation of variables*. (We are separating the time and space parts of the solution.)

It is important to realize that $y = X(x)T(t)$ is a less general solution than $y = f(x-ct)+g(x-ct)$. This latter class encompasses all separable solutions, in addition to all other possible solutions.

We substitute our expression for $y(x, t)$ in the wave equation

$$\frac{\partial^2 y}{\partial x^2} = \frac{1}{c^2}\frac{\partial^2 y}{\partial t^2}.$$

(11.24)

Take the left side first. We have

$$\frac{\partial^2 y}{\partial x^2} = \frac{\partial^2 X}{\partial x^2}T,$$

since $T(t)$ is constant for partial differentiation with respect to x, and similarly for the right hand side

$$\frac{\partial^2 y}{\partial t^2} = X\frac{d^2 T}{\partial t^2}.$$

Hence, substituting in equation (11.24), we require

$$T \frac{d^2 X}{dx^2} = \frac{X}{c^2} \frac{d^2 T}{dt^2}. \tag{11.25}$$

Divide through by $y = XT$:

$$\frac{1}{X} \frac{d^2 X}{dx^2} = \frac{1}{c^2 T} \frac{d^2 T}{dt^2}. \tag{11.26}$$

In equation (11.26) we have been able to write our equation in a very special form: the left side is a function of x only; the right side is a function of t only, not of x. Thus, a function of x on the left equals something that is independent of x on the right. The only way this can happen is if each side of the equation equals a constant. We have the freedom (at the moment) to choose this constant. For the applications we shall be making it will turn out that the best choice is for this constant to be negative. With hindsight we therefore write this separation constant as $-a^2$. (Writing $+a^2$ for the constant would lead a to be imaginary; this is not wrong but it is not so convenient.) Thus equation (11.4) gives

$$\frac{1}{X} \frac{d^2 X}{dx^2} = -a^2, \quad \frac{1}{T} \frac{d^2 T}{dt^2} = -a^2 c^2. \tag{11.27}$$

Thus we have separated out the x and t parts of the equation. The dependence on x and t is sometimes said to be decoupled. Equations (11.27) can be solved for X and T; possible solutions are

$$X = \cos(ax) \text{ or } \sin(ax),$$

and

$$T = \cos(act) \text{ or } \sin(act).$$

Since each of these four solutions are possible, we make linear combinations to form a solution:

$$\begin{aligned} y(x, t) &= [a \cos(ax) + b \sin(ax)] [c \cos(act) + d \sin(act)] \\ &= A \sin(ax + \psi) \sin(act + \phi). \end{aligned} \tag{11.28}$$

For the second line we have made use of the fact that

$$A\sin(z + \phi) = A \left[\sin(z) \cos(\phi) + \cos(z) \sin(\phi) \right]$$
$$= (A \sin(\phi)) \cos(z) + (A \cos(\phi)) \sin(z)$$
$$= a \cos(z) + b \sin(z).$$

The two constants on the left side (A, ϕ) are related to the two constants on the right side $(a$ and $b)$ by $a = A \sin(\phi)$ and $b = A \cos(\phi)$.

From the second form of equation (11.28) it is clearer that $y(x, t)$ involves three independent constants, not four, but the first form is often simpler in applications. Note that in a separable solution y varies in time in the same way (e.g. like $\sin(ct + \phi)$) at all points, x. Separable solutions are also referred to as *normal modes* of the string. Physically, these solutions are standing waves.

Exercise 11.19 Using the first form of the separable solutions above, show that this is consistent with the solution to Example 11.3.

Exercise 11.20 What are all the separable solutions of the wave equation obtained by taking $+a^2$ in equation (11.27)? Hence explain why the choice of $-a^2$ is made in applications involving the wave equation.

Exercise 11.21 For Laplace's equation in 2D,

$$\frac{\partial^2 \phi}{\partial x^2} + \frac{\partial^2 \phi}{\partial y^2} = 0,$$

use the method of separation of variables to find all solutions of the form $\phi(x, y) = X(x)Y(y)$ which vanish on $x = 0$ (for all y).

Exercise 11.22 Use the method of separation of variables to find all solutions of the form $\psi(x, t) = X(x)T(t)$ of Schrödinger's equation

$$\frac{\partial^2 \psi}{\partial x^2} = -i \frac{\partial \psi}{\partial t}$$

for which X is real.

Note that most equations do not have separable solutions; i.e. in general this method does not work because it cannot lead to an equation like equation (11.26) in which the variables have been separated! But those that do are among the most important in physical science.

11.12. EIGENVALUES

So far we have found separable solutions of the wave equation but we have not yet considered boundary conditions (e.g. at the fixed ends of the string) or initial conditions (e.g. how the string is set vibrating at time $t = 0$). The former will be considered here, the latter in Chapter 12. We shall illustrate the effect of boundary conditions by some examples. The result is always the same: the boundary conditions determine the possible values of the separation constant.

Possible values of a separation constant, giving rise to non-zero solutions satisfying the given boundary conditions, are called *eigenvalues*.

Example 11.10 Find the separable solutions of the wave equation on a string of length l subject to the conditions $y(0, t) = 0$, $y(l, t) = 0$ for all t (i.e. the string is fixed at its endpoints $x = 0$ and $x = l$).

First, write down the general form of the separable solutions:

$$y = A \sin(ax + \psi) \sin(act + \phi).$$

Next, impose the boundary conditions. For this problem these are given at $x = 0$ and $x = l$. At $x = 0$:

$$y(0, t) = A \sin(\psi) \sin(act + \phi) = 0$$

for all t. So $A\sin(\psi) = 0$, hence $\psi = 0$ (since $A \neq 0$). Similarly, at $x = l$:

$$y(l, t) = A\sin(al)\sin(act + \phi) = 0,$$

for all t. So it must be that $A\sin(al) = 0$. If

$$\sin(al) = 0$$

then al is an integer multiple of π, so we can put

$$a = a_n = n\pi/l, \text{ for } n = 1, 2, 3, \ldots.$$

These are all the solutions of $\sin(al) = 0$, and we see al must be an integer multiple of π. We have found an infinite number of separable solutions satisfying the given conditions. The separable solutions are

$$y(x, t) = A\sin(a_n x) \sin(a_n ct + \phi),$$

where $a_n = n\pi/l$ for $n = 1, 2 \ldots$. (See Figure 11.11.)

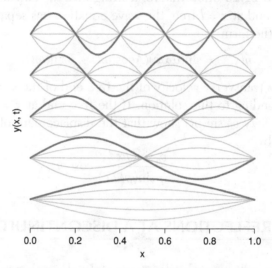

FIGURE 11.11: The first five eigenfunctions from Example 11.10. Bottom to top: $n = 1, 2, 3, 4, 5$.

The numbers a_n^2 are the eigenvalues of the problem; the corresponding solutions for y are called the *eigenfunctions* of the problem. So for the example above the eigenvalues are $(n\pi/l)^2$ and the eigenfunctions are

$$A\sin(n\pi x/l) \sin(n\pi ct/l + \phi),$$

or, expressed differently,

$$\sin(n\pi x/l)(a \cos(n\pi ct/l) + b \sin(n\pi ct/l)).$$

In the case of the wave equation we also refer to the eigenfunctions as the normal modes of the system and the eigenfrequencies as the normal frequencies or harmonics.

Exercise 11.23 Find the separable solutions of the wave equation on a string subject to the conditions $y = 0$ at $x = 0$ for all t and $\partial y/\partial x = 0$ at $x = l$ for all t. What are the eigenvalues and eigenfunctions? What are the normal modes of the string?

Exercise 11.24 Find the separable solutions of Laplace's equation (in 2D) subject to the conditions $\phi = 0$ on $x = 0$ and $x = l$, for all y. What are the eigenvalues and eigenfunctions?

Exercise 11.25 Show that for a string with the end at $x = 0$ fixed and the end at $x = l$ free, the wave equation has separable solutions of the form

$$y(x, t) = \sin(\omega x/c)(A\cos(\omega t) + B\sin(\omega t))$$

Find the transverse velocity of the free end of the string at $x = l$ corresponding to this solution. If the free end at $x = l$ is driven with velocity $v_0 \sin(pt)$ show that the displacement of the string is given by

$$y(x,t) = -\frac{v_0 \sin(px / c)}{p\sin(pl / c)}\cos(pt).$$

11.13. REFLECTIONS AT A DISCONTINUITY

We now look at two developments of the above theory. The first is to waves on a string which consists of two halves with different properties joined at the origin. The second is to a matter wave, satisfying the Schrödinger equation, incident on a potential step. The approach to these problems is similar. We consider a wave incident on the join in the strings, or the potential step, and determine what fraction of it is reflected and what fraction transmitted in order to satisfy certain conditions at the join. To solve these problems efficiently it is convenient to represent waves in exponential form. For example, the wave $A\cos(\omega t - kx)$ is represented as $\text{Re}(Ae^{i\omega t}e^{-ikx})$. The $\text{Re}(\ldots)$ is usually omitted, it being understood that the real part is intended.

Let a semi-infinite string $(x < 0)$, having density per unit length ρ, wavespeed c, be joined at $x = 0$ to a semi-infinite string $(x > 0)$ with density ρ', wavespeed c'. The tensions in the two strings are equal, from which it can be shown that $\rho'(c')^2 = \rho c^2 = F$, say. Let an incident wave of unit amplitude

$$y_i(x, t) = e^{i\omega t}e^{-ikx}$$

be incident from the left. (Note that the sign in the exponent is related to the direction of propagation in the usual way.) A wave

$$y_{\text{trans}}(x, t) = Be^{i\omega' t}e^{-ik'x}$$

is transmitted to the right in the region $x > 0$, and a wave

$$y_{\text{ref}}(x, t) = Ae^{i\omega t}e^{ikx}$$

is reflected to the left.

Thus, the total disturbance on the string in $x < 0$ is

$$y_1(x, t) = e^{i\omega t}(e^{ikx} + Ae^{-ikx}),$$

and for $x > 0$

$$y_2(x, t) = Be^{i\omega' t}e^{ik'x}.$$

Consider now the conditions at $x = 0$.

1. The string does not break, so y is continuous:

$$y_1(0, t) = y_2(0, t),$$
$$e^{i\omega t}(1 + A) = Be^{i\omega' t}.$$

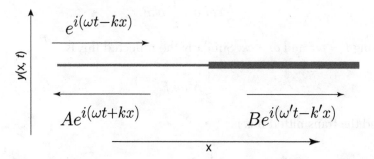

FIGURE 11.12: Wave incident on a join. In the text we show that $\omega = \omega'$ so this diagram is often drawn omitting the factors $e^{i\omega t}$.

For this to hold at all times we must have $\omega = \omega'$ and

$$1 + A = B. \tag{11.29}$$

2. The vertical components of the force at $x = 0$ must balance. Since the tension on each side of the origin is the same, this means the slope must be the same also. Thus

$$\frac{\partial y_1(0,t)}{\partial x} = \frac{\partial y_2(0,t)}{\partial x},$$

or

$$ik - ikA = ik'B. \tag{11.30}$$

We now have two equations (11.29) and (11.30) for A and B. Solving gives

$$A = \frac{k - k'}{k + k'}, \quad B = \frac{2k}{k + k'}.$$

The energy flux in a harmonic wave of amplitude a, frequency ω, on a string with density ρ, wavespeed c is known to be

$$\frac{1}{2}\rho\omega^2 a^2 c.$$

So the incident energy flux is

$$\frac{1}{2}\rho\omega^2 c = \frac{1}{2}\omega k F,$$

using $F = \rho c^2$ and $\omega = ck$. Similarly the reflected flux is

$$\frac{1}{2}A^2\omega k F$$

and the transmitted flux is

$$\frac{1}{2}B^2\omega k' F.$$

The reflection coefficient is defined as

$$R = \frac{\text{reflected flux}}{\text{incident flux}} = A^2$$

and the transmission coefficient as

$$T = \frac{\text{transmitted flux}}{\text{incident flux}} = \frac{k'}{k} B^2.$$

Notice that the transmission coefficient is not B^2 because the energy flux depends on wavenumber, which changes between the two strings.

▌ **Exercise 11.26** Show that $R + T = 1$.

▌ **Exercise 11.27** Show from first principles that the proportion of energy of a transverse wave which is reflected at a join between two different strings is given by

$$(z_1 - z_2)^2/(z_1 + z_2)^2$$

where z is the impedance of a string, $z = $ density × wavespeed.

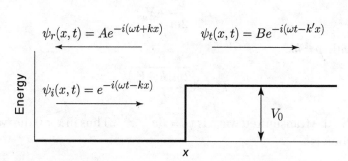

FIGURE 11.13: A matter wave incident from the left on a potential barrier V_0. The common exponential factors $e^{-i\omega t}$ are often omitted.

11.14. MATTER WAVES INCIDENT ON A POTENTIAL STEP

For matter waves incident on a small potential step the mathematics follows closely that of the previous section, although the interpretation is different.

The incident wave of unit amplitude is $\psi_i = e^{-i\omega t}e^{ikx}$, where $\omega = E/\hbar$ and $\kappa = p/\hbar$. The fact that this is a matter wave comes from the relation between ω and k: the Schrödinger equation for a free particle gives $E = p^2/2m$ or $\omega = \hbar k^2/2m$ and hence

$$k = \frac{\sqrt{2mE}}{\hbar}.$$

(Note the opposite signs in the exponentials in ψ_i relative to Section 11.13. This is dictated by the interpretation as matter waves. Solutions of the Schrödinger equation are proportional to $e^{-iEt/\hbar}$ and positive momentum means $e^{+ipx/\hbar}$. For waves on strings only the relative sign between t and kx matters.)

Similarly the reflected wave is $\psi_r = Ae^{-i\omega t}e^{-ikx}$.

To the right of the origin the Schrödinger equation requires

$$E = \frac{p^2}{2m} + V_0,$$

or, with $p = \hbar k'$,

$$k' = \frac{\sqrt{2m(E - V_0)}}{\hbar}.$$

So the transmitted wave is $\psi_t = Be^{-i\omega t}e^{ik'x}$. Thus in $x < 0$ the wavefunction is

$$\psi_1 = e^{-i\omega t}(e^{ikx} + Ae^{-ikx}),$$

and in $x > 0$

$$\psi_2 = Be^{-i\omega t}e^{ik'x}.$$

There are two cases to consider, depending on whether $E > V_0$ (k' real) or $E < V_0$ (k' imaginary).

(i) The small potential step. Assume first that $E > V_0$ so k' is real. In a steady state there can be no build up of probability at $x = 0$, so ψ must be continuous there:

$$\psi_1(0,t) = \psi_2(0,t),$$

or

$$1 + A = B. \tag{11.31}$$

Note that this holds for all time only if we can cancel the factors $e^{-i\omega t}$, which requires ω (or E) to be the same on both sides of the barrier at $x = 0$. This corresponds to *conservation of energy*. Similarly, the *conservation of probability* means that the probability current must be continuous across $x = 0$, so $\partial\psi/\partial x$ must be continuous. Thus

$$\frac{\partial\psi_1(0,t)}{\partial x} = \frac{\partial\psi_2(0,t)}{\partial x},$$

or

$$ik - ikA = -ik'B. \tag{11.32}$$

Equations (11.31) and (11.32) are the same as equations (11.29) and (11.30) and yield the solutions

$$A = \frac{k - k'}{k + k'}, \quad B = \frac{2k}{k + k'}.$$

The reflection and transmission coefficients are the same as for the string example (above). Again, we see an extra factor of k'/k in the transmission region to take account of the different speed of the wave there.

(ii) The large potential step. In this case $E < V_0$ so k' is imaginary. Thus A and B are complex and the definition of reflection and transmission coefficients requires a little more care. The flux of energy is proportional to

$$E \times \text{particle probability density} \times \text{speed} \propto E|\psi|^2 k.$$

414 • Mathematical Physics

Thus the reflection coefficient is not A^2 but $R = |A|^2$. In general (from the probability current)

$$R = |A|^2, \quad T = |B|^2 \frac{\mathrm{Re}(k')}{\mathrm{Re}(k)}$$

Exercise 11.28 Show for a large potential step the reflection coefficient is $R = 1$ and find the wavefunction in the region $x > 0$.

Exercise 11.29 From first principles derive the reflection and transmission coefficients at a small potential step in terms of the energy E of the incoming particle and the height of the barrier V_0.

Revision Notes

After completing this chapter you should be able to

- Write down the wave equation in one, two or three dimensions
- Understand the language used in the study of waves: frequency, wavenumber, wavespeed, phase, beats, phase and group velocity, eigenfunctions and eigenvalues, transmission and reflection coefficients
- Write down and derive the general solution of the wave equation in one space dimension
- Obtain the solution of the wave equation for an infinite string given its initial position and velocity
- Compute the group velocity of a wave given a relation between frequency and wavenumber
- Obtain separable solutions of simple linear partial differential equations and explain the terms *eigenvalue, eigenfunction, normal mode*
- Match solutions of the wave equation (and Schrödinger's equation) across a boundary and find the corresponding reflection and transmission coefficients

11.15. EXERCISES

1. Verify that $y = A\sin(ax)\cos(act)$ is a solution of the wave equation

$$\frac{\partial^2 y}{\partial x^2} = \frac{1}{c^2}\frac{\partial^2 y}{\partial t^2}$$

where A, a are constants. Hence find the solution satisfying $y(x,0) = 2\sin\left(\dfrac{\pi x}{l}\right)$ and $y(0, t) = y(l, t) = 0$ for all t.

2. Verify that $y = \dfrac{1}{2}\big[f(x - ct) + f(x + ct)\big]$ is the solution of the wave equation $\dfrac{\partial^2 y}{\partial x^2} = \dfrac{1}{c^2}\dfrac{\partial^2 y}{\partial t^2}$ satisfying $y(x, 0) = f(x)$, $\dfrac{\partial y}{\partial t}(x,0) = 0$.

3. Verify that $y(x,t) = \dfrac{1}{2c}\displaystyle\int_{x-ct}^{x+ct} G(s)\,ds$ is the solution of the wave equation $\dfrac{\partial^2 y}{\partial x^2} = \dfrac{1}{c^2}\dfrac{\partial^2 y}{\partial t^2}$ satisfying the initial conditions $y(x, 0) = 0$, $\dfrac{\partial y}{\partial t}(x,0) = G(x)$.

4. What is the general solution of $\dfrac{\partial^2 y(x,t)}{\partial x \partial t} = xt$?

5. If $y(x,t) = \exp[i(kx - \omega t)]$ satisfies $\dfrac{\partial^2 y}{\partial x^2} = \dfrac{1}{c^2}\dfrac{\partial^2 y}{\partial t^2}$ what is the relation between ω and k?

6. Write down the normal modes of a stretched string with wavespeed c which is of unit length fixed at its end-points. Such a string is given a displacement $y(x, 0) = 3\sin(2\pi x)$ and released from rest. Show that the subsequent displacement of the string is $y(x,t) = 3\sin(2\pi x)\cos(2\pi ct)$.

7. A semi-infinite string $0 \le x < \infty$ is fixed at $x = 0$. A wave $y = f(x + ct)$ impinges on $x = 0$ from the right. What is the reflected wave?

11.16.   PROBLEMS

1.   Verify that

$$y(x,t) = \frac{1}{2}\left[F(x-ct) + F(x+ct) + \frac{1}{c}\int_{x-ct}^{x+ct} G(s)ds \right]$$

is the solution of the wave equation $\dfrac{\partial^2 y}{\partial x^2} = \dfrac{1}{c^2}\dfrac{\partial^2 y}{\partial t^2}$ satisfying

the initial conditions $y(x, 0) = F(x)$, $\dfrac{\partial y}{\partial t}(x,0) = G(x)$.

2.   An infinite string is at rest along the x-axis. At time $t = 0$ it is struck in such a way that its velocity is non-zero only in the region $|x| \leq L$. Show that, after a time L/c, at the point $x = 0$ the string is again at rest. What parts of the string are at rest for times $t \geq L/c$?

3.   An infinite string is released from rest with the initial shape

$$y = \begin{cases} a(l+x) & \text{if } -l < x < 0 \\ a(l-x) & \text{if } 0 < x < l \\ 0 & \text{otherwise} \end{cases}$$

where a, l are constants. Show that after a time $t = l/c$ the string at the point $x = 0$ is again at rest. What parts of the string are at rest for $t > l/c$?

4.   An infinite string lies along the x-axis except for the region $-l < x < l$, where it is deformed from the axis. At $t = 0$ the string is released from rest. Show that, whatever the initial deformation, the string is again at rest at the origin (i.e. $y(0, t) = 0$) for $t > l/c$.

5.   Separate variables in the wave equation $\dfrac{\partial^2 y}{\partial x^2} = \dfrac{1}{c^2}\dfrac{\partial^2 y}{\partial t^2}$ and hence find the solution obeying the conditions

$$y(x,0) = B\sin\left(\frac{\pi x}{l}\right) \quad \text{and} \quad \frac{\partial y}{\partial t}(x,0) = v\sin\left(\frac{\pi x}{l}\right)$$

for $0 < x < \pi$, where B, l, v are constants. Verify that the combination

$$\left(\frac{\partial y}{\partial t}\right)^2 + \left(\frac{\pi c}{l}\right)^2 y^2$$

is constant in time at each x.

6.   By seeking solution of the form $y(x, t) = X(x)T(t)$ to the wave equation $\dfrac{\partial^2 y}{\partial x^2} = \dfrac{1}{c^2}\dfrac{\partial^2 y}{\partial t^2}$ find the solution obeying the conditions $y(x, 0) = 0$, $\dfrac{\partial y}{\partial t}(x,0) = \dfrac{2c\pi}{l}\sin\left(\dfrac{\pi x}{l}\right)$ and $y(0, t) = y(1, t) = 0$ for all t.

7.   By seeking solutions of the form $y(x, t) = X(x)T(t)$ to the wave equation $\dfrac{\partial^2 y}{\partial x^2} = \dfrac{1}{c^2}\dfrac{\partial^2 y}{\partial t^2}$ find the solution obeying the conditions $y(0, t) = y(1, t) = 0$ for all t and $\dfrac{\partial y}{\partial t}(x,0) = 0$ and $y(x,0) = a\sin\left(\dfrac{\pi x}{l}\right) + b\sin\left(\dfrac{3\pi x}{l}\right)$ with a and b given constants.

8.   Use separation of variables to find all the separable solutions of the equation

$$\frac{\partial^2 \phi}{\partial t^2} - c^2\frac{\partial^2 \phi}{\partial x^2} + \omega_0^2\phi = 0,$$

where c and ω_0 are constants.

9.   An infinite stretched string, with wavespeed c, is given an initial displacement in the form of a Gaussian curve $y(x, 0) = \exp(-x^2)$ and an initial velocity of the form $y_t(x, 0) = cx\exp(-x^2)$. Show that the subsequent displacement consists of a Gaussian wave of fixed shape with amplitude 3/4 moving to positive x and a Gaussian wave of amplitude 1/4 moving to negative x. If the two waves cannot be resolved while the displacement of the string has its greatest value at $x = 0$, show that the two waves will become apparent (i.e. can be resolved) after a time $\dfrac{1}{c}\left[\ln(\dfrac{4}{3})\right]^{1/2}$.

12

FOURIER SERIES

12.1. INTRODUCTION

Many problems in physics involve vibrations and oscillations. Often the oscillatory motion is simple (e.g. weights on springs, pendulums, harmonic waves, etc.) and can be represented as a single sine or cosine function. However, in many cases (electromagnetism, heat conduction, quantum theory, etc.) the waveforms are not simple and, unlike sines and cosines, can be difficult to treat analytically.

Fourier methods give us a set of powerful tools for representing any periodic function as a sum of sines and cosines. The *Fourier series* of a function, $f(x)$, with period $2L$ is

$$f(x) = \frac{1}{2}a_0 + \sum_{n=1}^{\infty}\left(a_n \cos\frac{n\pi x}{L} + b_n \sin\frac{n\pi x}{L}\right). \qquad (12.1)$$

To see how this works, let us consider a sound wave. The vibration of a tuning fork produces a sound wave of a given frequency. If we plot the pressure as a function of distance, x, or time, t, it looks like a single sine wave, or a "pure tone" (Figure 12.1: left). When a note is played on a flute, we get a more complex sound (Figure 12.1: center). The note that we get is made up from the sum of many pure tones: the fundamental and different harmonics with frequencies 2, 3, 4, . . . times the frequency of the fundamental (Figure 12.1: right). This series is the Fourier series representation of the complex waveform.

FIGURE 12.1: Left: a pure sine wave, sin(ωt). Center: example waveform, $f(t)$, from a flute. Right: the note from the flute is made up of the sum of the fundamental sine wave and a series of harmonics. In this example, $f(t)$ = sin(ωt) + sin($2\omega t$) + 0.2sin($3\omega t$) + 0.4sin($4\omega t$). This is the Fourier series.

Fourier methods are used very heavily in signal and data analysis. By Fourier analyzing a signal – essentially by expanding it in the form of equation (12.1) – we can immediately tell which harmonics are the important ones. For example, in the note from the flute (Figure 12.1), the harmonic at frequency 2ω has relatively large amplitude, while the harmonic at 3ω is small. If, for example, a poorly designed speaker filtered out the harmonic at 2ω it would greatly change the character of the sound, while filtering out the harmonic at 3ω would have a much less discernible effect.

Fourier methods are also commonly used in mathematical physics. In this chapter, we will focus on using them to solve differential equations, and the wave equation in particular. We will examine Fourier half range series and Fourier full range series, study some applications of Fourier series, then finish by introducing Fourier transforms and the convolution theorem.

12.2. FOURIER HALF RANGE SINE SERIES

In Chapter 11, we calculated the separable solutions for a wave on a string that is fixed at both ends, at $x = 0$ and at $x = L$. In general, the displacement of such a string is

$$y(x,t) = \sum_{n=1}^{\infty} \sin\frac{n\pi x}{L}\left(a_n \sin\frac{n\pi ct}{L} + b_n \cos\frac{n\pi ct}{L} \right), \quad (12.2)$$

where each of the a_n and b_n for $n = 1, 2, 3 \ldots$ is an arbitrary constant that we can set once we know the boundary conditions for any given problem. The range 0 to L is called the *half range* because it is half the maximum wavelength or spatial period.

Consider the case when the string is initially at rest and has initial displacement $y(x, 0) = f(x)$. Then, by substituting $t = 0$ into equation (12.2), we find

$$f(x) = \sum_{n=1}^{\infty} b_n \sin \frac{n\pi x}{L}. \tag{12.3}$$

Given any physically reasonable function, $f(x)$, can we find the coefficients, b_n, such that equation (12.3) is satisfied? Remarkably, yes! This is known as Fourier's theorem.

Equation (12.3) is the Fourier half range sine series of a function, $f(x)$. This is a very powerful result. It tells us that, within the range 0 to L, we can write *any* (physically reasonable) function as a sum of sine waves.

12.3. FOURIER SINE SERIES COEFFICIENTS

The Fourier half range sine series coefficients, b_n, are given by

$$b_n = \frac{2}{L} \int_0^L f(x) \sin \frac{n\pi x}{L} dx. \tag{12.4}$$

Derivation of the Fourier Sine Series Coefficients

The formula for the b_n can be derived directly from the Fourier series representation (Equation 12.3). First, multiply both sides of equation (12.3) by $\sin(m\pi x/L)$, then integrate from 0 to L. This gives

$$\int_0^L f(x) \sin \frac{m\pi x}{L} dx = \sum_{n=1}^{\infty} b_n \int_0^L \sin \frac{n\pi x}{L} \sin \frac{m\pi x}{L} dx.$$

The integral on the right hand side is a standard integral (equation 12.35) with result $(L/2)\delta nm$, where δnm is the Kronecker delta defined by $\delta nm = 1$ if $m = n$ and $\delta nm = 0$ otherwise. Essentially, when the two sine waves in the integral on the right have a different wavelength they interfere destructively and cancel to zero. We get a non-zero result for the integral only when the wavelengths are the same and hence when $n = m$.

Substituting in the result from equation (12.35), we have

$$\int_0^L f(x)\sin\frac{m\pi x}{L}dx = \sum_{n=1}^{\infty} b_n \frac{L}{2}\delta_{nm} = b_m \frac{L}{2}.$$

Finally, rearranging this equation and replacing the symbol m with n we find equation (12.4), the formula for the Fourier series sine coefficients.

12.4. USING THE FOURIER SERIES RESULTS

We can now use the results from equation (12.3) and equation (12.4) to find the Fourier half range sine series for any function, $f(x)$.

Example 12.1 Calculate the Fourier series representation of the function $f(x) = 1$ in $0 \leq x < L$.

We wish to represent the function $f(x)$ as a Fourier sine series,

$$f(x) = 1 = \sum_{n=1}^{\infty} b_n \sin\frac{n\pi x}{L}.$$

To do this, we simply need to calculate the appropriate Fourier coefficients b_n using equation (12.4):

$$b_n = \frac{2}{L}\int_0^L f(x)\sin\frac{n\pi x}{L}dx = \frac{2}{L}\int_0^L \sin\frac{n\pi x}{L}dx$$

$$= -\frac{2}{L}\frac{L}{n\pi}\left[\cos\frac{n\pi x}{L}\right]_0^L$$

$$= -\frac{2}{L}\frac{L}{n\pi}(\cos(n\pi) - 1).$$

We can simplify this equation using $\cos(n\pi) = (-1)^n$. Thus

$$b_n = \frac{2}{n\pi}\left(1 - (-1)^n\right)$$

$$= \begin{cases} 0 & \text{if } n \text{ is even} \\ \dfrac{4}{n\pi} & \text{if } n \text{ is odd.} \end{cases} \tag{12.5}$$

So the Fourier sine series of $f(x) = 1$ for $0 \le x < L$ is

$$1 = \sum_{n \text{ odd}} \frac{4}{n\pi} \sin \frac{n\pi x}{L}, \tag{12.6}$$

where the notation "n odd" simply means to take only the odd integer terms in the summation. We could also write this explicitly by defining a new integer counter, $m = 0, 1, 2, \ldots$, and setting $n = 2m + 1$ so that n is always odd:

$$1 = \sum_{m=0}^{\infty} \frac{4}{(2m+1)\pi} \sin \frac{(2m+1)\pi x}{L}.$$

Figure 12.2 illustrates how the representation of $f(x) = 1$ is built up by adding together sine waves from the series. Writing out the first few terms in equation (12.6) explicitly, we have

$$1 = \frac{4}{\pi}\left(\sin \frac{\pi x}{L} + \frac{1}{3} \sin \frac{3\pi x}{L} + \frac{1}{5} \sin \frac{5\pi x}{L} + \ldots \right).$$

As we add each successive sine term from the infinite series we get closer and closer to an exact representation of the function (Figure 12.2: right).

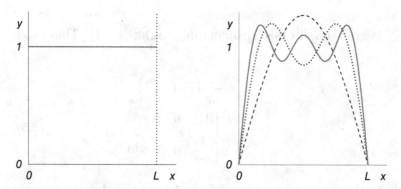

FIGURE 12.2: Left: the function $f(x) = 1$ in $0 \leq x < L$. Right: the first three partial sums of its Fourier sine series, dashed line $f_1 = 4\sin(\pi x/L)/\pi$, dotted line $f_2 = f_1 + 4\sin(3\pi x/L)/(3\pi)$, solid line $f_3 = f_2 + 4\sin(5\pi x/L)/(5\pi)$. As we add each successive term from the infinite series we get closer and closer to the exact representation of the function.

Exercise 12.1 If $f(x)$ is given by $f(x) = x$ in $0 \leq x < L$, show that the Fourier sine series coefficients are given by $b_n = 2L(-1)^{n+1}/(n\pi)$. Hence show that the Fourier sine series for $f(x)$ is

$$x = \frac{2L}{\pi}\left(\sin\frac{\pi x}{L} - \frac{1}{2}\sin\frac{2\pi x}{L} + \frac{1}{3}\sin\frac{3\pi x}{L} - \ldots\right). \qquad (12.7)$$

Figure 12.3 shows $f(x) = x$ and the first three partial sums of the Fourier sine series of $f(x)$.

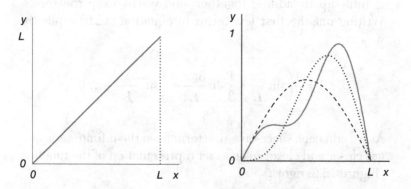

FIGURE 12.3: Left: the function $f(x) = x$ in $0 \leq x < L$. Right: the first three partial sums of its Fourier sine series, dashed line $f_1 = 2L\sin(\pi x/L)/\pi$, dotted line $f_2 = f_1 - 2L\sin(2\pi x/L)/(2\pi)$, solid line $f_3 = f_2 + 2L\sin(3\pi x/L)/(3\pi)$.

Exercise 12.2 A function $g(x)$ is defined by

$$g(x) = \begin{cases} x/L & \text{if } 0 \le x < L/2 \\ 1 - x/L & \text{if } L/2 \le x < L. \end{cases}$$

By expanding $g(x)$ as a Fourier sine series show that

$$g(x) = \sum_{n \text{ odd}} \frac{4(-1)^{\frac{n-1}{2}}}{n^2 \pi^2} \sin \frac{n\pi x}{L}.$$

Hint: the integral for b_n can be split into the sum of two parts of the form $\int_a^b f(x)dx = \int_a^c f(x)dx + \int_c^b f(x)dx$ where $a < c < b$. See Figure 12.4 for a depiction of $f(x)$ and the first three partial sums of its Fourier sine series.

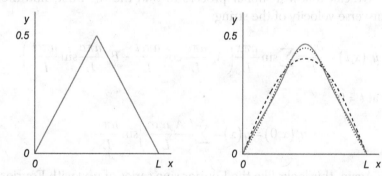

FIGURE 12.4: Left: the function $g(x)$ from exercise 12.2. Right: the first three partial sums of its Fourier sine series, dashed line $f_1 = 4\sin(\pi x/L)/\pi^2$, dotted line $f_2 = f_1 - 4\sin(3\pi x/L)/(9\pi^2)$, solid line $f_3 = f_2 + 4\sin(5\pi x/L)/(25\pi^2)$.

12.5. APPLICATION TO THE WAVE EQUATION

The separable solutions to the wave equation for a string fixed at $x = 0$ and $x = L$ are

$$y(x,t) = \sum_{n=1}^{\infty} \sin \frac{n\pi x}{L} \left(A_n \sin \frac{n\pi ct}{L} + B_n \cos \frac{n\pi ct}{L} \right). \qquad (12.8)$$

Armed with results 12.3 and 12.4, we can now find the coefficients A_n and B_n given a set of initial conditions.

Let us examine the general case when the string is given an initial displacement, $y(x, 0) = p(x)$, and an initial velocity, $y_t(x, 0) = q(x)$. By substituting $t = 0$ into equation (12.8) we immediately find that

$$y(x,0) = p(x) = \sum_{n=1}^{\infty} B_n \sin \frac{n\pi x}{L}.$$

This looks like the Fourier sine series of $p(x)$ with Fourier series coefficients B_n. So, to find the coefficients B_n we simply need to apply the formula in equation (12.4):

$$B_n = \frac{2}{L} \int_0^L p(x) \sin \frac{n\pi x}{L} dx. \tag{12.9}$$

We can follow a similar process to find the A_n. First, find the transverse velocity of the string:

$$y_t(x,t) = \frac{\partial y}{\partial t} = \sum_{n=1}^{\infty} \sin \frac{n\pi x}{L} \left(A_n \frac{n\pi c}{L} \cos \frac{n\pi ct}{L} - B_n \frac{n\pi c}{L} \sin \frac{n\pi ct}{L} \right).$$

So at $t = 0$

$$y_t(x,0) = q(x) = \sum_{n=1}^{\infty} \left(\frac{A_n n\pi c}{L} \right) \sin \frac{n\pi x}{L}.$$

Again, this looks like the Fourier sine series of $q(x)$ with Fourier series coefficients $A_n n\pi c/L$. So to find $A_n n\pi c/L$ we simply need to apply the formula in equation (12.4):

$$\left(\frac{A_n n\pi c}{L} \right) = \frac{2}{L} \int_0^L q(x) \sin \frac{n\pi x}{L} dx,$$

giving

$$A_n = \frac{2}{n\pi c} \int_0^L q(x) \sin \frac{n\pi x}{L} dx. \tag{12.10}$$

Example 12.2 A string fixed at $x = 0$ and at $x = L$ is given constant initial velocity, $y_t(x, 0) = v$, and zero initial displacement, $y(x, 0) = 0$. Find $y(x, t)$.

In general,

$$y(x,t) = \sum_{n=1}^{\infty} \sin\frac{n\pi x}{L}\left(A_n \sin\frac{n\pi ct}{L} + B_n \cos\frac{n\pi ct}{L} \right).$$

To find $y(x, t)$ given a set of initial conditions, substitute the initial conditions into the general solution, find equations involving the unknown coefficients A_n and B_n, then calculate A_n and B_n using the formula for the Fourier sine series coefficients, equation (12.4).

At $t = 0$ the displacement of the string is zero, so

$$0 = \sum_{n=1}^{\infty} B_n \sin\frac{n\pi x}{L}$$

giving $B_n = 0$.

At $t = 0$ the initial velocity is v, so

$$v = \sum_{n=1}^{\infty} \left(\frac{A_n n\pi c}{L} \right) \sin\frac{n\pi x}{L}.$$

This is just the Fourier series representation of a constant, v. So

$$\left(\frac{A_n n\pi c}{L} \right) = \frac{2}{L} \int_0^L v \sin\frac{n\pi x}{L} dx.$$

Then using the result from equation (12.5),

$$\left(\frac{A_n n\pi c}{L} \right) = \begin{cases} 0 & \text{if } n \text{ is even} \\ \dfrac{4v}{n\pi} & \text{if } n \text{ is odd} \end{cases}$$

from which

$$A_n = \begin{cases} 0 & \text{if } n \text{ is even} \\ 4vL/n^2\pi^2 c & \text{if } n \text{ is odd}. \end{cases}$$

Once we have calculated the A_n and B_n we can write down the full solution $y(x, t)$ that describes the displacement of the string as a function of x and t. In the case when the initial displacement is 0 and the initial velocity is v,

$$y(x,t) = \sum_{n \text{ odd}} \frac{4vL}{n^2\pi^2 c} \sin\frac{n\pi x}{L} \sin\frac{n\pi ct}{L}. \qquad (12.11)$$

Exercise 12.3 The initial displacement of a string of length L fixed at its end points $x = 0$ and $x = L$ is given by $y(x, 0) = ax$, where a is a constant. The initial velocity is zero. Find the solution for $y(x, t)$ as an infinite series.

Exercise 12.4 A string is fixed at its end-points at $x = 0$ and $x = L$. If the initial displacement is $y(x, 0) = \sin(\pi x/L)$ and the initial velocity is $y_t(x, 0) = x$, find the solution for $y(x, t)$ as an infinite series.

Exercise 12.5 A string fixed at its end-points $x = 0$ and $x = L$ is released from rest with initial displacement $y(x, 0) = \exp(-a^2 (x - L/2)^2)$ where $a \gg 1/L$. Find the displacement $y(x, t)$ at time t. You may assume that if aL is large then

$$\int_0^L e^{-a^2(x-L/2)^2} \sin\frac{n\pi x}{L} dx \approx \frac{\sqrt{\pi}}{a} e^{-n^2\pi^2/4L^2a^2} \sin\frac{n\pi}{2}.$$

Figure 12.5 shows the displacement of the string, $y(x, t)$, at four different times.

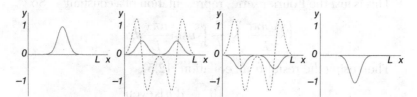

FIGURE 12.5: Wave from exercise 12.5. From left to right the panels show the displacement (solid line) and transverse velocity (dashed line) of the string at $t = 0$, $t = L/4c$, $t = 3L/4c$ and $t = L/c$.

12.6. APPLICATION TO OTHER DIFFERENTIAL EQUATIONS

We can use the techniques from Section 12.5 to find the solutions to other differential equations.

As a brief example, let us consider the solution to the Laplace equation for the electrostatic potential $\phi(x, y)$ on a metal plate,

$\phi_{xx} + \phi_{yy} = 0$. Suppose the potential $\phi(x, y) \rightarrow 0$ as $y \rightarrow \infty$ and is set to zero at $x = 0$ and $x = L$; then

$$\phi(x,y) = \sum_{n=1}^{\infty} B_n \sin\frac{n\pi x}{L} e^{-n\pi y/L}.$$

If the potential at $y = 0$ has the form $p(x)$, then

$$\phi(x,0) = p(x) = \sum_{n=1}^{\infty} B_n \sin\frac{n\pi x}{L},$$

and we can calculate the coefficients simply by applying the formula for the Fourier sine series coefficients,

$$B_n = \frac{2}{L} \int_0^L p(x)\sin\frac{n\pi x}{L} dx.$$

Figure 12.6 shows a solution to the Laplace equation, $\phi(x, y)$, satisfying the boundary equations in Exercise 12.6.

Exercise 12.6 The electric potential $\phi(x, y)$ has boundary conditions $\phi(x, 0) = x^2$, $\phi(0, y) = 0$, $\phi(L, y) = 0$ and $\phi(x, y) \rightarrow 0$ as $y \rightarrow \infty$. Show that

$$\phi(x,y) = \sum_{n=1}^{\infty} \left(\frac{2L^2}{n\pi}(-1)^{n+1} + \frac{4L^2}{n^3\pi^3}\left((-1)^n - 1\right) \right) \sin\frac{n\pi x}{L} e^{-n\pi y/L}. \quad (12.12)$$

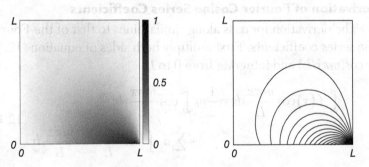

FIGURE 12.6: $\phi(x, y)$ from equation (12.12). Left: map of the electric potential. Right: equipotential lines.

12.7. FOURIER HALF RANGE COSINE SERIES

Another representation of a function defined in the range $0 \le x < L$ uses cosine instead of sine functions. This is equally valid: within the range $0 \le x < L$ both sines and cosines form mathematically *complete sets*. This means we can expand any function within this range in terms of either sines or cosines.

The cosine representation of a function $f(x)$ is

$$f(x) = \frac{1}{2}a_0 + \sum_{n=1}^{\infty} a_n \cos\frac{n\pi x}{L}. \tag{12.13}$$

The factor of $\frac{1}{2}$ in front of a_0 is simply for convenience in that it allows us to give a single formula for the a_n for all n.

The coefficients a_n, $n = 0, 1, 2, \ldots$ are given by

$$a_n = \frac{2}{L}\int_0^L f(x)\cos\frac{n\pi x}{L}dx. \tag{12.14}$$

It is often easier to perform the calculation for a_0 separately from all the other a_n. Then a_0 is simply

$$a_0 = \frac{2}{L}\int_0^L f(x)dx. \tag{12.15}$$

Note that $\frac{1}{2}a_0$ is simply the average of the function $f(x)$ in the range $0 \le x < L$.

Derivation of Fourier Cosine Series Coefficients

The derivation for a_n is along similar lines to that of the Fourier sine series coefficients. First multiply both sides of equation (12.13) by $\cos(m\pi x/L)$ and integrate from 0 to L:

$$\int_0^L f(x)\cos\frac{m\pi x}{L}dx = \frac{1}{2}a_0\int_0^L \cos\frac{m\pi x}{L}dx$$
$$+ \sum_{n=1}^{\infty} a_n\int_0^L \cos\frac{n\pi x}{L}\cos\frac{m\pi x}{L}dx. \tag{12.16}$$

From the standard integral (Equation 12.36) all the terms in the sum on the right of equation (12.16) are zero except for the one with $m = n$. Hence, if $m = n$,

$$\int_0^L f(x)\cos\frac{m\pi x}{L}\,dx = a_m \int_0^L \cos^2\frac{m\pi x}{L} = \frac{L}{2}a_m,$$

and by rearranging and replacing m by n we have the result in equation (12.4).

If we choose $m = 0$ in equation (12.16), then $\cos(m\pi x/L) = 1$ and all the terms in the sum go to zero. We are left with

$$\int_0^L f(x)\,dx = \frac{1}{2}a_0 \int_0^L dx = \frac{L}{2}a_0,$$

which gives the result for a_0 in equation (12.15).

Exercise 12.7 The function $f(x) = x$ in $0 \le x < \pi$ is expanded as a Fourier cosine series. Calculate the coefficients a_0 and a_n, and show that

$$x = \frac{\pi}{2} - \frac{4}{\pi}\sum_{n\,\text{odd}}\frac{1}{n^2}\cos nx. \tag{12.17}$$

The function $f(x) = x$ and the first three partial sums in its Fourier cosine series are shown in Figure 12.7.

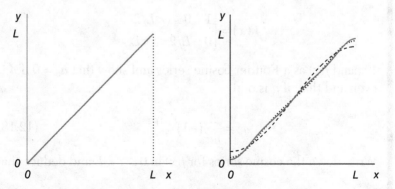

FIGURE 12.7: Left: the function $f(x) = x$ in $0 \le x < \pi$. Right: the first three partial sums of its Fourier cosine series, dashed line $f_1 = \pi/2 - 4\cos(x)/\pi$, dotted line $f_2 = f_1 - 4\cos(3x)/(9\pi)$, solid line $f_3 = f_2 - 4\cos(5x)/(25\pi)$. As we add each successive term from the infinite series we get closer and closer to the exact representation of the function $f(x) = x$.

12.8. NUMERICAL SERIES

Both sine and cosine Fourier representations can be used to derive interesting numerical series results for constants. Take, for example, the result from equation (12.17). Writing out the first few terms in the cosine series for x defined between 0 and π we have

$$x = \frac{\pi}{2} - \frac{4}{\pi}\left(\cos x + \frac{1}{9}\cos 3x + \frac{1}{25}\cos 5x + \dots\right).$$

This equation is true for any x where $0 \le x < \pi$. We can substitute a particular value for x into both the left and right sides and the equality will still hold. Let us choose, for example, $x = 0$. Then all the values $\cos nx = 1$ and so

$$0 = \frac{\pi}{2} - \frac{4}{\pi}\left(1 + \frac{1}{9} + \frac{1}{25} + \dots\right)$$

giving

$$\pi^2 = 8\left(1 + \frac{1}{9} + \frac{1}{25} + \dots\right),$$

which is a series expansion for π^2.

Exercise 12.8 A function $f(x)$ is defined by

$$f(x) = \begin{cases} 1 & 0 \le x < L/2 \\ 0 & L/2 < x < L. \end{cases}$$

Expand $f(x)$ as a Fourier cosine series and show that $a_n = 0$ if n is even and that, if n is odd,

$$a_n = \frac{2}{n\pi}(-1)^{(n-1)/2}. \tag{12.18}$$

Write down the cosine series for $f(x)$ in $0 \le x < L$ and deduce that

$$\frac{\pi}{4} = 1 - \frac{1}{3} + \frac{1}{5} - \frac{1}{7}.$$

12.9.   PERIODIC EXTENSION OF FOURIER SERIES

So far we have examined the sine and cosine Fourier series representations of functions within a limited range, $0 \leq x < L$. However, both sine and cosines repeat periodically. So, if we plot the Fourier series representations outside of this range we will get functions that repeat periodically with wavelength $2L$.

Outside the given finite range, the Fourier series of $f(x)$ represents a periodic extension of the function with $f(x + 2L) = f(x)$.

To understand the periodic extension of Fourier series it is important to first understand the symmetry of *even* and *odd* functions. There is a short section on even and odd functions in Chapter 3.

12.10.   EVEN AND ODD SYMMETRY OF PERIODIC FUNCTIONS

Sine waves are odd, so any Fourier sine series representation of a periodic function must have odd symmetry. Similarly, cosine waves are even, so any Fourier cosine series representation of a periodic function must have even symmetry.

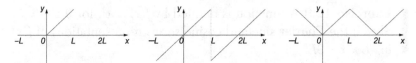

FIGURE 12.8: Left: Fourier sine or cosine representation of $f(x) = x$ within $0 \leq x < L$. Center: periodic extension of sine series representation of $f(x)$. Right: periodic extension of cosine series representation of $f(x)$.

Figure 12.8 shows sine and cosine representations of $f(x)$. Earlier we saw that within $0 \leq x < L$ we could expand the function $f(x) = x$ either as a sine series, *or* as a cosine series (Figure 12.8: left). However, because sines and cosines have different symmetry, when we expand the range we obtain different shape waveforms for each of the series: the "sawtooth" wave for the sine series (Figure 12.8: center) has odd symmetry, while the "triangle" wave for the cosine series (Figure 12.8: left) has even symmetry. For both periodic extensions, $f(x) = f(x + 2L)$. For the sine series we also have $f(x) = -f(-x)$ and, for the cosine series, $f(x) = f(-x)$.

Example 12.3 Sketch the Fourier cosine representation of $f(x) = x^3$ in the range $-L \leq x < 3L$.

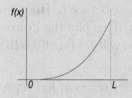

First sketch the function $f(x) = x^3$ between 0 and L.

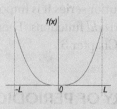

Next, we know that cosines are even functions, so the Fourier cosine series must be even. This allows us to sketch the Fourier cosine series between $-L$ and 0.

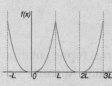

Cosines and hence the Fourier cosine series must repeat every $2L$. So the shape of the waveform between $-L$ and L simply repeats, and we can sketch the cosine series of $f(x) = x^3$ between $-L$ and $3L$.

Exercise 12.9 A function is defined by $f(x) = x^2$, for $0 \leq x < L$. Sketch the Fourier sine and cosine series representations of $f(x)$ in $-L < x < 3L$.

Exercise 12.10 A function $g(x)$ is defined by.

$$g(x) = \begin{cases} x/L & \text{if } 0 \leq x < L/2 \\ 1 - x/L & \text{if } L/2 \leq x < L. \end{cases}$$

Sketch the Fourier sine and Fourier cosine representations of $g(x)$ in $-L \leq x < 3L$.

Exercise 12.11 Within $0 \leq x < L$, $f(x) = x$ can be expanded as a sine series equation (12.7) or a cosine series equation (12.17). What functions do each of these series represent in the range $-L \leq x < 0$?

Figure 12.9 shows three examples of waveforms that are common in signal analysis. We can obtain a square wave by expanding the range of the Fourier half range sine series representation of $f(x) = 1$. We can similarly obtain a triangle wave from the Fourier half range cosine series representation of $f(x) = x$ from Exercise 12.7. But what happens if we consider a function, $f(x)$, defined in the *full range* from $x = -L$ to $x = L$? If $f(x)$ is neither even nor odd then we cannot expand it as only a sum of sine waves or only a sum of cosine waves. Instead we must use a Fourier full range series (Section 12.11).

The rectified half wave (Figure 12.9: right) is one such function. This is obtained from the full series representation of a function defined by $f(x) = 0$ where $-L \le x < 0$, $f(x) = \sin(\pi x/L)$ where $0 < x < L$.

FIGURE 12.9: Some common waveforms: square wave (left), triangle wave (center), rectified half wave (right).

12.11.   FOURIER FULL RANGE SERIES

The range 0 to $2L$ (or, alternatively, $-L$ to L) is called the *full range* because it contains a full wavelength of the periodic function. In this section we will mostly use the range $-L$ to L; however, the ranges 0 to $2L$ and $-L$ to L are exactly equivalent.

In the range $-L$ to L neither sine nor cosine waves form a complete set. If a function defined between $-L$ and L has odd symmetry it can be represented as a sine series. If a function has even symmetry it can be represented as a cosine series. However, in the general case, to represent a function of arbitrary symmetry, we need to include *both* sine and cosine terms in the representation.

The full range Fourier series for $f(x)$ in the range $-L < x < L$ is

$$f(x) = \frac{1}{2}a_0 + \sum_{n=1}^{\infty}\left(a_n \cos\frac{n\pi x}{L} + b_n \sin\frac{n\pi x}{L}\right). \qquad (12.19)$$

The formulae for the Fourier series full range coefficients can be derived in a similar way to the formulae for the sine and cosine half range coefficients. We have

$$a_0 = \frac{1}{L}\int_{-L}^{L} f(x)\,dx$$

$$a_n = \frac{1}{L}\int_{-L}^{L} f(x)\cos\frac{n\pi x}{L}\,dx \qquad (12.20)$$

$$b_n = \frac{1}{L}\int_{-L}^{L} f(x)\sin\frac{n\pi x}{L}\,dx.$$

Note that if the full range is defined to be between 0 to $2L$ the formulae remain the same except that the limits of the integration are from 0 to $2L$.

Exercise 12.12 A function $f(x) = 1 + x$ within $-\pi \le x < \pi$. Calculate the Fourier full range series of $f(x)$.

Exercise 12.13 Calculate the full range Fourier series for a "sawtooth" wave, $f(x) = x$, $-\pi < x < \pi$. Explain why the series is the same as the half range sine representation in Exercise 12.1. By writing out the result for an appropriately chosen value of x, show that

$$\sum_{m=0}^{\infty} \frac{(-1)^m}{2m+1} = \frac{\pi}{4}.$$

12.12.   COMPLEX FORM OF FOURIER SERIES

Instead of equation (12.19) we could equally well write the complex form

$$f(x) = \sum_{n=-\infty}^{\infty} c_n e^{in\pi x/L}, \qquad (12.21)$$

where

$$c_n = \frac{1}{2L}\int_{-L}^{L} f(x) e^{-in\pi x/L}\,dx. \qquad (12.22)$$

Sometimes this form is more convenient than the sine and cosine forms.

Derivation of Complex Series Results

Recalling that sines and cosines can be written in terms of complex exponentials we can obtain equation (12.21) directly from equation (12.19):

$$f(x) = a_0 + a_1 \cos\frac{\pi x}{L} + b_1 \sin\frac{\pi x}{L} + a_2 \cos\frac{2\pi x}{L} + b_2 \sin\frac{2\pi x}{L} + \ldots$$

$$= a_0 + a_1 \frac{e^{i\pi x/L} + e^{-i\pi x/L}}{2} + b_1 \frac{e^{i\pi x/L} - e^{-i\pi x/L}}{2i}$$

$$+ a_2 \frac{e^{i2\pi x/L} + e^{-i2\pi x/L}}{2} + b_2 \frac{e^{i2\pi x/L} - e^{-i2\pi x/L}}{2i} + \ldots.$$

Collecting together exponentials with the same powers we have

$$f(x) = \ldots + (a_2/2 - b_2/(2i)) e^{-i2\pi x/L} + (a_1/2 - b_1/(2i)) e^{-i\pi x/L} + a_0$$

$$+ (a_1/2 + b_1/(2i)) e^{i\pi x/L} + (a_2/2 + b_2/(2i)) e^{i2\pi x/L} + \ldots$$

$$= \ldots + c_{-2} e^{-i2\pi x/L} + c_{-1} e^{-i\pi x/L} + c_0 + c_1 e^{i\pi x/L} + c_2 e^{i2\pi x/L} + \ldots,$$

which is identical to equation (12.21).

To find the formula for the complex coefficients, c_n, multiply both sides of equation (12.21) by $e^{-i\pi m x/L}$ and integrate:

$$\int_{-L}^{L} f(x) e^{-im\pi x/L} dx = \sum_{n=-1}^{\infty} c_n \int_{-L}^{L} e^{i(n-m)\pi x/L} dx.$$

The integral on the right is a standard integral, equation (12.40). Using this result we find

$$\int_{-L}^{L} f(x) e^{-im\pi x/L} dx = \sum_{n=-\infty}^{\infty} c_n 2L\delta_{nm}. \qquad (12.23)$$

Then rearranging equation (12.23) for c_n we obtain the result in equation (12.22).

Exercise 12.14 If $f(x) = 1 + x$, $-L < x < L$, show that the complex Fourier series coefficients are given by

$$c_n = \begin{cases} 1 & n = 0 \\ (-1)^{n+1} L/(in\pi) & n \neq 0. \end{cases}$$

Exercise 12.15 The function $f(x) = \exp(px)$ in $-\pi \leq \pi < \pi$. Expand $f(x)$ as a sum of complex exponentials to show that

$$f(x) = \sum_{n=-\infty}^{\infty} \frac{(-1)^n}{\pi(p - in)} \sinh(p\pi) e^{inx}.$$

Example 12.4 Verify that the result for the complex Fourier series in Exercise 12.14 is equal to that obtained in Exercise 12.12.

The complex Fourier series for $f(x) = 1 + x$ in $-L < x < L$ is

$$f(x) = 1 + \sum_{n=-\infty}^{-1} \frac{(-1)^{n+1} L}{in\pi} e^{in\pi x/L} + \sum_{n=1}^{\infty} \frac{(-1)^{n+1} L}{in\pi} e^{in\pi x/L}.$$

Then replacing n by $-n$ in the first summation, and factorizing out $-L/\pi$, we have

$$f(x) = 1 + \frac{-L}{\pi} \sum_{n=1}^{\infty} \left[\frac{(-1)^{-n}}{i(-n)} e^{i(-n)\pi x/L} + \frac{(-1)^n}{in} e^{in\pi x/L} \right]$$

$$= 1 + \frac{-L}{\pi} \sum_{n=1}^{\infty} \frac{(-1)^n}{in} \left[e^{in\pi x/L} - e^{-in\pi x/L} \right]$$

$$= 1 + \sum_{n=1}^{\infty} \frac{2(-1)^{n+1} L}{n\pi} \sin \frac{n\pi x}{L}, \tag{12.24}$$

where, in the last step, we have used the fact that $e^{in\pi x/L} - e^{-in\pi x/L} = 2i \sin(n\pi x/L)$.

12.13. PROPERTIES OF FOURIER SERIES

Here we quote without proof some facts about Fourier series that you should know.

General Properties

a) Except for some pathological functions which do not occur in physical problems, we can *always* expand a function, $f(x)$, defined in a finite interval as a Fourier series which will converge with sum $f(x)$ at all points at which $f(x)$ is continuous.

b) If $f(x)$ has a discontinuity at $x = x_0$ its Fourier series will converge to the average of the limit from the left and the limit from the right. The Fourier series of a discontinuous function sums to a value midway along the discontinuity.

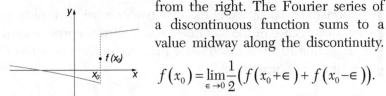

$$f(x_0) = \lim_{\epsilon \to 0} \frac{1}{2}\left(f(x_0 + \epsilon) + f(x_0 - \epsilon)\right).$$

c) A Fourier series can be integrated term by term: the resulting series always converges to $\int f(x)\,dx$.

d) Term by term differentiation of a Fourier series *may* produce a divergent series. If the series produced by differentiation does converge, then it is the Fourier series for $f'(x)$.

Convergence of Series

The properties above apply to the full Fourier series containing an infinite number of terms. In practice we often calculate the sum of only a finite number of the terms in the series which we use as an approximation. It is important to know when this is likely to give a good approximation.

a) If $f(x)$ or its periodic extension has *discontinuities* we expect a_n, b_n, c_n to be of order $1/n$ and the convergence is *slow*.

b) If $f(x)$ or its periodic extension is *continuous* we expect a_n, b_n, c_n to be of order $1/n^2$ and the convergence is *rapid*.

c) The Gibbs phenomenon. At a discontinuity (at x_0, say) convergence of a Fourier series is slow and a finite sum of N terms will persistently under- or over-estimate $f(x)$ near x_0. The size of the overshoot (or undershoot) does *not* tend to zero as $N \to \infty$, but the region of the overshoot (or undershoot) does become narrower as the series converges.

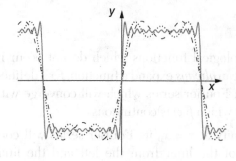

The figure on the left shows the partial Fourier sine series of $f(x) = 1$, $0 \le x < \pi$ with $N = 3$ (dashed), $N = 5$ (dotted) and $N = 25$ (solid line) terms.

Exercise 12.16 By using the results above briefly explain why the convergence of the half range cosine series representation of x (Equation 12.17) is much faster than that of the sine series representation (Equation 12.7).

Exercise 12.17 State whether each of the following functions, defined in the range $-\pi < x < \pi$, can be expanded as (a) a Fourier sine series, (b) a cosine series or (c) a Fourier series containing both sine and cosine terms: (i) x^3; (ii) $x^2 + x$; (iii) $\exp(x)$; (iv) $|x|$.

12.14. FOURIER TRANSFORMS AND FOURIER INTEGRALS

So far, we have seen that we can express an arbitrary periodic function as a Fourier series, a sum of harmonic (sine and cosine) waves.

The *Fourier transform* gives us an analogous way to represent a general function that is not periodic. Fourier transforms are used in an enormous range of pure and applied science, including information processing, electronics and communications.

The Fourier series of a function $f(x)$ with period $\lambda = 2L$ can be written as

$$f(x) = \sum_{n=-\infty}^{\infty} c_n e^{in\pi x/L} = \sum_{n=-\infty}^{\infty} c_n e^{2in\pi x/\lambda}. \tag{12.25}$$

So what happens if the function is non-periodic? A non-periodic function is equivalent to a periodic function in the limit that $\lambda \to \infty$. So, let us consider what happens to 12.25 when λ becomes large.

First we define a new variable $k = 2\pi n/\lambda$. Then we can think of the Fourier coefficients c_n as a function $\tilde{f}(k) = c_{k\lambda/2\pi}$ defined on a line of points in k-space. The distance between the points in k-space is $\Delta_k = 2\pi/\lambda$. Then we can rewrite the sum in equation (12.25) as

$$f(x) = \sum_{n=-\infty}^{\infty} c_n e^{ikx} = \sum_{n=-\infty}^{\infty} \tilde{f}(k) e^{ikx} = \frac{1}{2\pi} \sum_{n=-\infty}^{\infty} F(k) e^{ikx} \Delta_k$$

where $F(k) = \lambda \tilde{f}(k)$. In the limit $\lambda \to \infty$ the spacing between the points, Δ_k, tends to zero and the sum becomes an integral:

$$f(x) = \frac{1}{2\pi} \int_{-\infty}^{\infty} F(k) e^{ikx} dk. \tag{12.26}$$

This is a Fourier integral. It is a representation of an arbitrary (non-periodic) function, $f(x)$, in terms of simple harmonics.

12.15. FOURIER TRANSFORMS

To obtain a result for $F(k)$ in equation (12.26) we simply apply the formula (Equation 12.22) for the Fourier series coefficients, $c_n = c_{k\lambda/2\pi}$, in the limit as $\lambda \to \infty$,

$$F(k) = \lim_{\lambda \to \infty} \lambda c_{kL/2\pi} = \int_{-\infty}^{\infty} f(x) e^{-ikx} dx. \tag{12.27}$$

$F(k)$ is called the Fourier transform of $f(x)$. The two functions, $F(k)$ and $f(x)$, are called a Fourier transform pair; $f(x)$ is the inverse Fourier transform of $F(k)$.

Unfortunately there is no standard definition in the literature of what constitutes the transform and what constitutes the inverse transform. The only requirement is that one of equations (12.26) and (12.27) contains e^{-ikx} and one contains e^{ikx}. Similarly there is no set convention for the constant factors in front of these integrals. We have used a factor $1/2\pi$ on the inverse transform and 1 on the transform, but you will often see the opposite of this, or sometimes $1/\sqrt{2\pi}$ is used in front of both. This means that, when reading the literature, you should be careful to identify the conventions used.

The argument of the Fourier transform, k, has units that are the reciprocal of the dimensions of the variable x. We will often use x to denote position, in which case k is the wavenumber. Similarly, if the variable is a time, t, the transform variable will be a frequency, often denoted by ω. Physicists often talk about using Fourier transforms to transform from real space to reciprocal or k-space, or from the time-domain to the frequency-domain.

12.16. CALCULATING FOURIER TRANSFORMS

Fourier transforms and integrals are ordinary integrals that can be evaluated in the usual way. We will consider a specific example: the Fourier transform of a Gaussian.

Example 12.5 Find the Fourier transform of $f(x) = \exp(-x^2/\sigma^2)$. From equation (12.27) we have

$$F(k) = \int_{-\infty}^{\infty} e^{-x^2/\sigma^2} e^{-ikx} dx = \int_{-\infty}^{\infty} e^{-\left(x^2/\sigma^2 + ikx\right)} dx.$$

This integral is performed with a standard trick. First, we complete the square in the argument of the exponential,

$$x^2/\sigma^2 + ikx = \left(x/\sigma + ik\sigma/2\right)^2 + k^2\sigma^2/4.$$

Then

$$F(k) = e^{-k^2\sigma^2/4} \int_{-\infty}^{\infty} e^{-\left(x/\sigma + ik\sigma/2\right)^2} dx, \qquad (12.28)$$

$$= \sigma e^{-k^2\sigma^2/4} \int_{-\infty}^{\infty} e^{-x'^2} dx',$$

where we have changed variable from x to $x' = x/\sigma - ik\sigma/2$, so $dx' = dx/\sigma$. In this particular case, the change of variable leaves the limits on the integral unchanged at $-\infty$ and ∞. Then, using the fact that $\int_{-\infty}^{\infty}\exp\left(-x^2\right)dx = \sqrt{\pi}$, we find that the Fourier transform of the Gaussian is

$$F(k) = \sigma\sqrt{\pi}e^{-k^2\sigma^2/4}. \qquad (12.29)$$

So the Fourier transform of a Gaussian of half width σ is a Gaussian of half width $2/\sigma$; thus a wide Gaussian function in real space transforms to a narrow Gaussian function in k-space and vice-versa.

Exercise 12.18 Show that the Fourier transform of the function defined by $f(x) = a$ for $|x| \leq L$, $f(x) = 0$ for $|x| > L$ is

$$F(k) = \frac{2a\sin kL}{k}.$$

Exercise 12.19 Find the inverse Fourier transform of $F(k) = \exp(-k^2/4)$.

12.17. CONVOLUTIONS

Here we will briefly introduce a useful result: the convolution theorem.

A convolution integral has the general form

$$C(x) = \int_{-\infty}^{\infty} f(x-x')g(x')dx'. \qquad (12.30)$$

This integral relates an *output* function, $C(x)$, to the *input* $g(x)$ and the *response* $f(x)$. Figure 12.10 shows an example from imaging processing where we degrade an initially sharp image with a Gaussian "blur." The process is illustrated in one dimension in Figure 12.11: we convolve the input, $g(x)$, with a Gaussian response function, $f(x) = \exp(-x^2)$, to give the new blurred output

$C(x) = \int_{-\infty}^{\infty} \exp\left[-(x - x')^2\right] g(x') dx'$. You will often see a similar effect in experimental physics where some initially sharp "input" signal will suffer from Gaussian broadening as a result of an imprecise experimental response.

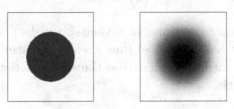

FIGURE 12.10: Image processing with convolutions. A sharp image of a circle (left) is convolved with a Gaussian response function to give the blurred image (right).

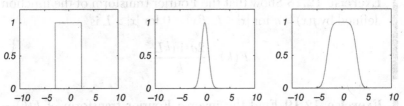

FIGURE 12.11: Example convolution. Left: sharp "input" signal, $g(x) = 1$, $|x| < 3$. Center: Gaussian response function, $f(x) = \exp(-x^2)$. Right: blurred output signal,

$$C(x) = \int_{-\infty}^{\infty} f(x - x') g(x') dx'.$$

12.18. CONVOLUTION THEOREM

The convolution theorem is a useful relation between the Fourier components of the input and output functions. According to the convolution theorem,

$$C(k) = F(k)G(k). \tag{12.31}$$

To prove this relation we simply calculate the Fourier transform of the output function $C(x)$,

$$C(k) = \int_{-\infty}^{\infty} C(x) e^{-ikx} dx = \int_{-\infty}^{\infty} \int_{-\infty}^{\infty} f(x - x') g(x') e^{-ikx} dx' dx.$$

Putting $u = x - x'$ and using u as the integration variable instead of x we have

$$C(k) = \int_{-\infty}^{\infty} \int_{-\infty}^{\infty} f(u)g(x')e^{-ik(x'+u)}dx'du$$
$$= \int_{-\infty}^{\infty} f(u)e^{-iku}du \int_{-\infty}^{\infty} g(x')e^{-ikx'}dx'$$
$$= F(k)G(k).$$

12.19. FOURIER DECONVOLUTION

The convolution theorem tells us that convolutions in real space are simply multiplications in reciprocal space. This is very useful if we want to *deconvolve* a signal.

For example, let us imagine that, in an experiment, we want to measure some input $g(x)$. However, because of our experimental response, $f(x)$, we actually measure a degraded output signal, $C(x)$. How do we get back to the actual input that we would like to measure? We could try to solve the integral equation (12.30) for $g(x)$ but it is much easier to make use of the convolution theorem.

In reciprocal space we can find $G(k)$ by dividing $C(k)$ by $F(k)$. Then we can easily obtain $g(x)$ with an inverse transform,

$$g(x) = \frac{1}{2\pi} \int_{-\infty}^{\infty} \frac{C(k)}{F(k)} e^{ikx}dk. \tag{12.32}$$

Figure 12.12 shows an illustration of this process. An initially blurred image is sharpened by deconvolving the image with a Gaussian response function.

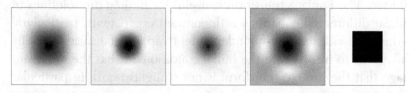

FIGURE 12.12: Sharpening of an image by deconvolution. From left to right: original blurred image $C(x, y)$; Fourier transform of blurred image, $C(k_x, k_y)$; response function in reciprocal space, $F(k_x, k_y)$; input function in reciprocal space, $G(k_x, k_y) = C(k_x, k_y)/F(k_x, k_y)$; deconvolved image, $g(x, y)$.

Exercise 12.20 Write down the analogous results for the convolution integral and convolution theorem, in the time and frequency domain.

Exercise 12.21 If $c(t) = \int_{-\infty}^{\infty} f(t-t')g(t')dt'$ show that $C(\omega) = F(\omega)G(\omega)$ where $C(\omega)$, $F(\omega)$ and $G(\omega)$ are the Fourier transforms of $c(t)$, $f(t)$ and $g(t)$, respectively.

Exercise 12.22 In reciprocal space an experimental signal that has been broadened with a Gaussian response function, $F(k) = \sqrt{\pi} \exp(-ak^2/4)$, has the form $C(k) = (2\sqrt{\pi}/k)\exp(-ak^2/4)\sin k$. Use the convolution theorem to write down the input signal, $G(k)$, in reciprocal space, then transform this into real space to find the original form of the input, $g(x)$. (Hint: you can often find a Fourier transform simply by recognizing that it is related to a known inverse transform.)

12.20. EXTENSION: LAPLACE TRANSFORM

We define the Laplace transform of the function $f(t)$ by

$$F(s) \neq \int_0^{\infty} e^{-st} f(t)dt.$$

Note how this differs from the Fourier transform: we replace the complex exponential by a real exponential and integrate only from 0 to ∞. The Laplace transform is useful for finding the solutions of linear differential equations with constant coefficients, particularly equations with time as the independent variable, because it readily allows us to incorporate the initial conditions. It has the disadvantage that the inverse transform is not straightfoward; in particular $f(t) \neq \int_0^{\infty} F(s)e^{-st}ds$ in general. In practice, one can look up tables of transform pairs.

To see how this works, consider the equation of a driven harmonic oscillator,

$$\ddot{x} + x = f(t). \tag{12.33}$$

We take the Laplace transform of both sides by multiplying by e^{-st} and integrating. We have

$$\int_0^\infty e^{-st}\ddot{x}\,dt = \left[e^{-st}\dot{x}(t) \right]_0^\infty + s\int_0^\infty e^{-st}\dot{x}(t)\,dt$$

$$= -\dot{x}(0) + s\left[e^{-st}x(t) \right]_0^\infty + s^2\int_0^\infty e^{-st}x(t)\,dt$$

$$= -\dot{x}(0) - sx(0) + s^2 X(s).$$

Thus the transform of equation (12.33) gives

$$X(s) = \frac{F(s) + sx(0) + \dot{x}(0)}{s^2 + 1}. \tag{12.34}$$

We can now find the solution of the differential equation by looking up the inverse Laplace transforms of $s/(s^2 + 1)$ and $1/(s^2 + 1)$ (which are $\sin(t)$ and $\cos(t)$, respectively) and, if it is known, $F(s)/(s^2 + 1)$ for the particular form of $F(s)$, the Laplace transform of $f(t)$. More complicated linear equations can be treated similarly, although explicit solutions will be obtained only where the inverse Laplace tranforms can be found.

You may be wondering what is special about the exponential here and in the Fourier transform. In fact, transforms can be and are defined for various classes of functions (besides the exponential) for which an inverse transform can be found. (A transform from which the original function cannot be retrieved is obviously losing information and is unlikely to be useful.)

Revision notes

After completing this chapter you should be able to

- Appreciate how an arbitrary periodic function can be represented by an infinite summation of sines and cosines, the *Fourier series* of a function

- Use the Fourier series representation to find the solution of partial differential equations, including the wave equation, satisfying given boundary conditions

- Distinguish between *full range* (both complex and real form) and *half range sine* and *cosine* Fourier representations of a given function

- Write down the formulae for the Fourier series coefficients in each case

- Find the Fourier half range and full range series representations of a given function in a finite interval

- Understand the periodic extensions of Fourier series

- Appreciate how an arbitrary non-periodic function can be represented by a *Fourier transform*

- Calculate the Fourier transform and inverse transform of a given function

- Prove the convolution theorem

12.A. APPENDIX: STANDARD INTEGRALS

In this appendix, we gather together some standard integrals that are useful for Fourier series.

$$\int_0^L \sin\frac{n\pi x}{L}\sin\frac{m\pi x}{L}\,dx = \begin{cases} 0 & \text{if } m = n = 0 \\ \dfrac{L}{2}\delta_{mn} & \text{otherwise} \end{cases} \tag{12.35}$$

$$\int_0^L \cos\frac{n\pi x}{L}\cos\frac{m\pi x}{L}\,dx = \begin{cases} L & \text{if } m = n = 0 \\ \dfrac{L}{2}\delta_{mn} & \text{otherwise} \end{cases} \tag{12.36}$$

$$\int_{-L}^L \sin\frac{n\pi x}{L}\sin\frac{m\pi x}{L}\,dx = \begin{cases} 0 & \text{if } m = n = 0 \\ L\delta_{mn} & \text{otherwise} \end{cases} \tag{12.37}$$

$$\int_{-L}^L \cos\frac{n\pi x}{L}\cos\frac{m\pi x}{L}\,dx = \begin{cases} 2L & \text{if } m = n = 0 \\ L\delta_{mn} & \text{otherwise} \end{cases} \tag{12.38}$$

$$\int_{-L}^L \sin\frac{n\pi x}{L}\cos\frac{m\pi x}{L}\,dx = 0 \tag{12.39}$$

$$\int_{-L}^L e^{i(n-m)\pi x/L}\,dx = 2L\delta_{nm}. \tag{12.40}$$

Derivation of Standard Integrals

The results for the standard integrals 12.35, 12.36, 12.37, 12.38, 12.39 can be found using the product formulae for trigonometric functions.

For example, to obtain equation (12.35) we write

$$\int_0^L \sin\frac{m\pi x}{L}\sin\frac{n\pi x}{L}\,dx = \frac{1}{2}\int_0^L\left(\cos\frac{(m-n)\pi x}{L} - \cos\frac{(n+m)\pi x}{L}\right)dx$$

$$= \frac{1}{2}\frac{L}{(m-n)\pi}\left[\sin\frac{(m-n)\pi x}{L}\right]_0^L$$

$$-\frac{1}{2}\frac{L}{(m+n)\pi}\left[\sin\frac{(m+n)\pi x}{L}\right]_0^L$$

$$= 0 \text{ if } m \neq n, (\text{all the sine terms are zero}).$$

If $m = n \neq 0$ then the integral becomes

$$\int_0^L \sin\frac{m\pi x}{L}\sin\frac{n\pi x}{L}dx = \int_0^L \sin^2\frac{n\pi x}{L}$$

$$= \frac{1}{2}\int_0^L\left(1 - \cos\frac{2n\pi x}{L}\right)dx$$

$$= \frac{L}{2}.$$

To obtain equation (12.40) we have, if $n \neq m$,

$$\int_{-L}^L \exp\left(\frac{i(n-m)\pi x}{L}\right)dx = \left[\frac{L}{i(n-m)\pi}\exp\left(\frac{i(n-m)\pi x}{L}\right)\right]_{-L}^L$$

$$= \frac{L}{i(n-m)}\left[e^{i(n-m)\pi} - e^{-i(n-m)\pi}\right]$$

$$= 0,$$

because $\exp(i(n-m)\pi) = \exp(-i(n-m)\pi) = \cos((n-m)\pi)$.
However, if $n = m$ then

$$\int_{-L}^L \exp\left(\frac{i(n-m)\pi x}{L}\right)dx = \int_{-L}^L dx$$

$$= 2L.$$

12.21. EXERCISES

1. A function $f(\theta)$, $-\pi \le \theta \le \pi$ is known to have a Fourier series of the form

$$f(\theta) = \frac{1}{2}a_0 + \sum_{n=0}^{\infty} a_n \cos(n\theta). \qquad (12.41)$$

Show that $f(\theta) = f(-\theta)$.

2. If the function $f(\theta)$ is given in the range $0 \le \theta \le \pi$ to be $f(\theta) = \theta$ sketch the extension of this function to the range $-\pi \le \theta \le \pi$ if it is to have a Fourier series of the form (12.41). Sketch also the function represented by the Fourier series outside this range.

3. The function $f(\theta) = \theta^3$, $0 \le \theta \le \pi$ is expanded in (a) a Fourier sine series, and (b) a Fourier cosine series. Sketch the forms of the functions represented by these series in the range $-3\pi \le \theta \le 3\pi$.

4. Starting from the assumption that any function defined between $-L$ and L can be expanded as a sum of sine and cosine functions,

$$f(x) = \frac{1}{2}a_0 + \sum_{n=1}^{\infty} \left(a_n \cos\frac{n\pi x}{L} + b_n \sin\frac{n\pi x}{L} \right),$$

derive the formulae for the Fourier coefficients, a_0, a_n and b_n.

5. The following functions are all defined in the range $-L < x < L$. For each of the functions, give the symmetry of the function and state whether the function can be expanded as a half range sine series, a half range cosine series or whether it *must* be expanded as a full range Fourier series.

   (i) $1/(|x| + d)$,

   (ii) $x^3 \sin(2x)$,

   (iii) $\exp(-4x)$,

   (iv) $x \exp(ax^2)$,

   (v) $ax^4 + bx^2 + c$.

12.22.   PROBLEMS

1.   Find the Fourier series for $f(\theta) = \theta^2$, $-\pi \le \theta < \pi$, and by setting $\theta = 0$, show that

$$\frac{\pi^2}{12} = \sum_{n=1}^{\infty} \frac{(-1)^{n+1}}{n^2}.$$

2.   Expand the function $f(x) = x$ as a half range Fourier cosine series in the range $0 \le x < L$.

By considering the series for a suitable value of x, show that

$$\sum_{n=0}^{\infty} \frac{1}{(2n+1)^2} = \frac{\pi^2}{8}.$$

What function does the series represent in the range $-L < x < L$?

3.   Expand the function $f(x) = \sin(\pi x)$ as a half range Fourier cosine series in the range $0 \le x \le 1$.

By considering the series for a suitable value of x show that

$$\sum_{p=1}^{\infty} \frac{(-1)^p}{4p^2 - 1} = \frac{1}{2} - \frac{\pi}{4}.$$

4.   If $f(x) = \frac{1}{2}\big(h(x) + h(-x)\big)$ and $g(x) = \frac{1}{2}\big(h(x) - h(-x)\big)$,

show that $f(x)$ is even and $g(x)$ is odd.

Derive the following results:

$$\int_{-a}^{a} h(x)\,dx = \begin{cases} 0 & \text{if } h(x) \text{ is odd} \\ 2\int_{0}^{a} h(x)\,dx & \text{if } h(x) \text{ is even.} \end{cases}$$

Hint: first split the integral into two regions, a part from $-a$ to 0 and a part from 0 to a, then change variable from x to $x' = -x$ in the region where $x < 0$.

5. Find the Fourier series for the function

$$f(\theta) = \begin{cases} a, & -\pi \le \theta < 0 \\ \beta, & 0 < \theta < \pi, \end{cases}$$

where a and β are constants. (Hint: the neatest approach is to write $f(\theta)$ as the sum of a symmetric function and an antisymmetric one.)

By setting $\theta = \pi/2$ show that

$$\sum_{n=0}^{\infty} \frac{(-1)^n}{2n+1} = \frac{\pi}{4}$$

6. Find the Fourier series for

$$f(t) = \begin{cases} -\sin t, & -\pi < t < 0 \\ \sin t, & 0 < t < \pi. \end{cases}$$

Hence show that

$$\sum_{n=2,4,6,\ldots}^{\infty} \frac{1}{n^2 - 1} = \frac{1}{2}.$$

7. Calculate the full range complex Fourier series representation of $f(x) = x + x^2$.

8. Show that the Fourier transform of the function defined by $f(x) = \exp(-a^2 x)$ for $x \ge 0$, $f(x) = 0$ otherwise, is $1/(ik + a^2)$.

9. Calculate the inverse Fourier transform of $F(k) = \exp(-k^2)$.

CHAPTER

13

INTRODUCTION TO VECTOR CALCULUS

In physics we are often concerned with physical properties that vary over some region of space, e.g. air temperature in a room, charge density through a solid body, velocity within a fluid. For these, we need to use the idea of a *field*. In this chapter we will discuss fields, and the most important *differential operators* that physicists use to work with fields. These are called *grad*, *div* and *curl*. Here we will study the mathematical definitions and some rules for manipulating fields and operations on fields, and we will begin to look into their geometrical interpretations. These will be crucial to your studies of physics, especially electromagnetic fields, but also fluid mechanics.

In this chapter, we will consider only three-dimensional space and use only Cartesian coordinate systems, i.e. using coordinates (x, y, z) and basis vectors **i, j, k.**

13.1. VECTOR FIELDS

A *scalar field* associates a scalar (a real or complex number) to each point in space. Let $f(P)$ be a function defined on the points P in some region of three-dimensional space. For example f might be the temperature at each point in a room. Then f is called a *scalar* field. In terms of the usual Cartesian coordinates a general point P is labeled by (x, y, z) and the scalar field f is the function $f(x, y, z)$. So f takes as input a point in three-dimensional space, or equivalently three numbers for the coordinate values, and returns as output a single number.

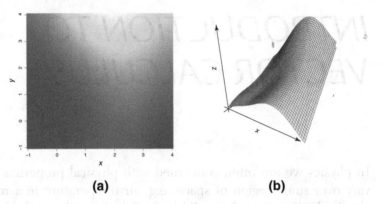

(a)   **(b)**

FIGURE 13.1: Two different ways to visualize a scalar field $f(x, y)$. (a) A heat-map or intensity image, where the shade represents the value of f. (b) A surface plot where $z = f(x, y)$ is the height of the surface.

You may wonder why, if f expressed in this way is just a familiar function, we go to the trouble of calling it a scalar field. The reason is to emphasize that a scalar field takes on values at the points in a region and these values do not depend on the coordinates used to label the points. For example, the temperature obviously depends on where you measure, but not on what system of coordinates you use to specify the point where you make the measurement. See Figure 13.1 for a visualization of a scalar field. Other examples of scalar fields are the air pressure in a room and the gravitational potential around a planet.

A *vector field* associates a vector to each point in space. Let $\mathbf{V}(P)$ be a vector defined on the points P in some region of space (see Figure 13.2); i.e. $\mathbf{V}$ associates a vector with each point. Then $\mathbf{V}$ is called a vector field. For example $\mathbf{V}$ might be the velocity at each point of a fluid. In terms of the usual Cartesian coordinates (x, y, z) the vector field has components $(f(x, y, z), g(x, y, z), h(x, y, z))$ for some functions f, g and h (each of which is a function of position). Instead of (f, g, h) we usually write

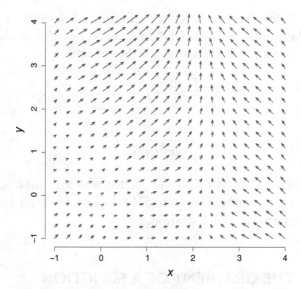

FIGURE 13.2: A vector field $\mathbf{V}(x, y)$: a vector is defined at each point (x, y). Here, vectors are drawn as arrows at grid points.

$$\mathbf{V} = \mathbf{V}(x, y, z) = (V_x(x, y, z), V_y(x, y, z), V_Z(x, y, z)), \quad (13.1)$$

the subscripts indicating that the components are along the x-, y- and z-axes (not to be confused with shorthand notation for the partial derivatives of a scalar $V(x, y, z)$. If there is any danger of confusion, the derivatives would be denoted by V_x etc. with the additional comma.)

A vector field is simply a vector-valued function, i.e. an assignment of a vector (three real numbers) to each point of a region of space. Examples of vector fields are the electric field in space and the velocity of fluid in a pipe.

Examples:

- $\mathbf{A} = (x, y, z)$ is a vector field with $A_x = x$, $A_y = y$, $A_z = z$ at the point (x, y, z).
- $\mathbf{B} = (x^3, x^2 y, y^2(z + 1))$ is a vector field.
- $\mathbf{V} = (0, 0, 2)$ is a vector field with $\mathbf{V}$ equal to this constant value throughout space.

A large part of this chapter is devoted to discussing differential operations on scalar fields and vector fields–in other words, different kinds of fields that can be constructed out of derivatives of other fields. Many of these involve the *del* operator or *nabla* symbol which is defined (in Cartesian coordinates) by

$$\nabla \equiv \mathbf{i}\frac{\partial}{\partial x} + \mathbf{j}\frac{\partial}{\partial y} + \mathbf{k}\frac{\partial}{\partial z}$$

$$\equiv \left(\frac{\partial}{\partial x}, \frac{\partial}{\partial y}, \frac{\partial}{\partial z}\right). \tag{13.2}$$

This is not a function; it is an operator, which means it is "waiting for a function" on which to operate. In the next few sections, we will see various ways in which it works.

13.2. THE GRADIENT OF A FUNCTION

Let $f(x, y, z)$ be a scalar field. We now define the gradient of f, which is a new vector field, by

$$\nabla f = \left(\frac{\partial f}{\partial x}, \frac{\partial f}{\partial y}, \frac{\partial f}{\partial z}\right), \tag{13.3}$$

or equivalently

$$\nabla f = \mathbf{i}\frac{\partial f}{\partial x} + \mathbf{j}\frac{\partial f}{\partial y} + \mathbf{k}\frac{\partial f}{\partial z}. \tag{13.4}$$

We take a scalar field f and make a vector field ∇f out of the partial derivatives of f. To emphasize that ∇f is a vector ∇ is often printed in bold or written as $\underline{\nabla}$. We read this as "grad f," and sometimes it is also written as grad f. To get equation (13.3) we inserted the scalar function f into the "slots" in the del operator, equation (13.2).

Example 13.1 Let $f(x, y, z) = xz$; find ∇f and its value at the point P with coordinates $(3, 2, 17)$.

According to the definition we have to work out the three partial derivatives. Recall that z is a constant for differentiation with respect to x etc.

Thus:

$$\frac{\partial f}{\partial x} = z, \quad \frac{\partial f}{\partial y} = 0, \quad \frac{\partial f}{\partial z} = x, \qquad (13.5)$$

and so we have

$$\nabla f = (z, 0, x). \qquad (13.6)$$

Substituting the values of x, y and z at $P = (3, 2, 17)$ gives

$$\nabla f = (17, 0, 3). \qquad (13.7)$$

Example 13.2 Let $f(x, y, z) = (x^2 + y^2 + z^2)^{1/2}$; find ∇f.

Once more, according to the definition, we have to work out the three partial derivatives of f. Using the chain rule (i.e "the function of a function" rule):

$$\frac{\partial f}{\partial x} = \frac{1}{2} \frac{1}{\left(x^2 + y^2 + z^2\right)^{1/2}} \times 2x = \frac{x}{\left(x^2 + y^2 + z^2\right)^{1/2}}. \qquad (13.8)$$

Note that there is no need to differentiate to get f_y and f_z in this problem. Since f is symmetrical in x, y and z, f_y is obtained from f_x by interchanging x and y in the numerator. (And similarly for f_z.) Hence

$$\frac{\partial f}{\partial y} = \frac{y}{\left(x^2 + y^2 + z^2\right)^{1/2}}, \quad \text{and} \quad \frac{\partial f}{\partial z} = \frac{z}{\left(x^2 + y^2 + z^2\right)^{1/2}}. \qquad (13.9)$$

Now, ∇f is a vector with three components:

$$\nabla f = \left(\frac{x}{\left(x^2 + y^2 + z^2\right)^{1/2}}, \frac{y}{\left(x^2 + y^2 + z^2\right)^{1/2}}, \frac{z}{\left(x^2 + y^2 + z^2\right)^{1/2}}\right). \qquad (13.10)$$

We can extract a common factor from each component:

$$\nabla f = (x^2 + y^2 + z^2)^{-1/2}(x, y, z). \qquad (13.11)$$

Exercise 13.1 Find the gradients of the following functions:

(i) $f(x, y, z) = (x^2 + y^2 + z^2)^{-1/2}$,

(ii) $f(x, y, z) = \exp(lx + my + nz)$ where l, m, n are constants.

13.3. MAXIMUM RATE OF CHANGE

The gradient of a scalar field, ∇f, has a simple geometrical inter-pretation. Its magnitude $|\nabla f|$ gives the maximum rate of change of the function f and the vector ∇f points in the direction of this maxi-mum rate of change.

Consider two neighboring points in three-dimensional space, with position vectors $\mathbf{r} = (x, y, z)$ and $\mathbf{r} + \delta\mathbf{r} = (x + \delta x, y + \delta y, z + \delta z)$, and let $f(x, y, z)$ be a function of position (a scalar field). We are going to calculate the change in the value of f as we pass from $\mathbf{r}$ to $\mathbf{r} + \delta\mathbf{r}$:

$$\delta f = f(\mathbf{r} + \delta\mathbf{r}) - f(\mathbf{r}). \tag{13.12}$$

To do this we add up the change in f as a result of first increasing x (to $x + \delta x$), then increasing y, and finally increasing z. (See Figure 13.3 and compare the chain rule in Section 10.15.) The increase in f is

$$\delta f = f(x + \delta x, y + \delta y, z + \delta z) - f(x, y, z)$$
$$\approx \frac{\partial f}{\partial x}\delta x + \frac{\partial f}{\partial y}\delta y + \frac{\partial f}{\partial z}\delta z. \tag{13.13}$$

The approximation in the last line is sometimes known as the *incremental approximation*. This consists of the first order terms of the Taylor series for a function of three variables (x, y, z). (See Section 10.11.)

Let $\delta\mathbf{r} = \mathbf{u}\delta s$ where $\mathbf{u}$ is a unit vector and $\delta s = |\delta\mathbf{r}|$. We can write equation (13.13), using the scalar product of vectors, as

$$\delta f = \left(\frac{\partial f}{\partial x}, \frac{\partial f}{\partial y}, \frac{\partial f}{\partial z}\right) \cdot (\delta x, \delta y, \delta z) = \nabla f \cdot \delta\mathbf{r} = (\nabla f \cdot \mathbf{u})\delta s.$$

$$\tag{13.14}$$

But the change in *f* is

$$\delta f = (\text{rate of change of } f \text{ in direction } \mathbf{u}) \times$$
$$(\text{distance between the two points}). \qquad (13.15)$$

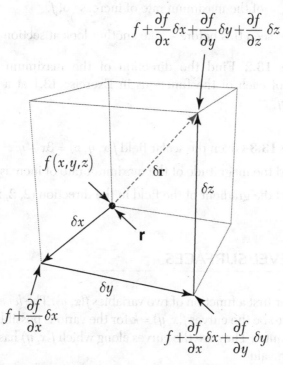

$$f + \frac{\partial f}{\partial x}\delta x + \frac{\partial f}{\partial y}\delta y + \frac{\partial f}{\partial z}\delta z$$

$f(x,y,z)$

$\delta \mathbf{r}$

δz

δx

$\mathbf{r}$

δy

$f + \dfrac{\partial f}{\partial x}\delta x$

$f + \dfrac{\partial f}{\partial x}\delta x + \dfrac{\partial f}{\partial y}\delta y$

FIGURE 13.3: Changes in *f*(x, y, z). We start at point **r** and end up at point **r** + δ**r** by moving first along the x-axis, then the y-axis, and then the z-axis.

Comparing equations (13.14) and (13.15), and noting that δs is the distance between the two points, we see that

$$(\text{rate of change of } f \text{ in direction } \mathbf{u}) = (\nabla f) \cdot \mathbf{u}. \qquad (13.16)$$

If *a* is the angle between the vectors ∇f and **u**, then equation (13.16) becomes (by the definition of the scalar product)

$$(\text{rate of change of } f \text{ in direction } \mathbf{u}) = |\nabla f| \cos(a), \qquad (13.17)$$

remembering that $|\mathbf{u}| = 1$ (as it is a unit vector). Since $0 \le |\cos(a)| \le 1$, from equation (13.17) we have

$$\text{(maximum rate of change of} f) = |\nabla f|, \qquad (13.18)$$

and this will occur when $a = 0$, i.e. when ∇f and $\mathbf{u}$ are parallel. Hence

$$\nabla f \text{ gives the magnitude and direction}$$
$$\text{of the maximum rate of increase of } f. \qquad (13.19)$$

You have seen this before – take another look at section 10.3.

Exercise 13.2 Find the direction of the maximum rate of increase of each of the functions in Exercise 13.1 at a general point (x, y, z).

Exercise 13.3 Given the scalar field $f(x, y, z) = 3x + y^2z^2$

(i) Find the magnitude of the maximum rate of increase of f,

(ii) Find the gradient of the field in the direction $(2, 2, 2)$.

13.4. LEVEL SURFACES

Consider first a function of two variables $f(x, y)$. The *level curves* are defined to be the curves $f(x, y) = k$ for the various possible values of the constant k. They are the curves along which $f(x, y)$ has a level, i.e. constant, value.

This idea is familiar from meteorology. If f = pressure, level curves are isobars; or if f = temperature, level curves are isotherms. And in geography, if f = height of land, level curves are contour lines.

The surface $f(x, y, z) = k$, with k a constant, is said to be a *level surface* of the function $f(x, y, z)$. A set of level surfaces is generated by taking a range of values for k. The notion of level surface captured by this definition is not the same as the non-technical use of "level" to mean a "flat" surface – it is the function that is "flat," not the surface.

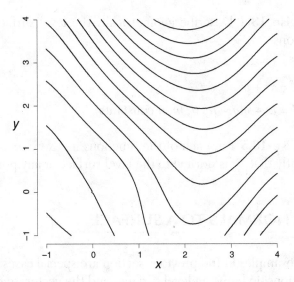

FIGURE 13.4: Level curves of of the function from Figure 13.1.

Example 13.3 Find level surfaces of the function $f(x, y, z)$ $= z$ and show that ∇f is normal (i.e. perpendicular) to these surfaces at any point.

The level surfaces are the planes obtained as the constant runs through a range of possible values. On a level surface, f = constant. As $f(x, y, z) = z$, the level surfaces z = constant are planes parallel to the (x, y) plane. We now find ∇f. Remember that

$$\frac{\partial z}{\partial x} \equiv \frac{\partial z}{\partial x}\bigg|_{y,z=\text{const}} = 0. \qquad (13.20)$$

Here we have that $f(x, y, z) = z$, and so

$$\nabla f = (f_x, f_y, f_z) = (0, 0, 1). \qquad (13.21)$$

The vector $(0, 0, 1)$ points along the z-axis, and hence is normal to the (x, y) plane.

Exercise 13.4 Describe geometrically the level surfaces of the functions

  (i) $f = (x^2 + y^2 + z^2)^{1/2}$,

  (ii) $f = x^2 + y^2$,

  (iii) $f = lx + my + nz$ (l, m, n constants).

Exercise 13.5 For each of the functions in Exercise 13.4, show explicitly that ∇f is normal to the level surface at any point.

13.5. NORMALS TO A SURFACE

The examples in the previous section are special cases of a general relationship between level surfaces and the vector gradient:

∇f is always in a direction normal to the surface f = constant.

$$(13.22)$$

To see this, consider a scalar function f and a level surface $f(x, y, z) = k$. Now let $\mathbf{r}$ and $\mathbf{r}+\delta\mathbf{r}$ be the position vectors of neighboring points in the level surface. As they are both in the level surface, $f(\mathbf{r}) = k$ and $f(\mathbf{r} + \delta\mathbf{r}) = k$, the difference in f at the two points is zero, i.e. $\delta f = 0$. Hence the directional derivative of f along the vector joining the points is $\delta f = \nabla f \cdot \delta\mathbf{r} = 0$. This means ∇f is normal (i.e. perpendicular) to $\delta\mathbf{r}$, hence is in the direction of the normal to the surface.

We have therefore shown that a unit normal to the surface $f(x, y, z) = k$ is

$$\mathbf{n} = \frac{\nabla f}{|\nabla f|}.$$

$$(13.23)$$

Note that $-\frac{\nabla f}{|\nabla f|}$ is also a unit normal to the surface (but it points in the opposite direction).

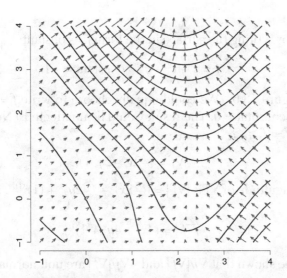

FIGURE 13.5: Level curves and gradient vectors of of the function from Figure 13.1. Notice how the gradient vectors are perpendicular to the level curves (contours). Also notice how the gradients are larger (longer arrows) when the contours are closer together.

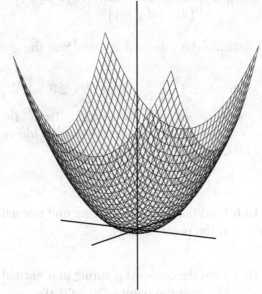

FIGURE 13.6: Surface defined by $z = x^2 + y^2 - 1$; see example 13.4.

Example 13.4 Find the outward pointing unit normal to the surface $z = x^2 + y^2 - 1$ at the point $(1, 1, 1)$.

Begin by writing the surface in the form f = constant:

$$f = z - x^2 - y^2 = \text{constant} = -1. \qquad (13.24)$$

See Figure 13.6. (We could equally choose to write $f = z - x^2 - y^2 + 1 = \text{constant} = 0$. The result would be unaffected.) Next, we find $\nabla f = (f_x, f_y, f_z)$:

$$\nabla f = (-2x, -2y, 1). \qquad (13.25)$$

The magnitude of ∇f is $|\nabla f| = [(-2x)^2 + (-2y)^2 + (1)^2]^{1/2}$, i.e.

$$|\nabla f| = (4x^2 + 4y^2 + 1)^{1/2}. \qquad (13.26)$$

We have shown that $\nabla f/|\nabla f|$ and $-\nabla f/|\nabla f|$ are unit normals, and so

$$\mathbf{n} = \pm \left(\frac{1}{\left(4x^2 + 4y^2 + 1\right)^{1/2}} \right)(-2x, -2y, 1). \qquad (13.27)$$

Next, we substitute the values of x, y and z at the given point $(1, 1, 1)$:

$$\mathbf{n} = (-2/3, -2/3, 1/3) \text{ or } (2/3, 2/3, -1/3). \qquad (13.28)$$

Since the surface expands outwards from the z-axis above $z = -1$, the outward normal points "down," so is $(2/3, 2/3, -1/3)$.

Exercise 13.6 Find the outward pointing unit normal to the surface $x^2 + y^2 = 8$ at the point $(2, 2, 1)$.

Exercise 13.7 Find the outward pointing unit normal to the surface $f(x, y, z) = x^2 - y^2$ at the point $(-20, -20, 0)$.

Exercise 13.8 Show that the unit normal to the surface $f = z - g(x, y) = 0$ is

$$\pm \left(\frac{\partial g}{\partial x}, \frac{\partial g}{\partial y}, -1 \right) \bigg/ \left(1 + \left(\frac{\partial g}{\partial x} \right)^2 + \left(\frac{\partial g}{\partial y} \right)^2 \right)^{1/2}. \tag{13.29}$$

13.6. THE DIVERGENCE OF A VECTOR FIELD

The *divergence* is another vector operator. The divergence operates on a vector field and produces a scalar field. (Compare *grad* which operates on a scalar field and produces a vector field.) The divergence of $\mathbf{A}(x, y, z)$, which is read and sometimes also written as "div A," is

$$\nabla \cdot \mathbf{A} = \frac{\partial A_x}{\partial x} + \frac{\partial A_y}{\partial y} + \frac{\partial A_z}{\partial z}, \tag{13.30}$$

where A_x, A_y and A_z are the components of $\mathbf{A}$, which are functions of position (x, y, z).

Look at this expression carefully. The x component of $\mathbf{A}$, A_x, is differentiated with respect to x, the y component with respect to y etc. and the results added together to give one function (i.e. a scalar).

There is a reason why we write the divergence using the nabla and dot product symbols. Let's write ∇ and $\mathbf{A}$ as vectors and multiply as if taking a dot product.

$$\nabla \cdot \mathbf{A} = \left(\mathbf{i} \frac{\partial}{\partial x} + \mathbf{j} \frac{\partial}{\partial y} + \mathbf{k} \frac{\partial}{\partial z} \right) \cdot \left(\mathbf{i} A_x + \mathbf{j} A_y + \mathbf{k} A_z \right)$$

$$= \frac{\partial A_x}{\partial x} + \frac{\partial A_y}{\partial y} + \frac{\partial A_z}{\partial z}. \tag{13.31}$$

We have used the results for the dot products of the basis vectors: $\mathbf{i} \cdot \mathbf{i} = 1$, $\mathbf{i} \cdot \mathbf{j} = 0$, and so on.

Example 13.5 Let $\mathbf{r} = (x, y, z)$. Find $\nabla \cdot \mathbf{r}$.

First, identify the three components of the vector field,

$$r_x = x, \; r_y = y, \; r_z = z. \tag{13.32}$$

Now work out the required partial derivatives:

$$\frac{\partial r_x}{\partial x} = 1, \quad \frac{\partial r_y}{\partial y} = 1, \quad \frac{\partial r_z}{\partial z} = 1. \tag{13.33}$$

Finally, substitute in the definition of $\nabla \cdot \mathbf{A}$:

$$\nabla \cdot \mathbf{r} = 1 + 1 + 1 = 3. \tag{13.34}$$

The answer is always a scalar function, in this case a constant. The vector field $\mathbf{r} = (x, y, z)$ is simply the radius vector at each point (Figure 13.7).

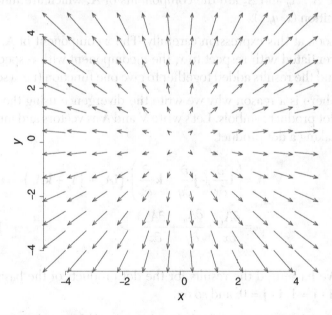

FIGURE 13.7: The vector field $\mathbf{r} = (x, y, z)$ shown in the plane $z = 0$. See Example 13.5.

Example 13.6 Let $\mathbf{A} = (x^3, x^2y, y^2(z + 1))$. Find $\nabla \cdot \mathbf{A}$ at the point $(1, 0, 0)$.

First we identify the three components of the vector field

$$A_x = x^3, A_y = x^2y, A_z = y^2(z + 1). \qquad (13.35)$$

Next, work out the required partial derivatives

$$\frac{\partial A_x}{\partial x} = 3x^2, \quad \frac{\partial A_y}{\partial y} = x^2, \quad \frac{\partial A_z}{\partial z} = y^2. \qquad (13.36)$$

Now we can determine the divergence of $\mathbf{A}$:

$$\nabla \cdot \mathbf{A} = 3x^2 + x^2 + y^2 = 4x^2 + y^2 \qquad (13.37)$$

and the answer is a single function. We now substitute for (x, y, z) at the given point; i.e. we let $(x, y, z) = (1, 0, 0)$ in $\nabla \cdot \mathbf{A}$ and so

$$\nabla \cdot \mathbf{A} = 4 \times 1^2 + 0^2 = 4. \qquad (13.38)$$

Exercise 13.9 Find the divergence of the following vector fields:

(i) $\mathbf{A} = (yz, xz, xy)$,

(ii) $\mathbf{B} = (x^2 + y^2 + z^2)^{n/2}(x, y, z)$ (i.e. $\mathbf{B} = r^n\mathbf{r}$),

(iii) $\mathbf{F} = \nabla \phi$ for some function $\phi(x, y, z)$.

We can think of divergence as representing the "spreading out" of a vector field. Negative divergence at a point means the vector field converges at that point. If $\mathbf{V}(x, y, z)$ is a vector field representing the velocity of a fluid, then its divergence at any point $\nabla \cdot \mathbf{V}$ is the outward flow of fluid (per unit volume) at that point (evaluated over an infinitesimal volume around the point). Negative divergence means there is a net inflow (convergence at the point), and zero divergence means the inflow and outflow balance.

13.7. CURL OF A VECTOR FIELD

Another vector operator is *curl*. The curl of a vector field is another vector field. The curl of a vector field **A**, which is read and sometimes written "curl **A**," is

$$\nabla \times \mathbf{A} = \left(\frac{\partial A_z}{\partial y} - \frac{\partial A_y}{\partial z}, \ \frac{\partial A_x}{\partial z} - \frac{\partial A_z}{\partial x}, \ \frac{\partial A_y}{\partial x} - \frac{\partial A_x}{\partial y} \right) \quad (13.39)$$

or equivalently

$$\nabla \times \mathbf{A} = \mathbf{i} \left(\frac{\partial A_z}{\partial y} - \frac{\partial A_y}{\partial z} \right) + \mathbf{j} \left(\frac{\partial A_x}{\partial z} - \frac{\partial A_z}{\partial x} \right) + \mathbf{k} \left(\frac{\partial A_y}{\partial x} - \frac{\partial A_x}{\partial y} \right).$$
$$(13.40)$$

Look at the definition carefully. The x component of the curl of **A** involves derivatives of the z and y components of **A** with respect to y and z, respectively, and so on cyclically. The result is a vector field (three components, each in general a function of x, y and z).

There is a reason we write curl using the nabla and vector (cross) product symbols. Let's write ∇ and **A** as vectors and work out the components of their cross product

$$\nabla \times \mathbf{A} = \left(\mathbf{i} \frac{\partial}{\partial x} + \mathbf{j} \frac{\partial}{\partial y} + \mathbf{k} \frac{\partial}{\partial z} \right) \times \left(\mathbf{i} A_x + \mathbf{j} A_y + \mathbf{k} A_z \right)$$

$$= \mathbf{i} \left(\frac{\partial A_z}{\partial y} - \frac{\partial A_y}{\partial z} \right) + \mathbf{j} \left(\frac{\partial A_x}{\partial z} - \frac{\partial A_z}{\partial x} \right) + \mathbf{k} \left(\frac{\partial A_y}{\partial x} - \frac{\partial A_x}{\partial y} \right).$$
$$(13.41)$$

using the usual rules for vector (cross) products (see Section 4.4). This is the same as equation (13.40).

Example 13.7 Let $\mathbf{A} = (3xz, 2yz, -z^2)$; find $\nabla \times \mathbf{A}$.

First, identify the components of $\mathbf{A}$:

$$A_x = 3xz,\ A_y = 2yz,\ A_z = -z^2. \qquad (13.42)$$

We now work out each of the components of $\nabla \times \mathbf{A}$. For the x component:

$$\left(\nabla \times \mathbf{A}\right)_x = \frac{\partial A_z}{\partial y} - \frac{\partial A_y}{\partial z} = 0 - 2y = -2y; \qquad (13.43)$$

while for the y component:

$$\left(\nabla \times \mathbf{A}\right)_y = \frac{\partial A_x}{\partial z} - \frac{\partial A_z}{\partial x} = 3x - 0 = 3x; \qquad (13.44)$$

and finally for the z component:

$$\left(\nabla \times \mathbf{A}\right)_z = \frac{\partial A_y}{\partial x} - \frac{\partial A_x}{\partial y} = 0 - 0 = 0. \qquad (13.45)$$

Thus we have that $\nabla \times \mathbf{A} = (-2y, 3x, 0)$. It is a vector field so has three components each of which is a function of x, y and z.

Exercise 13.10 Find the curls of the following vector fields. In each case begin by identifying explicitly the components of the vector for which the curl is required.

(i) $(1 + x^2, xy, y(z + 1))$,

(ii) $(\mathbf{a} \cdot \mathbf{r})\mathbf{a}$ where $\mathbf{r} = (x, y, z)$ and $\mathbf{a} = (a_x, a_y, a_z)$ is a constant vector,

(iii) $\omega \times \mathbf{r}$ where $\mathbf{r} = (x, y, z)$ and $\omega = (w_x, w_y, w_z)$ is a constant vector.

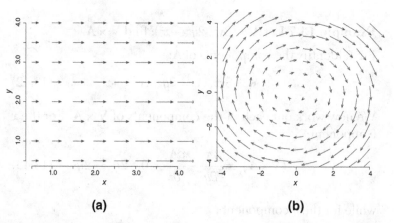

FIGURE 13.8: Two examples of vector fields plotted with the z-axis suppressed. (a) The vectors always lie along the x-axis with y-and z-components zero. The curl is therefore zero everywhere. If we placed a small paddle wheel in the field it would not be made to rotate. (b) The vectors change direction as we move around the space; if we placed a paddle wheel anywhere in the field it would begin to rotate clockwise. For example, at the point $(2, 0)$ there is a stronger "current" on the right than on the left side. This means there is a non-zero curl; at any point the curl vector here points into the page in agreement with the "right hand screw rule" (see last paragraph in Section 13.7).

Example 13.8 Let $\mathbf{A} = (3xz, 2yz, -z^2)$ as in example 13.7. Verify that $\nabla \cdot \nabla \times \mathbf{A}) = 0$.

We worked out $\nabla \times \mathbf{A}$ in example 13.7 so we just quote it here. We will call the resulting vector field $\mathbf{B}$:

$$\mathbf{B} = \nabla \times \mathbf{A} = (-2y, 3x, 0). \qquad (13.46)$$

As $\mathbf{B}$ is a vector field we can take its divergence,

$$\nabla \cdot (\nabla \times \mathbf{A}) = \nabla \cdot \mathbf{B} = \frac{\partial B_x}{\partial x} + \frac{\partial B_y}{\partial y} + \frac{\partial B_z}{\partial z} = 0 + 0 + 0 = 0. \qquad (13.47)$$

Exercise 13.11 Show that $\nabla \cdot (\nabla \times \mathbf{A}) = 0$ for any vector field $\mathbf{A}$.

In fact, although we shall not show it here, the converse also holds: if **B** is a vector field such that $\nabla \cdot \mathbf{B} = 0$, then it is always possible to find a vector field **A** such that $\mathbf{B} = \nabla \times \mathbf{A}$. This is an important result for theoretical physics.

❚ Exercise 13.12 Show that $\nabla \times \nabla \phi = \mathbf{0}$ for any scalar field ϕ.

A vector field with zero curl is said to be *irrotational* or *curl free*. The result of Exercise 13.12 shows that any vector field that is the gradient of a scalar field is irrotational. In fact, although we shall not show it here, the converse of this also holds: if **E** is a vector field such that $\nabla \times \mathbf{E} = 0$, then it is always possible to find a scalar field ϕ such that $\mathbf{E} = \nabla \phi$. This is another important result for theoretical physics.

We will return to the geometrical interpretation of curl a little later. But for now we can think of the curl of a vector field at a point as representing the amount of "circulation" of the field about that point. If $\mathbf{V}(x, y, z)$ is a vector field representing the velocity of a fluid, then its curl $\nabla \times \mathbf{V}$ at any point is a measure of the angular velocity of the fluid about that point. Imagine holding a tiny paddle wheel at a point in the fluid (Figure 13.8). If the curl is zero, the paddle wheel would not rotate at that point. If the paddle wheel rotates, the curl there is non-zero. The direction of the curl follows the "right hand screw rule": if you curl the fingers of your right hand in the direction of the circulation, your thumb will point in the direction of positive curl.

13.8. FLUX OF A VECTOR FIELD THROUGH A SURFACE

In this and the next section, we will introduce two important quantities associated with vector fields: *flux* and *circulation*. Here we will try to emphasize what these mean, and also show how they can be calculated for simple plane geometries.

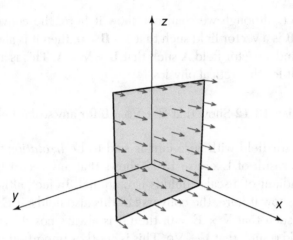

FIGURE 13.9: The flux of a field through a plane surface.

Flux[1] is the amount of "something" passing through a surface. If the vector field is the velocity of a fluid, then flux is the flow rate through the surface. We can also consider heat flux (the rate of heat energy transfer through a surface), and electric flux, which we can think of as the number of electric field lines passing through a surface (although notice that nothing is actually flowing in this case).

The flux depends on the vector field and the size, shape and location of the surface. If the vector field is stronger (faster moving fluid, stronger electric field etc.) then the flux will be higher; if the surface is made larger or oriented better to "catch" more of the field, then the flux will be higher. Imagine catching fish with a fishing net: a larger area fishing net will catch more fish, and moving the fishing net to a region with a higher density of fish will catch more fish.

First, a simple example. Let $\mathbf{B} = (b(x), 0, 0)$ be a vector field in the x-direction, with $b(x)$ depending only on x. Let S be a rectangular surface of area A at $x = x_0$, aligned normal to the positive x-axis, as in Figure 13.9. The flux of the vector field $\mathbf{B}$ through the surface S is

$$\text{(flux of } \mathbf{B} \text{ through } X) = \text{Area} \times \text{field strength} = Ab(x_0).$$
$$(13.48)$$

---

[1]The word *flux* is from the Latin *fluere* meaning "to flow" and was introduced into calculus by Newton's use of the word *fluxion* to mean derivative.

More generally, if

$$\mathbf{B}(x, y, z) = (B_x, B_y, B_z) \qquad (13.49)$$

where each of B_x, B_y and B_z depends on position, then the flux of **B** through X is defined as an integral over the surface X,

$$\left(\text{flux of } \mathbf{B} \text{ through } X\right) = \iint_X B_x\left(x_0, y, z\right) dy\, dz. \qquad (13.50)$$

The flux of a vector field through a surface is *defined* as the integral of the normal component of the field over the surface. Later we shall find an alternative way to *calculate* the flux through a closed surface.

Exercise 13.13 Write down the integral analogous to equation (13.50) defining the flux of the vector field $\mathbf{B}(x, y, z)$ through the rectangle Y normal to the y-axis, cutting the y-axis at $y = y_0$.

Note that to calculate the flux through a surface we need to be given the direction of the positive normal to the surface. In the examples below, we take this along the positive coordinate axes.

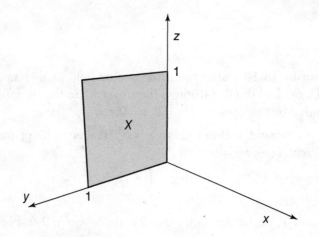

FIGURE 13.10: The surface X.

Example 13.9 Calculate the flux of the vector field $\mathbf{B} = ((1 + x)^{-1}, (y + x), 0)$ through the surface X, in the (y, z) plane bounded by the lines $y = 0$, $y = 1$, $z = 0$, $z = 1$.

When tackling problems like this, it is often helpful to draw the region of integration. This is done in Figure 13.10. We start from the definition. The component of $\mathbf{B}$ normal to the surface X is the component B_x. The element of area in X is $dydz$, so

$$\text{flux} = \iint_X B_x \, dydz. \tag{13.51}$$

For a general value of z in X, y runs between 0 and 1. Within X, z runs between 0 and 1. So these are the limits of integration:

$$\text{flux} = \int_0^1 \int_0^1 B_x \, dydz. \tag{13.52}$$

Next substitute for B_x. On the surface X we have $x = 0$, so $\mathbf{B} = (1, y, 0)$ on X. Thus $B_x = 1$. This leaves a simple double integral:

$$\text{flux} = \int_0^1 \int_0^1 1 \, dydz = \int_0^1 \left[y \right]_0^1 dz = \int_0^1 1 \, dz = \left[z \right]_0^1 = 1. \tag{13.53}$$

The field has 1 unit of flux through X.

Example 13.10 Using the same vector field as in Example 13.9, calculate the flux through the surface Y in the (x, z) plane bounded by the lines $x = 1$, $x = 2$, $z = 0$, $z = 1$.

The component of $\mathbf{B}$ normal to Y is B_y. The element of area in Y is $dxdz$. Thus we have that

$$\text{flux} = \iint_Y B_y \, dxdz. \tag{13.54}$$

To cover Y, x runs between 1 and 2 and z between 0 and 1, so

$$\text{flux} = \int_0^1 \int_1^2 B_y \, dxdz. \tag{13.55}$$

On the surface Y (in the (x, z) plane) $y = 0$, so $\mathbf{B} = (1/(1 + x), x, 0)$ and $B_y = x$. This leaves a standard double integral:

$$\text{flux} = \int_0^1 \int_1^2 x\, dx dz = \int_0^1 \left[\frac{x^2}{2} \right]_1^2 dz = \int_0^1 \frac{1}{2}(4-1) dz$$

$$= \int_0^1 \frac{3}{2} dz = \frac{3}{2}. \qquad (13.56)$$

The field has 3/2 units of flux through Y.

Example 13.11 Calculate the flux of the vector field $\mathbf{B} = (ye^{-x}, 1, 0)$ through the surface X in the (y, z) plane bounded by the lines $y = 0$, $y = 1$, $z = 0$, $z = 1$.

The component of $\mathbf{B}$ normal to the surface X is B_x, and so

$$\text{flux} = \iint_Y B_x \, dy dz. \qquad (13.57)$$

The surface of integration is the square $0 \le y \le 1$, $0 \le z \le 1$ so these are the limits of integration. The surface lies at $x = 0$, in the (y, z) plane, and so $B_x = ye^0 = y$. This gives a standard double integral:

$$\text{flux} = \int_0^1 \int_0^1 y\, dy dz = \int_0^1 \left[\frac{y^2}{2} \right]_0^1 dz = \int_0^1 \frac{1}{2} dz = \frac{1}{2}. \qquad (13.58)$$

Exercise 13.14 Calculate the flux of the vector field $(x^3 + 1, yz, 2)$ through

(i) the surface X_1, in the y-z plane bounded by the lines $y = 0$, $y = 1$, $z = 0$, $z = 1$,

(ii) the surface X_2, in the plane $x = 1$ bounded by the lines $y = 0$, $y = 1$, $z = 0$, $z = 1$.

(iii) Hence obtain the difference in the fluxes through X_2 and X_1.

13.9. CIRCULATION OF A VECTOR FIELD

The circulation of a vector field around a loop (closed curve) is defined as the integral of the tangential component of the vector field along the loop. If the vector field represents the force on a body, then the circulation represents the work done by the force in moving the body around the loop. There are many other physical applications in electro-magnetism and fluid mechanics.

Let's see an example. We have a vector field $\mathbf{V} = (u(y), v(x), 0)$ where, for the moment, u depends on y only and v depends on x only. Let $PQRS$ be a rectangle in the (x, y) plane; sides PS and QR have length $(x_2 - x_1)$ and PQ and SR have length $(y_2 - y_1)$ (Figure 13.11). We are going to define the circulation of $\mathbf{V}$ around the rectangle $PQRS$ in the counterclockwise direction.[2]

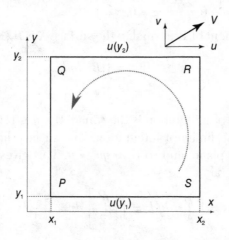

FIGURE 13.11: The rectangle PQRS. Above is shown the vector **V** at a point, and its components in the x-and y-directions.

We define the contribution from the side RQ to the circulation of the vector field $\mathbf{V}$ around the rectangle to be

(length of RQ) × (tangential component of $\mathbf{V}$ along $\overrightarrow{RQ}$)

$$= (x_2 - x_1) \times (-u(y_2)), \qquad (13.59)$$

---

[2]The direction is chosen to follow the right hand screw convention with respect to the (x, y, z) coordinate system.

where the minus sign for u arises because the tangential component of $\mathbf{V}$ is positive in the direction QR, opposite to RQ. Similarly, for the contribution from PS,

(length of PS) × (tangential component of $\mathbf{V}$ along $\overrightarrow{PS}$)
$$= (x_2 - x_1) \times u(y_1).$$

The net contribution of the pair of sides QR and SP is

$$(x_2 - x_1)(u(y_1) - u(y_2)). \tag{13.60}$$

Similarly, from the pair of sides QP and SR we get a net contribution

$$(y_2 - y_1)(-v(x_1) + v(x_2)).$$

The circulation of $\mathbf{V}$ around $PQRS$ (i.e. around the sides in the order RQ, QP, PS, SR) is the sum of these net contributions:

$$\text{circulation} = (x_2 - x_1)(u(y_1) - u(y_2)) + (y_2 - y_1)(v(x_2) - v(x_1))$$
$$= (x_2 - x_1)u(y_1) + (y_2 - y_1)v(x_2) - (x_2 - x_1)u(y_2) - (y_2 - y_1)v(x_1). \tag{13.61}$$

To extend this to more general vector fields with tangential components that vary along an arc we replace the products (length) × (tangential component) by an integral.

For a vector field $\mathbf{V} = (V_x, V_y, V_z)$ we define the contribution to the circulation from RQ to be

$$\int_{x_2}^{x_1} V_x(x, y_2, 0)\,dx = -\int_{x_1}^{x_2} V_x(x, y_2, 0)\,dx \tag{13.62}$$

(replacing Equation 13.59). We have put $z = 0$ in V_x because we have assumed above that $PQRS$ lies in the $z = 0$ plane. The contribution to the circulation from PS is

$$\int_{x_1}^{x_2} V_x(x, y_2, 0)\,dx. \tag{13.63}$$

The net contribution from PQ and RS is

$$\int_{x_1}^{x_2}\left(-V_x\left(x,\,y_2,\,0\right)+V_x\left(x,\,y_1,\,0\right)\right)dx \tag{13.64}$$

(replacing Equation 13.60). The total circulation of **V** around $PQRS$ is therefore

$$\text{circulation} = \int_{x_1}^{x_2}\left(-V_x\left(x,y_2,0\right)+V_x\left(x,y_1,0\right)\right)dx$$
$$+ \int_{y_1}^{y_2}\left(V_y\left(x_2,y,0\right)-V_y\left(x_1,y,0\right)\right)dx. \tag{13.65}$$

Example 13.12 Calculate the circulation of the vector field $\mathbf{V} = (1+x^2,\ xy,\ y(z+1))$ around the unit square in the first quadrant of the $(x,\ y)$ plane having one vertex at the origin (Figure 13.12).

The circuit $PQRS$ is in the $(x,\ y)$ plane, meaning $z = 0$. The simplest way to get the signs right is to let the limits take care of them.

On RQ the tangential component of **V** is $V_x = (1 + x^2)$, and x runs between 1 (lower limit) and 0 (upper limit). So, the contribution to the circulation from RQ is

$$\text{circulation from } RQ = \int_{1}^{0}\left(1+x^2\right)dx = -\int_{0}^{1}\left(1+x^2\right)dx.$$

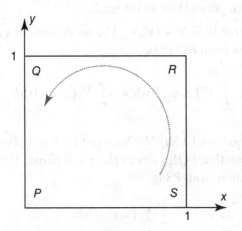

FIGURE 13.12: The square PQRS (see Example 13.12).

On QP the tangential component of $\mathbf{V}$ is $V_y = xy = 0$ since $x = 0$, so

$$\text{circulation from } QP = \int_1^0 0.y\,dy = 0.$$

On PS the tangential component of $\mathbf{V}$ is $V_x = (1 + x^2)$, so

$$\text{circulation from } PS = \int_0^1 \left(1 + x^2\right) dx. \qquad (13.66)$$

On SR the tangential component of $\mathbf{V}$ is $V_y = xy = y$ since $x = 1$ on SR; y runs between 0 and 1 and so

$$\text{circulation from } SR = \int_0^1 y\,dy.$$

The net circulation from $PQRS$ is the sum of these contributions around the loop,

$$\text{circulation } PQRS = -\int_0^1 \left(1 + x^2\right) dx + \int_0^1 y\,dy + \int_0^1 \left(1 + x^2\right) dx.$$

The contributions from QR and SP cancel since V_x is not a function of y, and so

$$\text{circulation } PQRS = \int_0^1 y\,dy = \left[\frac{y^2}{2}\right]_0^1 = \frac{1}{2}.$$

Exercise 13.15 Find the (counterclockwise) circulation of the vector field

$$\mathbf{V}(x, y, z) = (1 + x^2, xy, y(z + 1))$$

around the unit square in the (x, y) plane, centered on the origin with sides parallel to the axes.

13.10. DIVERGENCE THEOREM (GAUSS'S THEOREM)

We have defined two physically important quantities associated with vector fields–the divergence and the flux through a surface. For a special type of surface, namely a *closed surface*, these two are related. The relation between them is called the *divergence theorem,* also known as *Gauss's theorem.* The aim of this section is to derive this theorem for some special cases.

We start by considering a vector field **B** and a closed surface S enclosing a volume V. The divergence theorem says the flux of **B** (outwards) through the closed surface is equal to the integral of the net divergence inside the volume

$$\Delta \text{flux} = \int_V \nabla \cdot \mathbf{B} \, dV. \tag{13.67}$$

Example 13.13 Verify the divergence theorem for a simplified case in which **B** = $(b(x), 0, 0)$ is a vector field in the x-direction, with $b(x)$ depending only on x, and V is a cuboid with faces X_1 and X_2 of areas A perpendicular to **B**.

The normal component of B on X_2 is $B_x = b(x_2)$. The flux of **B** through X_2 is

$$\iint_{X_2} b(x_2) \, dy dz = b(x_2) \iint_{X^2} dy dz = Ab(x_2). \tag{13.68}$$

since B_x is constant on X_2. Similarly the flux of **B** through X_1 into V is (area)×(normal component) since the normal component B_x is constant on X_1, thus

$$\text{flux through } X_1 = Ab(x_1). \tag{13.69}$$

The net flux out of the volume is the sum of the net flux out of each face. As the vector field is along the x-direction, crossing only the surfaces X_1 and X_2, these are the only faces with non-zero flux. Therefore the net flux is the difference of the flux out of X_2 and the flux into X_1, i.e.

$$\Delta\text{flux} = A(b(x_2) - b(x_1)).\qquad(13.70)$$

We want to relate the expression for Δflux to $\nabla\cdot\mathbf{B}$, so we bring in derivatives of $b(x)$:

$$\Delta\text{flux} = A \int_{x_1}^{x_2} \frac{\partial b}{\partial x}\, dx.\qquad(13.71)$$

However, since

$$\Delta\cdot\mathbf{B} = \frac{\partial B_x}{\partial x} + \frac{\partial B_y}{\partial y} + \frac{\partial B_z}{\partial z}.\qquad(13.72)$$

where $B_x = b(x)$, $B_y = 0$ and $B_z = 0$, so

$$\Delta\cdot\mathbf{B} = \frac{\partial b}{\partial x}.\qquad(13.73)$$

We can write

$$A = \int_{z_1}^{z_2}\int_{y_1}^{y_2} dy\, dz.\qquad(13.74)$$

Then

$$\Delta\text{flux} = \left(\int_{z_1}^{z_2}\int_{y_1}^{y_2} dy\, dz\right)\left(\int_{x_1}^{x_2}\nabla\cdot\mathbf{B}\, dx\right),\qquad(13.75)$$

or

$$\Delta\text{flux} = \int_{z_1}^{z_2}\int_{y_1}^{y_2}\int_{x_1}^{x_2}\nabla\cdot\mathbf{B}\, dx\, dy\, dz = \int_V \nabla\cdot\mathbf{B}\, dV.\qquad(13.76)$$

This example shows us the physical meaning of the divergence of a vector field. The divergence of a field gives the rate per unit volume at which flux is being gained or lost from a point. If we think of a vector field as a fluid flowing through space (the arrows representing the velocity at each point) then we gain or lose flux where there are *sources* or *sinks* of the fluid. At a point where $\nabla\cdot\mathbf{B} = 0$ there are no sources or sinks. And if there are no sources or sinks inside a volume, then $\nabla\cdot\mathbf{B} = 0$ everywhere inside, and the divergence theorem tells us that the net flux through the surrounding surface is also zero.

More generally, if $\mathbf{B} = (b(x, y, z), 0, 0)$ the definition of the flux of $\mathbf{B}$ through the rectangle X_2 gives

$$\text{flux of } \mathbf{B} \text{ through the plane surface } X_2 = \int_{z_1}^{z_2} \int_{y_1}^{y_2} b(x_2, y, z)\, dy\, dz$$

(13.77)

and similarly for X_1. Now the difference in flux between X_2 and X_1 is

$$\Delta\text{flux} = \int_{z_1}^{z_2} \int_{y_1}^{y_2} \left[b(x_2, y, z) - b(x_1, y, z) \right] dy\, dz. \qquad (13.78)$$

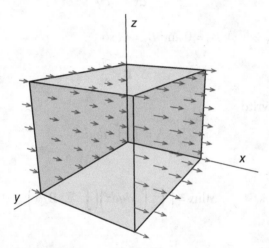

FIGURE 13.13: Flux through a closed surface.

Example 13.14 Verify the divergence theorem for the case in which $\mathbf{B} = (B_x, B_y, B_z)$ is a general vector field (with components in all three directions each of which is a function of x, y and z), and V is a cuboid with faces X_1, X_2, Y_1, Y_2, Z_1, Z_2 perpendicular to the x-, y- and z-axes.

We calculate the flux through each of the six faces of the rectangular parallelepiped (cuboid) in Figure 13.13. For each pair of faces we write down an equation like equation (13.78). So, for faces X_1 and X_2,

$$(\Delta\text{flux})_x = \int_{z_1}^{z_2} \int_{y_1}^{y_2} \left[B_x\left(x_2,y,z\right) - B_x\left(x_1,y,z\right)\right] dy dz$$

$$= \int_{z_1}^{z_2} \int_{y_1}^{y_2} \left(\int_{x_1}^{x_2} \frac{\partial B_x}{\partial x} dx \right) dy dz. \tag{13.79}$$

Similarly

$$(\Delta\text{flux})_y = \int_{z_1}^{z_2} \int_{x_1}^{x_2} \int_{y_1}^{y_2} \frac{\partial B_y}{\partial y} dy dx dz. \tag{13.80}$$

and

$$(\Delta\text{flux})_z = \int_{y_1}^{y_2} \int_{x_1}^{x_2} \int_{z_1}^{z_2} \frac{\partial B_z}{\partial z} dz dx dy. \tag{13.81}$$

The flux out of the volume V is the sum of the net flux through each pair of faces:

$$\left(\text{flux out of } V\right) = (\Delta\text{flux})_x + (\Delta\text{flux})_y + (\Delta\text{flux})_z,$$

$$\tag{13.82}$$

$$= \int_{z_1}^{z_2} \int_{x_1}^{x_2} \int_{y_1}^{y_2} \left(\frac{\partial B_x}{\partial x} + \frac{\partial B_y}{\partial y} + \frac{\partial B_z}{\partial z} \right) dx dy dz$$

$$= \int_V \nabla \cdot \mathbf{B} dV, \tag{13.83}$$

where we change the order of integration in equations (13.80) and (13.81).

We have therefore obtained the divergence theorem for a general vector field, but only for a cuboidal volume. In fact, although we shall not prove it, the theorem holds for general vector fields and general volumes. It is usual to write the flux through a surface S in a notation that indicates its origin as an integral of the vector field over an area.

In this notation the divergence theorem becomes

$$\int_S \mathbf{B} \cdot d\mathbf{S} = \int_V \nabla \cdot \mathbf{B} \, dV, \qquad (18.84)$$

where S is a surface enclosing the volume V.

The divergence theorem therefore tells us that we can calculate the flux of a vector field through a closed surface if we know what it is throughout the whole interior volume. It relates a surface (two-dimensional) integral to a volume (three-dimensional) integral. It says that if we know the flux of a field through a closed surface, which can be obtained from the value of the field only on the surface, we can deduce the net sources and sinks of the field within the enclosed volume.

> **Exercise 13.16** Evaluate the flux of the vector field $\mathbf{F} = (x^3, x^2 y, y^2(z + 1))$ through the closed surface of the unit cube bounded by the six planes $x = 0, x = 1, y = 0, y = 1, z = 0$ and $z = 1$
>
> (i) directly by evaluating the six surface integrals,
>
> (ii) by integrating $(\nabla \cdot \mathbf{F})$ over the volume of the cube. Hence verify the divergence theorem in this case.

> **Exercise 13.17** Use the divergence theorem to calculate the flux of the vector field
>
> $$\mathbf{F}(x, y, z) = (xz, y^3, xy)$$
>
> through the closed surface of the unit cube bounded by the six planes $x = -1, x = 0, y = -1, y = 0, z = -1$ and $z = 0$.

13.11. STOKES'S THEOREM

We have defined two other physically important quantities associated with vector fields the curl of the field and its circulation around a closed loop. These two are also related to each other. The relation between them is called *Stokes's theorem*. The aim of this section is to derive this theorem for some special cases.

Example 13.15 Let S be a closed rectangle in the (x, y) plane with sides parallel to the axes and perimeter C. Show that the circulation of a vector field **V** around C is given by the flux of $\nabla \times \mathbf{V}$ through S.

The circulation around $C \equiv PSRQ$ is found from the sum of the contributions from the four sides,

$$PQRS = \int_R^Q V_x\, dx + \int_Q^P V_y\, dy + \int_P^S V_x\, dx + \int_S^R V_y\, dy$$

$$= \int_{x_1}^{x_2} \left(V_x\left(x, y_2, 0\right) - V_x\left(x, y_1, 0\right)\right) dx$$

$$+ \int_{y_1}^{y_2} \left(V_y\left(x_2, y, 0\right) - V_y\left(x_1, y, 0\right)\right) dy.$$

Now

$$V_x\left(x, y_2, 0\right) - V_x\left(x, y_1, 0\right) = \int_{y_2}^{y_1} \frac{\partial V_x}{\partial y}(x, y, 0)\, dy$$

and

$$V_y\left(x_2, y, 0\right) - V_y\left(x_1, y, 0\right) = \int_{x_1}^{x_2} \frac{\partial V_y}{\partial x}(x, y, 0)\, dx$$

and so

$$\text{circulation} = -\int_{x_1}^{x_2}\int_{y_1}^{y_2} \frac{\partial V_x}{\partial y} dy dx + \int_{y_1}^{y_2}\int_{x_1}^{x_2} \frac{\partial V_y}{\partial x} dx dy$$

$$= \int_{x_1}^{x_2}\int_{y_1}^{y_2} \left(\frac{\partial V_y}{\partial x} - \frac{\partial V_x}{\partial y} \right) dx dy.$$

Making use of the fact that the integrand is the same as the z-component of the curl $\nabla \times \mathbf{V}$ (see Equation 13.39) we have that

$$\text{Circulation of } \mathbf{V} \text{ round } C = \int_S \left(\nabla \times \mathbf{V} \right)_z dA. \quad (13.85)$$

The integration in the final expression is over the area of the plane rectangle S.

The equality (13.85) is known as *Stokes's theorem* (for a plane area). Stokes's theorem enables us to express the circulation of a vector field around a closed curve as an integral of the curl of the vector field over the area enclosed by the curve.

$$\oint_C \mathbf{V} \cdot d\mathbf{l} = \int_S (\nabla \times \mathbf{V}) \cdot d\mathbf{S}. \qquad (13.86)$$

The physical meaning of the curl of a vector field is therefore that it gives the rate per unit area at which the circulation changes (roughly speaking, the "vorticity" of the vector field). We have not here proved the result for a general vector field, only for a plane rectangle. But in fact, the theorem holds for most open surfaces[3] one encounters in physics, even surfaces with holes. It is usual to write the theorem in a way that emphasizes this:

$$\text{circulation} = \int_S (\nabla \times \mathbf{V})_z \, dS. \qquad (13.87)$$

On the right side we have the flux (of $\nabla \times \mathbf{V}$) through a general surface, which involves the integral of the normal component of $\nabla \times \mathbf{V}$ through a general surface S that has unit normal **n**. On the left side the circulation is written as the integral of the tangential component (the scalar product of **V** with a element of length $d\mathbf{l}$ tangent to the curve C) around the closed curve C that is the boundary of S. The circle on the integral symbol is the conventional way of indicating that the curve C must be closed (otherwise the equation is false).

Example 13.16 By using Stokes's theorem, compute the circulation of the vector field $\mathbf{V} = (1 + x^2, xy, y(z + 1))$ around the unit square S in the first quadrant of the (x, y) plane having one vertex at the origin.

By Stokes's theorem we can work out the circulation as an integral of the curl of **A** over S:

---

[3]Provided they are *orientable*–roughly speaking, that they have two sides and so one can define a consistent choice of normal vectors to the surface at every point, unlike a Möbius strip.

$$\text{circulation} = \int_S (\nabla \times \mathbf{V})_z \, dS.$$

We have already worked out $\nabla \times \mathbf{V}$ in Exercise 13.10:

$$\nabla \times \mathbf{V} = (z + 1, 0, y).$$

The integrand is the z–component of $\nabla \times \mathbf{V}$ on the surface S:

$$(\nabla \times \mathbf{V})_z = y.$$

So we have

$$\text{circulation} = \int_0^1 \int_0^1 (\nabla \times V)_z \, dx dy = \int_0^1 \int_0^1 y \, dx dy,$$

and so

$$\text{circulation} = \int_0^1 dx \int_0^1 y \, dy = 1 \times \frac{1}{2} = \frac{1}{2}.$$

This agrees with the direct calculation of the circulation in Exercise 13.12.

Exercise 13.18 Use Stokes's theorem to show that the circulation of the vector field

$$\mathbf{V}(x, y, z) = (1 + x^2, xy, y(z + 1))$$

around the unit square in the (x, y) plane, centered on the origin, is zero.

Exercise 13.19

(i) The steady drift of a straight section of river can be described by the velocity field

$$\mathbf{v} = v\mathbf{i}. \tag{13.88}$$

A lobster pot consisting of a cube of side 1m (designed by a physicist!) is immersed in the water with all its sides parallel or perpendicular to the flow. Write down the flux of $\mathbf{v}$ into each of the sixsides and hence the net flux into the cube. Verify this result using Gauss's divergence theorem.

(ii) The pot is now moved to a section of the river approaching a weir where the water is speeding up with a constant acceleration a, so the velocity field becomes

$$\mathbf{v} = (v_0 + 2ax)^{1/2}\mathbf{i}, \tag{13.89}$$

where v_0 is the velocity at $x = 0$. If the pot is placed in the same orientation as before with its center at $x = 10.5$m, write down the flux into each face and the net flux into the cube. Again verify this result using Gauss's theorem.

(iii) What is the circulation of $\mathbf{v}$ around the top face of the pot in the two locations? Verify these results using Stokes's theorem.

(iv) The pot is placed near an outcrop where the flow is disturbed and the velocity field is given by

$$\mathbf{v} = (v + t_1 y)\mathbf{i} + t_2 y\mathbf{j}, \tag{13.90}$$

where $y = 0$ corresponds to the riverbank. If the pot is placed in the same orientation as before with its center 1.5m away from the riverbank, work out the circulation of $\mathbf{v}$ around the top face. Verify this result using Stokes's theorem.

13.12.   EXTENSIONS: INDEX NOTATION

We can write compact expressions for the vector operators using the index notation introduced in Section 4.15. Thus

$$(\nabla\phi)_i = \frac{\partial\phi}{\partial x_i}$$

gives the components of the gradient of the function $\phi(x_1, x_2, x_3)$. The divergence and curl of a vector field $\mathbf{A}(x_1, x_2, x_3)$ are given by

$$\nabla \cdot \mathbf{A} = \frac{\partial A_i}{\partial x_i}$$

$$(\nabla \times \mathbf{A})_i = \epsilon_{ijk} \frac{\partial A_k}{\partial x_j}$$

using the notation of Section 4.15. Once one gets the hang of this notation the derivation of vector identities becomes a mechanical process.

13.13. EXTENSIONS: OTHER COORDINATES

In Cartesian coordinates, the distances between neighboring points along the coordinate axes are dx, dy and dz along the x-, y- and z-axes, respectively. In other coordinate systems the distances between points are more complicated. For example, in spherical polar coordinates the physical distance on the sphere of radius r between points at the same latitude θ separated by a coordinate increment $d\phi$ is not $d\phi$ (which is not a distance) but $r \sin(\theta)d\phi$. This leads to more complicated expressions for the gradient, divergence and curl of a vector field in non–Cartesian coordinates.

In a general orthogonal coordinate system (ξ_1, ξ_2, ξ_3) we let the distances between neighboring points be $ds_1 = h_1 d\xi_1$, $ds_2 = h_2 d\xi_2$ and $ds_3 = h_3 d\xi_3$. This defines (h_1, h_2, h_3). (So, for example, in spherical polars $h_3 = r \sin(\theta)$.) Then

$$\nabla \psi = \left(\frac{1}{h_1} \frac{\partial \psi}{\partial \xi_1}, \frac{1}{h_2} \frac{\partial \psi}{\partial \xi_2}, \frac{1}{h_3} \frac{\partial \psi}{\partial \xi_3} \right),$$

$$\nabla \cdot \mathbf{A} = \frac{1}{h_1} \frac{\partial}{\partial \xi_1} (h_2 h_3 A_1) + \frac{1}{h_2} \frac{\partial}{\partial \xi_2} (h_1 h_3 A_2) + \frac{1}{h_3} \frac{\partial}{\partial \xi_3} (h_1 h_2 A_3),$$

$$\nabla \times \mathbf{A} = \left(\frac{1}{h_2 h_3} \left(\frac{\partial (h_3 A_3)}{\partial \xi_2} - \frac{\partial (h_2 A_2)}{\partial \xi_3} \right), \frac{1}{h_1 h_3} \left(\frac{\partial (h_1 A_1)}{\partial \xi_3} - \frac{\partial (h_3 A_3)}{\partial \xi_1} \right), \right.$$

$$\left. \frac{1}{h_1 h_2} \left(\frac{\partial (h_2 A_2)}{\partial \xi_1} - \frac{\partial (h_1 A_1)}{\partial \xi_2} \right) \right).$$

In particular, in cylindrical coordinates (ρ, θ, z) we have $(h_1, h_2, h_3) = (1, \rho, 1)$ and in spherical polars (r, θ, ϕ), $(h_1, h_2, h_3) = (1, r, r\sin(\theta))$.

Revision Notes

After completing this chapter you should be able to

- Understand the difference between scalar fields $f(x, y, z)$ and vector fields $\mathbf{A}(x, y, z) = (A_x, A_y, A_z)$

- Calculate the *gradient*

$$\nabla f = (\partial f/\partial x, \partial f/\partial y, \partial f/\partial z)$$

  of a scalar field $f(x, y, z)$ and use it to find the rate of change, $\mathbf{u} \cdot \nabla f$, of a field f in a given direction $\mathbf{u}$, and the unit normal, $\mathbf{n} = \pm \nabla f / |\nabla f|$, to the surface $f = $ constant

- Calculate the *divergence* of $\mathbf{A}$, a scalar, given by

$$\nabla \cdot \mathbf{A} = \frac{\partial A_x}{\partial x} + \frac{\partial A_y}{\partial y} + \frac{\partial A_z}{\partial z}$$

- Calculate the *curl* of $\mathbf{A}$, a vector field,

$$\nabla \times \mathbf{A} \left(\frac{\partial A_z}{\partial z} - \frac{\partial A_y}{\partial y}, \frac{\partial A_x}{\partial x} - \frac{\partial A_z}{\partial z}, \frac{\partial A_y}{\partial y} - \frac{\partial A_x}{\partial x} \right)$$

- Calculate, for a closed rectangular box S, the *flux* of $\mathbf{A}$ out of S

$$\int A \cdot d\mathbf{S} = \text{sum of fluxes of } \mathbf{A} \text{ out of each face}$$

- State the *divergence theorem* (Gauss's theorem):

$$\int_S \mathbf{A} \cdot d\mathbf{S} = \int_V \nabla \cdot \mathbf{A}\, dV$$

where V is the volume interior to S, and explain the meaning of the terms with reference to simple examples

- Compute, for a closed rectangular curve C, the circulation of **A** around C:

$$\int_C \mathbf{A} \cdot d\mathbf{l} = \text{sum of integrals along each edge of } C.$$

- State *Stokes's theorem*:

$$\int_C \mathbf{A} \cdot d\mathbf{l} = \int_S (\nabla \times \mathbf{A}) \cdot d\mathbf{S}$$

= flux of $\nabla \times \mathbf{A}$ through interior of the closed curve C

and explain the meaning of the terms with reference to simple examples.

13.14. EXERCISES

1. Find the gradient of the function $f(x, y, z) = x^2 + 2xyz + z^2$ in the direction parallel to the vector $(1, -1, 2)$ at the point $(0, 2, 3)$.

2. Find expressions for the divergence and the curl of the vector function

$$\mathbf{A}(x, y, x) = ze^{xy}\mathbf{i} + xe^{xy}\mathbf{j} + ye^{xy}\mathbf{k}.$$

3. Calculate the flux of the vector field $\mathbf{E} = (y^3, x^2y, y^2z)$ through a rectangular surface with corners at $(0, 0, 0)$, $(0, 0, -2)$, $(0, 1, 0)$ and $(0, 1, -2)$.

4. Find the divergence of the vector field

    $$\mathbf{A} = x^2\mathbf{i} + 2xy\mathbf{j} - \frac{z^2 y}{x}\mathbf{k}$$

    at the point $(1, 2, 3)$.

5. A curve C in the form of a closed unit square spans a plane area S which has a unit normal $\hat{\mathbf{n}}$ (so $\hat{\mathbf{n}}\,dS = \hat{\mathbf{n}}dS$). Evaluate (a) $\oint_C d\mathbf{r}$, (b) $\iint_S d\mathbf{S}$, (c) $\iint_S dS$ and (d) $\int_C dr$.

6. Show that $\mathrm{curl}(x\mathbf{a}) = \mathbf{i} \times \mathbf{a}$ where $\mathbf{a}$ is a constant vector.

7. Calculate the divergence of the vector field $3r^2\mathbf{r}$ at the point $\mathbf{r} = (0, 2, 1)$.

8. Find the divergence of the vector field (yz, xy^2z, xyz^2).

9. Find the curl of the vector field (x, xz, xy).

10. Show that $\nabla(|\mathbf{r}|2)^n = 2n(|\mathbf{r}|2)^{n-1}\mathbf{r}$ where $\mathbf{r} = x\mathbf{i} + y\mathbf{j} + z\mathbf{k}$.

11. If $\mathbf{A} = (x^2 - y^2, y^2 - z^2, z^2 - x^2)$ and $Q = x + y + z$, find $\nabla \cdot (Q\mathbf{A})$ (the divergence of $Q\mathbf{A}$) at $(1, 0, 1)$.

12. Show that the *curl* of $\mathbf{k}\sin(\mathbf{k} \cdot \mathbf{r})$ is identically zero if $\mathbf{k}$ is a constant vector.

13.15. PROBLEMS

1. State Stokes's theorem and use it to show that

    $$\oint_C (\mathbf{A} \times \mathbf{r}) \cdot d\mathbf{r} = 2\pi A,$$

    where C is a unit circle spanned by a surface normal to the constant vector $\mathbf{A}$.

Use the above result to interpret the geometrical meaning of the integral

$$\frac{1}{2}\oint_C \mathbf{r} \times d\mathbf{r}.$$

2. Calculate the integral over the surface of a cube whose corners are located at $(0, 0, 0)$, $(0, 2, 0)$, $(0, 0, 2)$, $(2, 0, 0)$, $(0, 2, 2)$, $(2, 0, 2)$, $(2, 2, 0)$ and $(2, 2, 2)$ of the vector field

$$\mathbf{B} = (x^2 + y^2, y^2 + z^2, z^2 + x^2).$$

Calculate the divergence of **B** and verify that its integral over the volume enclosed by the cube above has the same value as the flux through the surface.

3. State Gauss's theorem and use it to evaluate

$$\int_V \mathrm{div}\left(r^2 \mathbf{r}\right) dV,$$

where $\mathbf{r} = (x, y, z)$ and the integral is taken over a unit cube with its center at the origin and sides parallel to the axes.

4. State Gauss's theorem and verify it for the vector field (x^n, y^n, z^n) for a unit cube whose surface normals are parallel to the x-, y- or z- axes and with its center at the origin. (Assume n is an odd positive integer.)

5. Prove that the volume of a solid is given by

$$\frac{1}{3}\int \mathbf{r} \cdot d\mathbf{S},$$

where **r** is the position vector of any point on its surface and $d\mathbf{S}$ is the vector element of area of the surface. Show that the expression is correct for a cube of side a, with its center at the origin and sides parallel to the axes.

A

PREREQUISITES

This appendix contains some basic facts and methods that will be review for most students, as well as some useful reference material. Consult it as necessary.

A.1. NUMBERS, FUNCTIONS, AND PROOFS

Real Numbers

Real numbers are values that represent a quantity along a line. Imagine each real number as a point on a line with the negatives to the left of zero and the positives to the right. There are two types of points: rational numbers that can be represented as fractions or as either finite or recurring decimals and irrational numbers that are not fractions and are represented by an unending and non-repeating decimal expansion.

There is one point about the decimal expansion of real numbers that you may need reminding of: the recurring expansion $0.99999\overline{9}$ (the bar shows that the 9 repeats forever) is the same as $1.0000\overline{0}$, and similarly for other numbers with recurring 9s. (You can see why, once you try adding something to $0.99999\overline{9}$, such as 0.00001 or 0.0000001. There is nothing you can add that does not give a number that exceeds 1.)

Functions

A function of a real variable is a rule that assigns a real number to each permitted value of the variable. For example the square function x^2 assigns the square of x for any real number x. We often write $y = f(x)$ for a function of x. Here x is the *independent variable* and y the *dependent variable*.

Proofs

You need to be quite clear about the notion of proof in mathematics. A finite number of verifiable instances of a general result does not constitute a proof. Take the following example, the expansion of

$$(1 + x)^n = 1 + nx + \ldots,$$

where we want to prove that the coefficient of x in the expansion is n. It is obviously true for $n = 1$. For $n = 2$ we can get the result by explicitly multiplying out $(1 + x)(1 + x) = 1 + 2x + x^2$. You might see how to go on to verify the result for $n = 3$. But we cannot prove that the result is true in general by enumerating instances, because we might eventually come across a value of n for which the result does not hold. For example, take the (incorrect) statement that every even number, $2n$, is followed by a prime number. This is true for $n = 1, 2$ and 3 where, $2n = 2,4$ and 6, which are followed by $3, 5$, and 7, but it is clearly not true for $n = 4$ since $2n = 8$ is followed by 9, which is not prime.

There are various valid methods of proof in mathematics. One example is proof by mathematical induction.

Example A.1 Show that $(1 + x)^n = 1 + nx + \ldots$.

We want to show that

$$(1 + x)^n = 1 + nx + \ldots,$$

for $n = 1,2,3,4, \ldots$ This result is clearly true for $n = 1$. We shall assume that it holds for $n = 2, 3, \ldots$ up to some general n, then consider $(1 + x)^{n+1}$.

$$(1+ x)^{n+1} = (1+ x)(1 + x)^n,$$
$$= (1 + x)(1 + nx + \ldots).$$

Then, multiplying out the parentheticals and collecting like terms we find

$$(1 + x)^{n+1} = (1 + nx + \dots + x + nx^2 + \dots),$$
$$= 1 + (n + 1)x + \dots.$$

So if $(1 + x)^n = 1 + nx + \dots$ is true for n we have shown that it is true for $n + 1$. But we know that it is true for $n = 1$. So it must be true also for $n = 2, 3, \dots$ and so on for all n.

This method (assuming the result for n, deduce it for $n + 1$; check it for $n = 1$) is called *proof by induction*.

A.2. NECESSARY AND SUFFICIENT CONDITIONS

The distinction between what is necessary for a result to be true and what is sufficient is sometimes important. It is best approached through some examples:

- For $x^2 > 1$ to be true it is sufficient that $x > 1$ but not necessary (x could also be < -1).

- For $x^2 < 1$ to be true it is necessary that $x < 1$, but it is not sufficient (x must also be > -1).

- For $x^{1/2} > 1$ it is both necessary and sufficient for $x > 1$.

Suppose that $x > 1$; then it follows from the first of the above statements that $x^2 > 1$.

Suppose $x < 1$; then it does not follow that $x^2 < 1$ (because according to the second statement $x < 1$ is necessary but not sufficient). Another way of thinking about this is in terms of the direction of implication:

- $x > 1$ *implies* $x^2 > 1$ so $x > 1$ is a sufficient condition ("sufficient" to imply the result).

- $x < 1$ is *implied by* $x^2 < 1$ so the result $x < 1$ is a necessary consequence.

A.3. LOGARITHMS

We define the function $y = \log_a x$ ("the logarithm to the base a of x") for $x > 0$ and $a > 0$, as the number, y, satisfying the relation $x = a^y$. i.e. the log of a positive number is the power to which the base must be raised to get that number.

$$\text{For } x > 0, y = \log_a(x) \text{ if } x = a^y.$$

Example A.2 For example,

    (i) $\log_{10}(10) = 1$ (put $a = 10, x = 10$, then $10 = 10^1$ so $y = 1$) and similarly $\log_a a = 1$,

    (ii) $\log_{10}(10^n) = n$,

    (iii) $\log_{10}(0.1) = -1$ (because $0.1 = 1/10 = 10^{-1}$).

To find values of $\log_a(x)$ for more general values of x you will need to use a calculator. Usually calculators have "log" meaning $\log_{10}$, and "ln" meaning $\log_e$ (see Chapter 2) but you can get from these to any base (see below). In this section, for formulae that are true in any base we will omit explicit reference to the base and just write $\log x$.

Example A.3 Show that $\log(x_1 x_2) = \log(x_1) + \log(x_2)$, and deduce that $\log(1/x) = -\log(x)$.

Let $\log(x_1) = y_1$, $\log(x_2) = y_2$ and $\log(x_1 x_2) = y$. Then, using the definition of *log*, we can write

$$x_1 = a^{y_1}, x_2 = a^{y_2} \text{ and } x_1 x_2 = a^y.$$

Then

$$x_1 x_2 = a^{y_1} a^{y_2} = a^{y_1 + y_2} = a^y.$$

So, by examining the exponents we have that

$$y = y_1 + y_2.$$

Finally, substituting for y, y_1 and y_2 we obtain the required result that

$$\log(x_1 x_2) = \log(x_1) + \log(x_2).$$

Now for the second part. We know $a^0 = 1$ and taking logs of both sides we get $0 = \log(1)$. Therefore

$$\log(x \times (1/x)) = \log(1) = 0.$$

Using the result that $\log(x_1 x_2) = \log(x_1) + \log(x_2)$ we can then therefore write that

$$\log(x) + \log(1/x) = 0$$

and hence,

$$\log(1/x) = -\log(x).$$

This is a useful result that is worth remembering.

The formula for the logarithm of a product is true in any base. Obviously it follows from Example A.3 that $\log(x^2) = 2\log(x)$ and similarly

$$\log(x^a) = a\log(x)$$

for any a.

Example A.4 *Change of Base*

Show that $\log_a(x) = (\log_a b)(\log_b x)$ and that $\log_a(x) = \dfrac{\log_b x}{\log_b a}$.

First, we let $y = \log_b x$ and $w = \log_a b$, then

$$x = b^y,$$
$$b = a^w.$$

Next, substituting for $b = a^w$, we have

$$x = (a^w)^y,$$
$$= a^{wy}.$$

Taking logs of both sides gives

$$\log_a x = wy.$$

Then, substituting for y and w we find the required result that

$$\log_a x = \log_a b \times \log_b x.$$

For the second part, we use the fact that $b = a^w$, so

$$a = b^{1/w}.$$

Taking $\log_b$ of both sides gives

$$\log_b a = 1/w,$$
$$= 1/\log_a b.$$

Then, substituting this into the result for the first part we have

$$\log_a x = \log_a b \log_b x = \frac{\log_b x}{\log_b a}.$$

Exercise A.1 By using logs on your calculator find

(i) $\sqrt[3]{5}$

(ii) 1.254×0.378

(iii) $1.254/0.378$.

Exercise A.2

(i) If a plot of $\log(y)$ against $\log(x)$ is a straight line of slope k show that $y = Ax^k$ for some constant A.

(ii) Express $\log(x^x)$ in terms of x and $\log(x)$.

(iii) Find $\log_3(5.2)$.

A.4. TRIGONOMETRIC FUNCTIONS

FIGURE A.1: Right angle triangle; h is the length of the hypotenuse, a the adjacent side and o the opposite side with respect to the angle θ.

The elementary definition of the functions sin, cos, and tan gives them in terms of the ratios of sides of a right-angled triangle,

$$\sin(\theta) = \frac{o}{h}, \quad \cos(\theta) = \frac{a}{h} \quad \text{and} \quad \tan(\theta) = \frac{o}{a}.$$

From this definition various relations can be deduced. For example,

$$\tan(\theta) = \frac{\sin(\theta)}{\cos(\theta)} \quad \text{and} \quad \sin^2(\theta) + \cos^2(\theta) = 1.$$

However, the elementary definition in terms of a simple right angle triangle has limitations – and more convenient and technical definitions exist. We will not go through these here, but simply supply some relevant and general results.

First, the extension to all angles ($0° \leq \theta \leq 360°$) is most easily represented by showing the signs of the sides of the triangle in each of the quadrants (see Figure A.2). These follow the obvious pattern of coordinate axes. So the overall signs of each of the functions are as in Figure A.3.

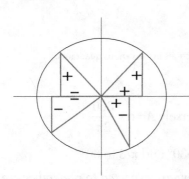

FIGURE A.2: The signs of the sides of the triangles for construction of the trigonometric functions follow the pattern of the coordinate axes.

For example, $\sin(135°) = \sin(45°)$, $\sin(192°) = -\sin(12°)$ and $\sin(307°) = -\sin(53°)$. Also, $\cos(135°) = -\cos(45°)$, $\cos(192°) = -\cos(12°)$ and $\cos(307°) = \cos(53°)$.

In general,

$$\sin(-\theta) = -\sin(\theta),$$
$$\cos(-\theta) = \cos(\theta).$$

Radians and Degrees

Note that in general mathematical relations are given with the *angles measured in radians*, not in degrees. The relation is that 2π radians is equivalent to $360°$. Thus to convert an angle A in degrees to the same

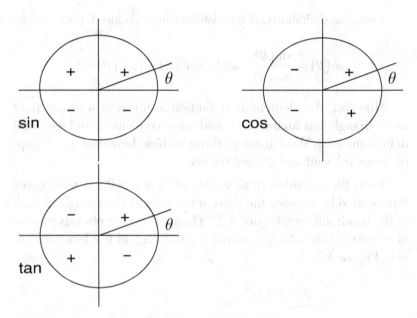

FIGURE A.3: The signs of the trigonometric functions in the various quadrants.

angle a in radians we have

$$a = A \times \frac{2\pi}{360} \text{ and conversely } A = \alpha \times \frac{360}{2\pi}.$$

Thus $\pi/4$ corresponds to $45°$, $\pi/6$ to $30°$ and $\pi/3$ to $60°$.

Some very useful exact values *you are expected to remember* are

$$\sin(0) = 0, \quad \sin\left(\frac{\pi}{6}\right) = \frac{1}{2}, \quad \sin\left(\frac{\pi}{4}\right) = \frac{1}{\sqrt{2}}, \quad \sin\left(\frac{\pi}{3}\right) = \frac{\sqrt{3}}{2}, \quad \sin\left(\frac{\pi}{2}\right) = 1,$$

$$\cos(0) = 1, \quad \cos\left(\frac{\pi}{6}\right) = \frac{\sqrt{3}}{2}, \quad \cos\left(\frac{\pi}{4}\right) = \frac{1}{\sqrt{2}}, \quad \cos\left(\frac{\pi}{3}\right) = \frac{1}{2}, \quad \cos\left(\frac{\pi}{2}\right) = 0,$$

and, for any angle θ,

$$\sin(\theta) = \cos\left(\frac{\pi}{2} - \theta\right).$$

A graphical representation of these functions is given in Figures A.4 (a) and (b).

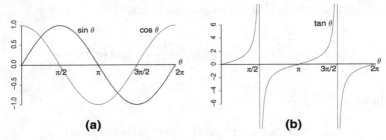

FIGURE A.4: (a) The sin and cos functions. (b) The tan function.

Combining Angles

We can combine angles:

$$\sin(\alpha + \beta) = \sin(\alpha)\cos(\beta) + \cos(\alpha)\sin(\beta),$$
$$\cos(\alpha + \beta) = \cos(\alpha)\cos(\beta) - \sin(\alpha)\sin(\beta),$$

and

$$\sin(\alpha) + \sin(\beta) = 2\sin\left(\frac{\alpha + \beta}{2}\right)\cos\left(\frac{\alpha - \beta}{2}\right),$$
$$\cos(\alpha) + \cos(\beta) = 2\cos\left(\frac{\alpha + \beta}{2}\right)\cos\left(\frac{\alpha - \beta}{2}\right).$$

And also split them up:

$$\sin(2\alpha) = 2\sin(\alpha)\cos(\alpha),$$
$$\cos(2\alpha) = 2\cos^2(\alpha) - 1 = 1 - 2\sin^2(\alpha).$$

Exercise A.3 Show that

(i) $\sin(\pi + \alpha) = -\sin(\alpha)$,

(ii) $\sin(\pi - \alpha) = \sin(\alpha)$.

It is not overly important to remember these formulae exactly, although it is often convenient to have done so. However, it is absolutely vital that you know these formulae exist and know their general form.

Related Functions

Here are some related trigonometrical functions and the relations between them that are important to know:

$$\sec(a) = \frac{1}{\cos(a)}, \quad \mathrm{cosec}(a) = \frac{1}{\sin(a)}, \quad \cot(a) = \frac{1}{\tan(a)},$$

where

$$1 + \tan^2(a) = \sec^2(a), \quad 1 + \cot^2(a) = \mathrm{cosec}^2(a).$$

Finally, another useful result (see Exercise A5) to remember is

$$a\sin(a) + b\cos(a) = A\sin(a + \phi),$$

where $A = (a^2 + b^2)^{1/2}$ and $\phi = \tan^{-1}(b/a)$.

Exercise A.4 Use your calculator to find

(i) $\sin(135°)$

(ii) $\tan(15\pi/4)$

(iii) $\mathrm{cosec}(11\pi/8)$.

Exercise A.5 By using the identity for $\sin(a + \phi)$ find A and ϕ such that $a\sin(a) + b\cos(a) = A\sin(a + \phi)$.

Exercise A.6 Using the identities given, prove that

(i) $\tan(2a) = \dfrac{2\tan(a)}{1 - \tan^2(a)}$,

(ii) $\sin(a) + \sin(\beta) \le 2\sin\left(\dfrac{1}{2}(a + \beta)\right)$ for $(0 \le a + \beta \le \pi)$.

A.5.   DIVISION OF POLYNOMIALS

There are several ways to divide a polynomial of higher degree by one of lesser degree. Here we cover one simple example but you will find many other methods and examples in different textbooks.

Example A.5 Simplify $\dfrac{2x^3 + 9x^2 + 4x - 20}{2x + 5}$.

One basic way to divide a polynomial is to guess the form of the answer and compare coefficients. In this question the answer must start with an x^2 and hence will be of the form $ax^2 + bx + c + R(x)$ where the remainder, $R(x)$, will be of the form $d/(2x + 5)$.

We can therefore write

$$\frac{2x^3 + 9x^2 + 4x - 20}{2x + 5} = \left(ax^2 + bx + c\right) + \frac{d}{2x + 5}$$

so

$$2x^3 + 9x^2 + 4x - 20 = (2x + 5)(ax^2 + bx + c) + d$$
$$= 2ax^3 + (5a + 2b)x^2 + (5b + 2c)x + 5c + d.$$

Now we can equate the coefficients, starting with the highest power of x. This allows the unknowns to be read off directly: equating the coefficients of x^3 gives $a = 1$; equating the coefficients of x^2 gives $b = (9 - 5 \times 1)/2 = 2$; then, similarly, $c = (4 - 5 \times 2)/2 = -3$ and $d = (-20 - 5 \times (-3)) = -5$. Therefore

$$\frac{2x^3 + 9x^2 + 4x - 20}{2x + 5} = x^2 + 2x - 3 - \frac{5}{2x + 5}.$$

Exercise A.7 Using the method above, or your own preferred method, obtain the following results:

(i) $\dfrac{3x^4 + x^3 - 14x^2 + 9x + 20}{3x + 7} = x^3 - 2x^2 + 3\dfrac{1}{3x + 7}$,

(ii) $\dfrac{x^2 + 3x + 5}{x^2 + 4} = 1 + \dfrac{3x + 1}{x^2 + 4}$,

(iii) $\dfrac{x^3 + 4x^2 + x - 1}{x^2 + 3x - 2} = x + 1 + \dfrac{1}{x^2 + 3x - 2}.$

Exercise A.8 Find $\dfrac{x^4 + 5x^3 + 11x^2 + 13x + 7}{x^2 + 2x + 3}$.

A.6. PARTIAL FRACTIONS

A *rational function* is a polynomial divided by another polynomial, i.e. a fraction with polynomials as numerator and denominator. To find the sum (or difference) of two rational functions we put them over a common denominator. For example:

$$\frac{1}{(x-2)} - \frac{1}{(x+3)} = \frac{(x+3)-(x-2)}{(x-2)(x+3)} = \frac{5}{(x-2)(x+3)}.$$

The reverse process is called expressing a rational function in partial fractions.

Example A.6 Express $\dfrac{1}{(x-2)(x+3)}$ in partial fractions.

We begin with

$$\frac{1}{(x-2)(x+3)} = \frac{a}{(x-2)} + \frac{b}{(x+3)}.$$

To find a multiply through by the parenthetical $(x-2)$,

$$\frac{1}{(x+3)} = a + \frac{(x-2)b}{(x+3)}.$$

The two sides should be equal for all x, including $x = 2$ which will get rid of the messy term in b. By inserting $x = 2$ we find $a = 1/5$. Similarly, to find b we multiply by $(x+3)$ instead:

$$\frac{1}{(x-2)} = \frac{(x+3)a}{(x-2)} + b.$$

We choose $x = -3$ to get rid of the term in a. This gives $b = -1/5$.

Hence we have the solution

$$\frac{1}{(x-2)(x+3)} = \frac{1}{5(x-2)} - \frac{1}{5(x+3)}.$$

This is the easiest way to obtain the solution. Alternatively, if we had chosen different values of x to substitute in when finding a and b we would have obtained a pair of simultaneous equations for a and b, which we could then solve to find a and b.

Exercise A.9 Express $\dfrac{3}{(x+4)(x-1)}$ in partial fractions.

Exactly the same method works if the numerator is any polynomial of degree less than that of the denominator:

Exercise A.10 Use the method of Example A.6 to express $\dfrac{x}{(x+4)(x+1)}$ in partial fractions.

If the numerator is a polynomial of degree greater than or equal to that of the denominator, then a rearrangement equivalent to long division with remainder must first be carried out.

Example A.7 Express $\dfrac{x^2+6x+4}{(x+4)(x+1)}$ in partial fractions.

We have that

$$\frac{x^2+6x+4}{(x+4)(x+1)} = \frac{x^2+6x+4}{x^2+5x+4},$$

$$= \frac{(x^2+5x+4)+x}{x^2+5x+4},$$

$$= 1 + \frac{x}{(x+4)(x+1)},$$

$$= 1 + \frac{4}{3(x+4)} - \frac{1}{3(x+1)}.$$

A.7. SERIES

Arithmetic Series

An arithmetic sequence (or arithmetic progression) is a succession of terms, each of which differs from the previous one by adding a constant.

For example

$$1, 3, 5, 7, 9,... . \tag{A.1}$$

The sequence may be finite or infinite. If the sequence starts at a and each term differs from the previous one by a constant d then the nth term is

$$a_n = a + (n - 1)d,$$

giving $a_1 = a$, $a_2 = a + d$, $a_3 = a + 2d$, and so on. The unevaluated sum of the terms of a sequence is called a *series*. For example, the series obtained from the sequence (A.1) is

$$1 + 3 + 5 + 7 + 9 + ... \tag{A.2}$$

The sum of the first N terms of an arithmetic sequence is

$$S_N = \sum_{n=1}^{N} a_n = \sum_{n=1}^{N} \left(a + (n-1)d \right).$$

We can evaluate the sum of the first N terms as follows. Write out the sum in two different ways, first with the terms in ascending order, then in descending order.

$$S_N = a + (a + d) + (a + 2d) + ... + (a + (N - 2)d) + (a + (N - 1)d),$$
$$S_N = (a + (N - 1)d) + (a + (N - 2)d) + ... + (a + 2d) + (a + d) + a,$$

where $a_N = a + (N - 1)d$ is the final term in the sequence. If we add these two together we get

$$2S_N = (2a + (N - 1)d) + (2a + (N - 1)d) + ... + (2a + (N - 1)d)$$
$$= N (a + a_N).$$

So

$$S_N = \frac{N}{2}\left(a + a_N\right) = \frac{N}{2} \text{ (first term + last term)}.$$

For example, in the case of the series (A.2) we have $a = 1$ and $d = 2$. Therefore $a_{10} = 19$ and $S_{10} = 100$.

If there are an infinite number of terms the series will diverge. If d is positive then S_N increases without limit as N increases; if d is negative S_N will decrease without limit.

Geometric Series

A geometric sequence (or geometric progression) is a succession of terms in which each term differs from the previous one by a multiplicative constant. For example

$$1, (1/2), (1/4), (1/8),(1/16),\ldots .\tag{A.3}$$

The sequence may be finite or infinite. If the sequence starts at a and each term differs from the previous one by a constant r then the nth term is

$$a_n = ar^{n-1}$$

giving $a_1 = a$, $a_2 = ar$, $a_3 = ar^2$, and so on. The sum of the terms is called a *geometric series*. The series obtained from the sequence (A.3) is

$$1 + (1/2) + (1/4) + (1/8) + (1/16) + \ldots .\tag{A.4}$$

The sum of the first N terms of a geometric sequence is

$$S_N = \sum_{n=1}^{N} a_n = \sum_{n=1}^{N} ar^{n-1}.$$

We can evaluate the sum of the first N terms as follows. Write out the sum S_N and also rS_N,

$$S_N = a + ar + ar^2 + ar^3 \cdots + ar^{N-1},$$
$$rS_N = ar + ar^2 + ar^3 + ar^4 \cdots + ar^{N},$$

and then subtract the second from the first. The only terms that do not cancel are the first term of the first equation and the last term of the second equation

$$(1 - r)S_N = a - ar^N.$$

The sum of the first N terms can therefore be written as

$$S_N = \frac{a\left(1-r^N\right)}{1-r}.$$

For an infinite sequence, if $|r| < 1$ we have $\lim_{N\to\infty} r^N = 0$, and so the sum tends to

$$S = \sum_{n=1}^{\infty} ar^n = \frac{a}{1-r}. \qquad (A.5)$$

For example, in the case of the series (A.4) we have $r = 1/2$ and $a = 1$. Therefore $S = 2$. For $|r| \geq 1$ the sum S_N either diverges with N or it oscillates with increasing amplitude.

B

THE GREEK ALPHABET

alpha	α	A		nu	ν	N
beta	β	B		xi	ξ	Ξ
gamma	γ	Γ		omicron	o	O
delta	δ	Δ		pi	π	Π
epsilon	ε	E		rho	ρ	P
zeta	ζ	Z		sigma	σ	Σ
eta	η	H		tau	τ	T
theta	θ	Θ		upsilon	τ	Υ
iota	ι	I		phi	ϕ	Φ
kappa	κ	K		chi	χ	X
lambda	λ	Λ		psi	ψ	ψ
mu	μ	M		omega	ω	Ω

INDEX